Data Mining and Mathematical Programming

Volume 45

CRM PROCEEDINGS & LECTURE NOTES

Centre de Recherches Mathématiques
Montréal

Data Mining and Mathematical Programming

Panos M. Pardalos
Pierre Hansen
Editors

The Centre de Recherches Mathématiques (CRM) of the Université de Montréal was created in 1968 to promote research in pure and applied mathematics and related disciplines. Among its activities are special theme years, summer schools, workshops, postdoctoral programs, and publishing. The CRM is supported by the Université de Montréal, the Province of Québec (FCAR), and the Natural Sciences and Engineering Research Council of Canada. It is affiliated with the Institut des Sciences Mathématiques (ISM) of Montréal, whose constituent members are Concordia University, McGill University, the Université de Montréal, the Université du Québec à Montréal, and the École Polytechnique. The CRM may be reached on the Web at www.crm.umontreal.ca.

American Mathematical Society
Providence, Rhode Island USA

The production of this volume was supported in part by the Fonds pour la Formation de Chercheurs et l'Aide à la Recherche (Fonds FCAR) and the Natural Sciences and Engineering Research Council of Canada (NSERC).

2000 *Mathematics Subject Classification.* Primary 68R01, 68P01, 68P10, 90–08, 90C09, 90C27.

Library of Congress Cataloging-in-Publication Data

Data mining and mathematical programming / Panos M. Pardalos, Pierre Hansen, editors.
p. cm. — (CRM proceedings & lecture notes, ISSN 1065-8580 ; v. 45)
Includes bibliographical references and index.
ISBN 978-0-8218-4352-9 (alk. paper)
1. Programming (Mathematics) 2. Data mining. I. Pardalos, P. M. (Panos M.), 1954–
II. Hansen, P. (Pierre) III. Title. IV. Series.

QA402.5 .D268 2008
519.7—dc22 2008002772

∞ The paper used in this book is acid-free and falls within the guidelines
established to ensure permanence and durability.
This volume was submitted to the American Mathematical Society
in camera ready form by the Centre de Recherches Mathématiques.
Visit the AMS home page at http://www.ams.org/

10 9 8 7 6 5 4 3 2 1 13 12 11 10 09 08

Contents

Preface

Data mining aims at finding interesting, useful or profitable information in very large databases. The enormous increase in the size of available scientific and commercial databases (data avalanche) as well as the continuing and exponential growth in performance of present day computers make data mining a very active field. Its main tasks, i.e., discrimination, clustering, and relation finding have already been partially explored in Data Analysis, Pattern Recognition, Artificial Intelligence and other fields. The challenge, however, resides in the much increased size of problems considered. Indeed, instead of problem instances with a small number of entities, current ones often have thousands, and sometimes even several million entities (massive data sets). In many cases, the burgeoning volume of data sets has grown so large that it threatens to overwhelm rather than enlighten scientists. Therefore, traditional methods are revised and streamlined, complemented by many new methods to address challenging new problems.

Mathematical Programming plays a key role in this endeavor. It helps us to formulate precise objectives (e.g., a clustering criterion or a measure of discrimination) as well as the constraints imposed on the solution (e.g., find a partition, a covering or a hierarchy in clustering). It also provides powerful mathematical tools to build highly performing exact or approximate algorithms.

The chapters in this book are based on lectures presented at the workshop on "Data Mining and Mathematical Programming" (October 10–13, 2006, Montreal) and mainly focus on three basic areas of data mining:

- *Discrimination or supervised classification*: build a function that best discriminates between good and bad entities of a given set, and classifies as correctly as possible new entities.

- *Clustering or unsupervised classification*: find subsets of a given set of entities that are homogeneous and well-separated.

- *Relation finding*: given a set of entities, together with measurements or observations on them, find relations satisfied by all, or by most, of them.

This book will be a valuable scientific source of information to faculty, students, and researchers in optimization, data analysis and data mining, as well as people working in computer science, engineering and applied mathematics.

We would like to take the opportunity to thank the authors of the papers, the anonymous referees, the Centre de Recherches Mathématiques Montréal (Québec) Canada, the Center for Applied Optimization (University of Florida), and all sponsors for supporting this effort. Finally, we would like to thank Louise Letendre and the American Mathematical Society for their assistance in publishing this book.

Panos M. Pardalos (Florida) and Pierre Hansen (HEC Montreal).

Centre de Recherches Mathématiques
CRM Proceedings and Lecture Notes
Volume **45**, 2008

Support Vector Machines and Distance Minimization

Emilio Carrizosa

ABSTRACT. Support Vector Machine is a powerful classification tool of increasing popularity in Data Mining. When, as customary in practice, the sets to be classified are not separable, the so-called soft-margin approach is often used as a surrogate of the maximization of the margin.

In this note we show that different soft-margin problems already studied in the literature, as well as some of their extensions, can be formulated as minimum-distance problems. Using the machinery of Convex Analysis, optimality conditions are derived for such minimum-distance problems, yielding, in particular, a characterization for degeneracy.

1. Introduction

Support Vector Machines (SVM) is a popular methodology for supervised classification, with excellent theoretical and practical behavior. See e.g. [2, 3, 5, 6, 11] for an introduction.

A non-empty index set I, a non-trivial partition of I as $I = I_+ \cup I_-$, and subsets $\{x_i : i \in I_+\}$, $\{x_i : i \in I_-\}$ in $\mathbb{R}^N$ are given. The aim is to separate the two groups, labelled as I_+ and I_-, via an affine function, i.e., $\varphi_{\omega,\beta}(x) = \omega^\top x + \beta$ is sought satisfying

$$\begin{aligned} &\omega^\top x_i + \beta > 0, \quad \forall i \in I_+, \\ &\omega^\top x_i + \beta < 0, \quad \forall i \in I_-. \end{aligned} \tag{1.1}$$

Observe that, since we assume that the partition of I is non-trivial, i.e., $I_+ \neq \emptyset$, $I_- \neq \emptyset$, any (ω, β) satisfying (1.1) has $\omega \neq 0$, and thus the locus $\mathcal{H}(\omega, \beta) := \{x \in \mathbb{R}^N : \varphi_{\omega,\beta}(x) = 0\}$ is a hyperplane.

The coefficients ω, β will be used to label as members of I_+ or I_- future entries x, according to the rule

$$\begin{aligned} &\text{Assign } x \text{ to } I_+, \quad \text{if } \omega^\top x + \beta > 0, \\ &\text{Assign } x \text{ to } I_-, \quad \text{if } \omega^\top x + \beta < 0, \end{aligned} \tag{1.2}$$

remaining the points on the hyperplane $\mathcal{H}(\omega, \beta)$ unclassified.

2000 *Mathematics Subject Classification.* Primary: 90C25; Secondary: 62H30.

Partially supported by grants MTM2005-09362-103-01 of MEC, Spain, and FQM-329 of Junta de Andalucía, Spain.

This is the final form of the paper.

A *hard-margin* SVM hyperplane is defined as a hyperplane of the form $\mathcal{H}(\omega,\beta)$ for which (1.1) holds, and the minimum $\|\cdot\|$-distance from $\{x_i : i \in I\}$ to $\mathcal{H}(\omega,\beta)$ is maximized. Here, $\|\cdot\|$ is an arbitrary norm in $\mathbb{R}^N$. The usual choice in the literature is the ℓ_2 norm, although other choices, such as the ℓ_1, have also been analyzed, e.g. [1, 4, 10, 13].

We assume throughout this paper that the sets are not linearly separable, i.e., the convex hulls of $\{x_i : i \in I_+\}$ and $\{x_i : i \in I_-\}$ intersect. Under this assumption, system (1.1) is unfeasible. Different strategies have been proposed in the literature to face this situation. For instance, the hyperplane minimizing the number of misclassified points is suggested e.g. in [9]; the hyperplane minimizing the sum of its $\|\cdot\|$-distances to misclassified points has been proposed by [8]. However, the most popular extension to the case of non-separable groups is the so-called *soft-margin* SVM, which has been proposed in different versions, formulated as the linearly-constrained convex problem

$$\begin{aligned} &\min \omega^\top \omega + C\sum_{i\in I}\xi_i^p \\ &\text{s.t. } \omega^\top x_i + \beta + \xi_i \geq 1, \quad \forall i \in I_+, \\ &\quad \omega^\top x_i + \beta - \xi_i \leq -1, \quad \forall i \in I_-, \\ &\quad \xi_i \geq 0, \quad \forall i \in I, \end{aligned} \tag{1.3}$$

[5], where $C > 0, p \geq 1$, (typically $p = 1$ or $p = 2$) or

$$\begin{aligned} &\min \sum_{j=1}^N (\omega_j^+ + \omega_j^-) + C\sum_{i\in I}\xi_i \\ &\text{s.t. } (\omega^+ - \omega^-)^\top x_i + \beta + \xi_i \geq 1, \quad \forall i \in I_+, \\ &\quad (\omega^+ - \omega^-)^\top x_i + \beta - \xi_i \leq -1, \quad \forall i \in I_-, \\ &\quad \omega^+, \omega^- \geq 0, \\ &\quad \xi_i \geq 0 \quad \forall i \in I, \end{aligned} \tag{1.4}$$

where $C > 0$. Problem (1.4) can also be written as

$$\begin{aligned} &\min \sum_{j=1}^N (\omega_j^+ + \omega_j^-) + C\sum_{i\in I}|\xi_i|, \\ &\text{s.t. } (\omega^+ - \omega^-)^\top x_i + \beta + \xi_i \geq 1, \quad \forall i \in I_+, \\ &\quad (\omega^+ - \omega^-)^\top x_i + \beta - \xi_i \leq -1, \quad \forall i \in I_-, \\ &\quad \omega^+, \omega^- \geq 0, \end{aligned}$$

and as the minimum-distance problem

$$\begin{aligned} &\min \|\omega\|_1 + C\|\xi\|_1, \\ &\text{s.t. } \omega^\top x_i + \beta + \xi_i \geq 1, \quad \forall i \in I_+, \\ &\quad \omega^\top x_i + \beta - \xi_i \leq -1, \quad \forall i \in I_-, \end{aligned} \tag{1.5}$$

and thus can be seen as the problem of finding the point in the polyhedron $\mathcal{X}$,

$$\begin{aligned}\mathcal{X} = \{(\omega,\xi) \in \mathbb{R}^N \times \mathbb{R}^I :& \\ &\omega^\top x_i + \beta + \xi_i \geq 1, \quad \forall i \in I_+, \\ &\omega^\top x_i + \beta - \xi_i \leq -1, \quad \forall i \in I_-, \\ &\text{for some } \beta \in \mathbb{R}\}\end{aligned} \tag{1.6}$$

with minimum $\|\cdot\|$-norm, where $\|(\omega,\xi)\| = \|\omega\|_1 + C\|\xi\|_1$.

The purpose of this note is to show that Problem (1.3) is also equivalent to minimum-distance problems on $\mathcal{X}$ for certain norms. This interpretation enables us to present a common framework for these different soft-margin approaches. Within this general framework, we derive simple optimality conditions, and a characterization of *degeneracy,* i.e., of those cases in which, for any choice of the parameter C, an optimal solution of (1.3) and (1.4) has the form $(\omega^*,\beta^*) = (0,0)$, and is thus useless for classification purposes according to (1.2).

The remainder of the note is structured as follows. In Section 2 a minimum-distance problem is introduced, and related with (1.3) under the assumption that the sets $\{x_i : i \in I_+\}$ and $\{x_i : i \in I_-\}$ are not linearly separable. Optimality conditions for such minimum-distance problem are derived in Section 3, leading, in particular, to a characterization of degeneracy, as detailed in Section 4.

2. Minimum-distance Problems

Let γ_1, γ_2 be norms in $\mathbb{R}^N$ and $\mathbb{R}^I$, respectively, with γ_2 strictly monotonic in $\mathbb{R}^I_+$, i.e.,

$$(0 \leq \xi^1 \leq \xi^2, \xi^1 \neq \xi^2) \implies (\gamma_2(\xi^1) < \gamma_2(\xi^2)).$$

Observe that, in particular,

$$\gamma_2\big((\xi_i)_{i\in I}\big) = \gamma_2\big((|\xi_i|)_{i\in I}\big), \tag{2.1}$$

see [7].

For any $\mu > 0$, consider the optimization problem

$$\begin{aligned}&\min \gamma_1(\omega) + \mu\gamma_2(\xi), \\ &\text{s.t. } \omega^\top x_i + \beta + \xi_i \geq 1, \qquad \forall i \in I_+, \\ &\quad\ \omega^\top x_i + \beta - \xi_i \leq -1, \qquad \forall i \in I_-, \\ &\quad\ \omega \in \mathbb{R}^N,\ \beta \in \mathbb{R},\ \xi \in \mathbb{R}^I\end{aligned} \tag{P_μ}$$

which can also be written as the minimum-distance problem

$$\begin{aligned}&\min \|(\omega,\xi)\|, \\ &\text{s.t. } (\omega,\xi) \in \mathcal{X},\end{aligned} \tag{2.2}$$

where $\|\cdot\|$ is the norm in $\mathbb{R}^N \times \mathbb{R}^I$ defined as

$$\|(\omega,\xi)\| = \gamma_1(\omega) + \mu\gamma_2(\xi), \tag{2.3}$$

and $\mathcal{X}$ is the polyhedron defined in (1.6).

Observe that (1.5) corresponds to (P_μ) choosing γ_1 and γ_2 as the ℓ_1 norm in $\mathbb{R}^N$ and $\mathbb{R}^I$ respectively, and $\mu = C$.

We will show that, for a parameter C fixed, the optimal solutions of problems (1.3) are also optimal solutions of problems of type (P_μ) for appropriate choices of γ_1, γ_2 and μ. Moreover, the parametric problems (1.3) (C being the parameter, varying in $(0,+\infty)$) are equivalent to the parametric problem (P_μ), with μ varying also in $(0,+\infty)$.

Proposition 2.1. *For $\mu > 0$, (P_μ), has an optimal solution. Moreover, if γ_1 is a strictly convex norm, then any pair (ω^1,β^1,ξ^1), (ω^2,β^2,ξ^2) of optimal solutions to (P_μ) generate hyperplanes $\mathcal{H}(\omega^1,\beta^1)$ and $\mathcal{H}(\omega^2,\beta^2)$ which are parallel.*

PROOF. An optimal solution exists since (P_μ) consists in finding the point in the polyhedron $\mathcal{X}$ closest to the origin, where closeness is measured by the norm $\|\cdot\|$ defined in (2.3).

Suppose now given (ω^1,β^1,ξ^1), (ω^2,β^2,ξ^2), both optimal solutions to (P_μ). In particular, (ω^1,ξ^1) and (ω^2,ξ^2) are both optimal for (2.2). Moreover, the convexity of $\mathcal{X}$ and the strict convexity of γ_1 imply that ω^1 and ω^2 are proportional. In other words, the corresponding hyperplanes $\mathcal{H}(\omega^1,\beta^1)$ and $\mathcal{H}(\omega^2,\beta^2)$ are parallel, as asserted. □

Lemma 2.1. *For any choice of γ_1, γ_2 and $\mu > 0$, any (ω,β,ξ), optimal to (P_μ), satisfies*

$$\xi \geq 0, \quad \xi \neq 0.$$

PROOF. Any (ω,β,ξ) feasible for (P_μ), must have $\xi \neq 0$, since, else, there would exist ω,β satisfying

$$\begin{aligned} \omega^\top x_i + \beta &\geq 1, \quad \forall i \in I_+, \\ \omega^\top x_i + \beta &\leq -1, \quad \forall i \in I_-, \end{aligned}$$

and thus the sets $\{x_i : i \in I_+\}$ and $\{x_i : i \in I_-\}$ would be linearly separable, which is a contradiction. Moreover, for any (ω,β,ξ), feasible for (P_μ), define $(\xi)^+ \in \mathbb{R}^I_+$ as

$$(\xi)^+_i = \max\{\xi_i, 0\}, \quad i \in I.$$

One has that $(\omega,\beta,(\xi)^+)$, is also feasible for (P_μ), and, by the assumption of strict monotonicity of γ_2 and (2.1), it has a strictly lower objective value unless $\xi \geq 0$. □

Lemma 2.2. *For any $\alpha_1, \alpha_2 \geq 1$ and $C > 0$, consider the problem*

$$(Q_{\alpha_1,\alpha_2,C}) \qquad \begin{aligned} &\min \gamma_1^{\alpha_1}(\omega) + C\gamma_2^{\alpha_2}(\xi), \\ &s.t.\ \omega^\top x_i + \beta + \xi_i \geq 1, \qquad \forall i \in I_+, \\ &\quad\ \omega^\top x_i + \beta - \xi_i \leq -1, \qquad \forall i \in I_-, \\ &\quad\ \omega \in \mathbb{R}^N,\ \beta \in \mathbb{R},\ \xi \in \mathbb{R}^I. \end{aligned}$$

Given (ω^*,β^*,ξ^*), $\omega^* \neq 0$, the following statements are equivalent:

1. *There exists $\mu > 0$ such that (ω^*,β^*,ξ^*) is optimal for (P_μ).*
2. *There exists $C > 0$ such that (ω^*,β^*,ξ^*) is optimal for $(Q_{\alpha_1,\alpha_2,C})$.*

PROOF. Let (ω^*,β^*,ξ^*), be optimal for (P_{μ^*}) for some $\mu^* > 0$.

Let $\mathcal{S}$ denote the feasible region of all (P_μ) and $(Q_{\alpha_1,\alpha_2,C})$. Since (ω^*,β^*,ξ^*) is optimal for (P_{μ^*}), one has that

$$(\omega^*,\beta^*,\xi^*) \in \mathcal{S},$$

and there exist $p_1 \in \partial\gamma_1(\omega^*)$, $p_2 \in \partial\gamma_2(\xi^*)$, such that

$$(p_1, 0, p_2) \in -N_{\mathcal{S}}(\omega^*,\beta^*,\xi^*), \tag{2.4}$$

where $N_{\mathcal{S}}(s)$ denotes the normal cone of $\mathcal{S}$ at s.

Since $(\omega^*, \beta^*, \xi^*)$ is optimal for (P_{μ^*}), we have by Lemma 2.1 that $\xi^* \neq 0$. Moreover, since $-N_{\mathcal{S}}(\omega^*, \beta^*, \xi^*)$ is a cone, (2.4) implies that

$$\left((\alpha_1 \gamma_1^{\alpha_1 - 1}(\omega^*)) p_1, 0, \left(\frac{\alpha_1 \gamma_1^{\alpha_1 - 1}(\omega^*) \mu^*}{\alpha_2 \gamma_2^{\alpha_2 - 1}(\xi^*)} \alpha_2 \gamma_2^{\alpha_2 - 1}(\xi^*) \right) p_2 \right) \in -N_{\mathcal{S}}(\omega^*, \beta^*, \xi^*).$$

Hence, defining C^* as

$$C^* = \frac{\alpha_1 \gamma_1 (\omega^*)^{\alpha_1 - 1} \mu^*}{\alpha_2 \gamma_2 (\xi^*)^{\alpha_2 - 1}},$$

we have that $C^* > 0$ (else, $\gamma_1(\omega^*) = 0$, thus $\omega^* = 0$, which contradicts the assumptions.) Moreover,

$$((\alpha_1 \gamma_1 (\omega^*)^{\alpha_1 - 1}) p_1, 0, (\alpha_2 \gamma_2 (\xi^*)^{\alpha_2 - 1}) C^* p_2) \in -N_{\mathcal{S}}(\omega^*, \beta^*, \xi^*),$$

and thus $(\omega^*, \beta^*, \xi^*)$ satisfies the optimality conditions of $(Q_{\alpha_1, \alpha_2, C^*})$, as asserted.

Conversely, let $(\omega^*, \beta^*, \xi^*)$ be optimal to $(Q_{\alpha_1, \alpha 2, C^*})$ for some $C^* > 0$.

The optimality conditions of $(\omega^*, \beta^*, \xi^*)$ for $(Q_{\alpha_1, \alpha_2, C^*})$ imply the existence of $p_1 \in \partial \gamma_1(\omega^*)$, $p_2 \in \partial \gamma_2(\xi^*)$ such that

$$(\alpha_1 \gamma_1 (\omega^*)^{\alpha_1 - 1} p_1, 0, \alpha_2 \gamma_2 (\xi^*)^{\alpha_2 - 1} C^* p_2) \in -N_{\mathcal{S}}(\omega^*, \beta^*, \xi^*). \tag{2.5}$$

Since, by assumption, $\omega^* \neq 0$, (2.5) implies that

$$\left(p_1, 0, \frac{\alpha_2 \gamma_2 (\xi^*)^{\alpha_2 - 1} C^*}{\alpha_1 \gamma_1 (\omega^*)^{\alpha_1 - 1}} p_2 \right) \in -N_{\mathcal{S}}(\omega^*, \beta^*, \xi^*),$$

thus $(\omega^*, \beta^*, \xi^*)$ is optimal for (P_{μ^*}) with μ^* given by

$$\mu^* = \frac{\alpha_2 \gamma_2 (\xi^*)^{\alpha_2 - 1} C^*}{\alpha_1 \gamma_1 (\omega^*)^{\alpha_1 - 1}}.$$ □

As a direct consequence of these Lemmata we obtain the following

Proposition 2.2. *Let $(\omega^*, \beta^*, \xi^*) \in \mathbb{R}^N \times \mathbb{R} \times \mathbb{R}^I$, $\omega^* \neq 0$, and $p \geq 1$. The following statements are equivalent:*

(1) *There exists $C > 0$ such that $(\omega^*, \beta^*, \xi^*)$ is optimal to (1.3).*
(2) *For $\gamma_1 = \|\cdot\|_2$, $\gamma_2 = \|\cdot\|_p$, there exists $\mu > 0$ such that $(\omega^*, \beta^*, \xi^*)$ is optimal to (P_μ).*

In other words, we can consider both (1.3) and (1.4) as particular instances of (P_μ).

3. Optimality Conditions

Since (P_μ) is a linearly-constrained convex problem, optimality conditions for (P_μ) can easily be derived as follows

Proposition 3.1. *Let $\mu > 0$ and let $(\omega^*, \beta^*, \xi^*)$ be feasible for (P_μ). The following statements are equivalent:*

(1) *$(\omega^*, \beta^*, \xi^*)$ is optimal for (P_μ).*

(2) *There exists* $\lambda \in \mathbb{R}^I$ *satisfying*

$$\begin{aligned} &\lambda \geq 0,\\ &\sum_{i \in I_+} \lambda_i x_i - \sum_{i \in I_-} \lambda_i x_i \in \partial\gamma_1(\omega^*),\\ &\sum_{i \in I_+} \lambda_i - \sum_{i \in I_-} \lambda_i = 0,\\ &\lambda \in \mu\partial\gamma_2(\xi^*),\\ &\lambda_i({\omega^*}^\top x_i + \beta^* + \xi_i^* - 1) = 0, \quad \forall i \in I_+,\\ &\lambda_i({\omega^*}^\top x_i + \beta^* - \xi_i^* + 1) = 0, \quad \forall i \in I_-. \end{aligned} \tag{3.1}$$

PROOF. Let $N_S(\omega^*, \beta^*, \xi^*)$ denote the normal cone of the feasible region at the feasible point $(\omega^*, \beta^*, \xi^*)$, i.e.,

$$N_S(\omega^*, \beta^*, \xi^*) = -\left\{ \left(\sum_{i \in I_+} \lambda_i x_i - \sum_{i \in I_-} \lambda_i x_i, \sum_{i \in I_+} \lambda_i - \sum_{i \in I_-} \lambda_i, \lambda \right) : \right.$$
$$\lambda \geq 0 \lambda_i({\omega^*}^\top x_i + \beta^* + \xi_i^* - 1) = 0, \quad \forall i \in I_+,$$
$$\left. \lambda_i({\omega^*}^\top x_i + \beta^* - \xi_i^* + 1) = 0, \quad \forall i \in I_- \right\}$$

$(\omega^*, \beta^*, \xi^*)$ is optimal for (P_μ) iff there exist $p \in \partial\gamma_1(\omega^*)$, $q \in \partial\gamma_2(\lambda)$ such that

$$(p, 0, \mu q) \in -N_S(\omega^*, \beta^*, \xi^*),$$

which is equivalent to (3.1). □

By Proposition 2.1, (P_μ) may have multiple optimal solutions. By convexity, the set of optimal solutions is closed and convex. Moreover, such set is contained in the intersection of two hyperplanes, as shown in the following result.

Proposition 3.2. *Let* $\mu > 0$. *There exist* $i_+ \in I_+$, $i_- \in I_-$ *such that any* (ω, β, ξ), *optimal to* (P_μ) *satisfies*

$$\begin{aligned} \omega^\top x_{i_+} + \beta + \xi_{i_+} &= 1,\\ \omega^\top x_{i_-} + \beta - \xi_{i_-} &= -1 \end{aligned} \tag{3.2}$$

PROOF. Let $(\omega^*, \beta^*, \xi^*)$ be optimal for (P_μ). By Lemma 2.1, $\xi^* \neq 0$. By Proposition 3.1, $\lambda \in \mu\partial\gamma_2(\xi^*)$, and thus $\lambda \neq 0$. Moreover, since $\sum_{i \in I_+} \lambda_i - \sum_{i \in I_-} \lambda_i = 0$, we have

$$\sum_{i \in I_+} \lambda_i \neq 0,$$
$$\sum_{i \in I_-} \lambda_i \neq 0.$$

Hence, there exist $i_+ \in I_+$, $i \in I_-$ such that

$$\lambda_{i_+} \neq 0, \quad \lambda_{i_-} \neq 0.$$

By the complementary slackness conditions,

$$\begin{aligned} {\omega^*}^\top x_{i_+} + \beta^* + \xi_{i_+}^* &= 1,\\ {\omega^*}^\top x_{i_-} + \beta^* - \xi_{i_-}^* &= -1. \end{aligned} \tag{3.3}$$

Hence, we have shown that, for any $(\omega^*, \beta^*, \xi^*)$, optimal solution to (P_μ), there exist $i_+ \in I_+$, $i_- \in I_-$ satisfying (3.3).

Let us show by contradiction that a common pair (i_+, i_-) can be chosen for *all* optimal solutions to (P_μ). Suppose for each $i \in I_+$ there exists $(\omega^i, \beta^i, \xi^i)$, optimal to (P_μ), such that

$$\omega^{i\top} x_i + \beta^i + \xi^i_i > 1,$$

and, since $(\omega^i, \beta^i, \xi^i)$ is feasible for (P_μ),

$$\omega^{i\top} x_k + \beta^i + \xi^i_k \geq 1, \quad \forall k \in I_+,$$
$$\omega^{i\top} x_k + \beta^i - \xi^i_k \leq -1, \quad \forall k \in I_-.$$

Then, the feasible solution $(\omega^*, \beta^*, \xi^*)$ defined as

$$(\omega^*, \beta^*, \xi^*) = \frac{1}{|I_+|} \sum_{i \in I_+} (\omega^i, \beta^i, \xi^i)$$

is also optimal to (P_μ), and violates (3.3). This is a contradiction, and hence, a common $i_+ \in I_+$ is active for all optimal solutions to (P_μ). A completely analogous argument can be used to show that a common $i_- \in I_-$ is active for all optimal solutions to (P_μ), and thus the result follows. □

The name Support Vector Machines refers to the fact that just a few points *support* the solution, in the sense that they determine the maximal margin hyperplane. The next Proposition shows that this result also holds for arbitrary norms: an optimal solution exists which is determined by at most $N+1$ points.

Proposition 3.3. *Let* $(\omega^*, \beta^*, \xi^*)$ *be optimal for* (P_μ). *Then, there exists* $\lambda^* \in \mathbb{R}^I$ *satisfying* (3.1), *such that the set* $I(\lambda^*) := \{i \in I : \lambda^*_i > 0\}$ *has cardinality at most* $N+1$.

PROOF. By Proposition 3.1, there λ^* satisfying (3.1) exists. Let $p \in \partial\gamma_1(\omega^*)^\circ$, defined as

$$p = \sum_{i \in I_+} \lambda^*_i x_i - \sum_{i \in I_-} \lambda^*_i x_i,$$

and consider the set $\mathcal{E} \subset \mathbb{R}^I$ of solutions of the system

$$\begin{aligned}
&\lambda \geq 0,\\
&\sum_{i \in I_+} \lambda_i - \sum_{i \in I_-} \lambda_i = 0,\\
&\sum_{i \in I_+} \lambda_i x_i - \sum_{i \in I_-} \lambda_i x_i = p,\\
&\lambda_i = 0, \quad \forall i \in I_+ \text{ s.t. } (x^*_i \omega^* + \beta^*) > 1,\\
&\lambda_i = 0, \quad \forall i \in I_- \text{ s.t. } (x^*_i \omega^* + \beta^*) < -1.
\end{aligned} \tag{3.4}$$

It follows that $\mathcal{E}$ is a nonempty polyhedron which contains no lines. Hence, its set $\text{ext}(\mathcal{E})$ of extreme points is nonempty.

Since the system of equations

$$\sum_{i \in I_+} \lambda_i - \sum_{i \in I_-} \lambda_i = 0,$$
$$\sum_{i \in I_+} \lambda_i x_i - \sum_{i \in I_-} \lambda_i x_i = p$$

has rank not greater than $N+1$, one has that any extreme point of $\mathcal{E}$ has, at most, $N+1$ coordinates at a strictly positive value. In other words,

$$|I(\lambda)|+ \leq N+1, \quad \forall \lambda \in \operatorname{ext}(\mathcal{E}),$$

and thus any $\lambda \in \operatorname{ext}(\mathcal{E})$ satisfies the conditions. □

We show now that a maximal margin solution is also optimal if we restrict the data set to the support vectors. Since the vector ξ in this reduced problem is in a different (smaller) dimension, a *reduced* norm $\overline{\gamma_2}$ is to be defined. Some technical results are needed first.

Lemma 3.1. *Let $\lambda, \xi \in \mathbb{R}^I_+$ such that $\lambda \in \partial\gamma_2(\xi)$. Then,*

$$\xi_i = 0, \quad \forall i \in I \text{ such that } \lambda_i = 0.$$

PROOF. By definition of subgradient,

$$\gamma_2(\eta) - \gamma_2(\xi) \geq \sum_{i \in I} \lambda_i(\eta_i - \xi_i), \quad \text{for all } \eta \in \mathbb{R}^I.$$

In particular, for ξ^*, defined as

$$\xi_i^* = \begin{cases} \xi_i, & \text{if } \lambda_i > 0, \\ 0, & \text{if } \lambda_i = 0, \end{cases}$$

one has that $0 \leq \xi^* \leq \xi$, thus

$$0 \geq \gamma_2(\xi^*) - \gamma_2(\xi) \geq 0.$$

If $\xi^* \neq \xi$, the first inequality above would be strict, which is a contradiction. Hence, $\xi^* = \xi$, and thus, $\xi = 0$ for all $i \in I$ with $\lambda_i = 0$. □

Lemma 3.2. *Given a norm γ in $\mathbb{R}^{p+q}$, define the norm in $\mathbb{R}^p$ $\overline{\gamma}$ as*

$$\overline{\gamma}(x) = \gamma(x, 0).$$

For any $u \in \mathbb{R}^p$ such that $(u,0) \in \partial\gamma(x,0)$, one has that $u \in \partial\overline{\gamma}(x)$.

For $J \subset I$, $J \neq \emptyset$, define $g_J : \mathbb{R}^J \to \mathbb{R}^I$,

$$\left(g_J(u)\right)_i = \begin{cases} u_i, & \text{if } i \in J, \\ 0, & \text{if } i \in I \setminus J \end{cases}$$

and the projection $\pi_J : \mathbb{R}^I \to \mathbb{R}^J$,

$$\left(\pi_J(u)\right)_i = u_i, \quad \forall i \in J.$$

With this notation, we have

Proposition 3.4. *Let* $(\omega^*, \beta^*, \xi^*)$ *be optimal for* (P_μ)*, and let* λ^* *satisfy the optimality conditions* (3.1)*. Let* $I^* = \{i \in I : \lambda_i^* > 0\} \neq \emptyset$*. Define the norm* $\overline{\gamma}_2$ *in* $\mathbb{R}^{I^*}$ *as*

$$\overline{\gamma}_2(u) = \gamma_2\big(g_{I^*}(u)\big), \quad \forall \in \mathbb{R}^{I^*}.$$

Then, $\big(\omega^*, \beta^*, \pi_{I^*}(\xi^*)\big)$ *is optimal for*

$$\begin{aligned} &\min \gamma_1(\omega) + \mu\overline{\gamma}_2(\xi) \\ &s.t.\ \omega^\top x_i + \beta + \xi_i \geq 1, \qquad \forall i \in I^* \cap I_+, \\ &\quad\ \ \omega^\top x_i + \beta - \xi_i \leq -1, \qquad \forall i \in I^* \cap I_- \end{aligned} \tag{3.5}$$

PROOF. Let $\overline{\lambda} = \pi_{I^*}(\lambda^*)$. By the optimality conditions at λ^*, we have:

$$\begin{aligned} \overline{\lambda} &\geq 0, \\ \sum_{i \in I^* \cap I_+} \overline{\lambda}_i x_i - \sum_{i \in I^* \cap I_-} \overline{\lambda}_i x_i &\in \partial\gamma_1(\omega^*), \\ \sum_{i \in I^* \cap I_+} \overline{\lambda}_i - \sum_{i \in I^* \cap I_-} \overline{\lambda}_i &= 0, \\ \overline{\lambda}_i({\omega^*}^\top x_i + \beta^* + \xi_i^* - 1) = 0, &\quad \forall i \in I^* \cap I_+, \\ \overline{\lambda}_i({\omega^*}^\top x_i + \beta^* - \xi_i^* + 1) = 0, &\quad \forall i \in I^* \cap I_-. \end{aligned} \tag{3.6}$$

Moreover, since

$$\lambda^* \in \mu\partial\gamma_2(\xi^*),$$

and $\lambda^*, \xi^* \geq 0$, Lemma 3.1 implies that

$$\xi_i^* = 0, \quad \forall i \in I \setminus I^*.$$

In other words,

$$\xi^* = g_{I^*}\big(\pi_{I^*}(\xi^*)\big),$$

and thus, by Lemma 3.2,

$$\frac{1}{\mu}\overline{\lambda} \in \partial\overline{\gamma}_2\big(\pi_{I^*}(\xi^*)\big),$$

and thus the optimality conditions for Problem (3.5) hold, and the result follows. □

Proposition 3.5. *Let* $(\omega^*, \beta^*, \xi^*)$ *satisfy*

$$\begin{aligned} \omega^\top x_i + \beta + \xi_i \geq 1, &\quad \forall i \in I_+, \\ \omega^\top x_i + \beta - \xi_i \leq -1, &\quad \forall i \in I_-. \end{aligned}$$

The following statements are equivalent:

(1) *There exists* $\mu > 0$ *such that* $(\omega^*, \beta^*, \xi^*)$ *is optimal for* (P_μ)*.*

(2) *There exists* $\lambda \in \mathbb{R}^I$ *satisfying*

$$
\begin{aligned}
\sum_{i \in I_+} \lambda_i x_i - \sum_{i \in I_-} \lambda_i x_i \in \partial \gamma_1(\omega^*), \\
\sum_{i \in I_+} \lambda_i - \sum_{i \in I_-} \lambda_i = 0, \\
\gamma_2(\xi^*) \leq \frac{\lambda^\top \xi^*}{\gamma_2^\circ(\lambda)}, \\
\lambda_i({\omega^*}^\top x_i + \beta^* + \xi_i^* - 1) = 0, \quad \forall i \in I_+, \\
\lambda_i({\omega^*}^\top x_i + \beta^* - \xi_i^* + 1) = 0, \quad \forall i \in I_-, \\
\lambda \geq 0.
\end{aligned}
\tag{3.7}
$$

PROOF. Assume 1 holds. Hence, $(\omega^*, \beta^*, \xi^*)$ is optimal for some (P_{μ^*}), and thus $(\omega^*, \beta^*, \xi^*)$ satisfies (3.1) for such μ^*. Hence,

$$\frac{1}{\mu^*}\lambda \in \partial \gamma_2(\xi^*),$$

thus

$$\gamma_2\left(\frac{1}{\mu^*}\lambda\right)^\circ \leq 1, \tag{3.8}$$

and

$$\gamma_2(\xi^*) = \left(\frac{1}{\mu^*}\lambda\right)^\top \xi^*,$$

and thus, by (3.8),

$$1 \geq \gamma_2\left(\frac{1}{\mu^*}\lambda\right)^\circ = \gamma_2\left(\frac{\gamma_2(\xi^*)}{\lambda^\top \xi^*}\lambda\right)^\circ,$$

thus

$$\gamma_2(\xi^*) \leq \frac{\lambda^\top \xi^*}{\gamma_2(\lambda)^\circ}.$$

The remainder conditions are those of (3.1), and thus part 2 follows.

Conversely, let $(\omega^*, \beta^*, \xi^*)$ satisfy (3.7). Defining μ^* as

$$\mu^* = \frac{\lambda^\top \xi^*}{\gamma_2(\xi^*)},$$

one has that (3.1) follows, thus $(\omega^*, \beta^*, \xi^*)$ is optimal for (P_{μ^*}). □

4. A Characterization of Degeneracy

The coefficients $(\omega^*, \beta^*, \xi^*)$ will be used to classify future entries, given by $x \in \mathbb{R}^N$. In other words, we solve (P_μ) for a certain μ, obtain then $(\omega^*, \beta^*, \xi^*)$, and from its (ω^*, β^*) components we classify x according to (1.2).

Although, by Lemma 2.1, $(\omega^*, \beta^*, \xi^*) = (0, 0, 0)$, cannot be optimal (in fact it cannot be feasible due to the non-separability assumption), it is possible to come obtain $\omega^* = 0, \beta^* = 0, \xi^* \neq 0$, as optimal solution of (P_μ). Such solution does not allow us to classify future entries x according to (1.2). We will call this the *degenerate case*, discussed in [12] for more particular instances.

Lemma 4.1. *If (P_μ) has an optimal solution $(\omega^*, \beta^*, \xi^*)$ with $\omega^* = 0, \beta^* = 0$, then (ω^*, β^*, e) is also optimal for (P_μ), where $e \in \mathbb{R}^I$ is the vector with all its components equal to 1.*

PROOF. Consider the optimization problem

$$\begin{aligned} &\min \gamma_2(\xi) \\ &\text{s.t. } \eta_i \geq 1, \quad \forall i \in I. \end{aligned} \tag{4.1}$$

Since, by assumption, γ_2 is monotonic, $e \in \mathbb{R}^I$ is optimal to (4.1), thus

$$\gamma_1(0) + \mu\gamma_2(e) \leq \gamma_1(0) + \mu\gamma_2(\xi^*),$$

and, since $(0, 0, e)$ is also feasible for (P_μ), we obtain that it is also optimal for (P_μ). □

Optimality conditions at degenerate solutions follow from the previous results:

Proposition 4.1. *Given $\mu > 0$, the following statements are equivalent:*

(1) *There exists $\xi^* \in \mathbb{R}^I$ such that $(0, 0, \xi^*)$ is optimal for (P_μ).*
(2) *There exists $\lambda \in \mathbb{R}^I$ such that*

$$\begin{gathered} \lambda \geq 0, \\ \sum_{i \in I_+} \lambda_i - \sum_{i \in I_-} \lambda_i = 0, \\ \lambda \in \partial\gamma_2(e), \\ \gamma_1^\circ\left(\sum_{i \in I_+} \lambda_i x_i - \sum_{i \in I_-} \lambda_i x_i\right) \leq \frac{1}{\mu}. \end{gathered}$$

PROOF. By Lemma 4.1, 1 is equivalent to state that $(0, 0, e)$ is optimal for (P_μ). Optimality conditions at $(0, 0, e)$ given in (3.1) are easily seen to be equivalent to those given in condition 2. □

Proposition 4.2. *The following statements are equivalent:*

(1) *For all $\mu > 0$ there exists $\xi \in \mathbb{R}^I$ such that $(0, 0, \xi)$ is an optimal solution for (P_μ)*
(2) *There exists $\lambda \in \mathbb{R}^I$ satisfying*

$$\begin{gathered} \lambda \geq 0, \\ \sum_{i \in I_+} \lambda_i - \sum_{i \in I_-} \lambda_i = 0, \\ \lambda \in \partial\gamma_2(e), \\ \sum_{i \in I_+} \lambda_i x_i - \sum_{i \in I_-} \lambda_i x_i = 0. \end{gathered}$$

PROOF. By Proposition 4.1, 2 implies 1. To show the converse, consider the problem

$$(4.2)\qquad \begin{aligned} &\min \gamma_1^\circ\left(\sum_{i\in I_+}\lambda_i x_i - \sum_{i\in I_-}\lambda_i x_i\right)\\ &\text{s.t. } \lambda\in\partial\gamma_2(e)\\ &\quad\sum_{i\in I_+}\lambda_i - \sum_{i\in I_-}\lambda_i = 0\\ &\quad\lambda\geq 0. \end{aligned}$$

Problem (4.2) is feasible by assumption. Let μ^* be its optimal value, and let us show that $\mu^* = 0$. Indeed, if it were not the case, for any μ, $0 < \mu < \mu^*$, we would have that the system

$$\begin{gathered} \gamma_1\left(\sum_{i\in I_+}\lambda_i x_i - \sum_{i\in I_-}\lambda_i x_i\right) \leq \mu,\\ \lambda\in\partial\gamma_2(e),\\ \sum_{i\in I_+}\lambda_i - \sum_{i\in I_-}\lambda_i = 0,\\ \lambda\geq 0 \end{gathered}$$

is unfeasible, thus $(0, 0, e)$ could not be optimal to $(P_{1/\mu})$, contradicting 1. Hence, $\mu^* = 0$. Since the feasible set of (4.2) is compact, μ^* is attained at some λ^*, and thus such λ^* satisfies the assumptions. □

A natural assumption for γ_2 is *symmetry*, i.e.,

$$\gamma\big((\xi_i)_{i\in I}\big) = \gamma\big((\xi_{\sigma(i)})_{i\in I}\big) \text{ for any } \sigma, \text{ permutation of } I.$$

Under this assumption, a more detailed characterization of degeneracy is possible. Indeed, one has

Lemma 4.2. *Let γ_2 by symmetric. Then, there exists $\tau\in\mathbb{R}$ such that $\tau e\in\partial\gamma_2(e)$.*

PROOF. Let $p = (p_i)_{i\in I}\in\partial\gamma_2(e)$. For any σ, permutation in I, we have for each $u\in I$,

$$\begin{aligned} \gamma_2(u) - \gamma_2(e) &= \gamma_2\big((u_{\sigma^{-1}(i)})_{i\in I}\big) - \gamma_2(e)\\ &\geq \sum_{i\in I} p_i(u_{\sigma^{-1}(i)} - 1)\\ &= \sum_{i\in I} p_{\sigma(i)}(u_i - 1), \end{aligned}$$

thus the vector $p^\sigma = (p_{\sigma(i)})_{i\in I}$ is also subgradient of γ_2 at e. By convexity of the subdifferential, the vector $p^* := \frac{1}{|I|!}\sum_\sigma p^\sigma\in\partial\gamma_2(e)$, and, by construction, $p^* = \tau e$ for some τ. Hence, p^* satisfies the conditions. □

Proposition 4.3. *Let γ_2 be differentiable at e and symmetric. The following statements are equivalent:*

(1) *For all $\mu > 0$ there exists $\xi\in\mathbb{R}^I$ such that $(0, 0, \xi)$ is optimal for (P_μ).*

(2) $|I_+| = |I_-|$ *and* $\sum_{i \in I_+} x_i = \sum_{i \in I_-} x_i$.

PROOF. By Proposition 4.2, statement 1 is equivalent to feasibility of the following system

$$\begin{aligned} \lambda &\geq 0, \\ \sum_{i \in I_+} \lambda_i - \sum_{i \in I_-} \lambda_i &= 0, \\ \lambda &= \nabla \gamma_2(e), \\ \sum_{i \in I_+} \lambda_i x_i - \sum_{i \in I_-} \lambda_i x_i &= 0. \end{aligned} \tag{4.3}$$

By Lemma 4.2, $\nabla \gamma_2(e) = \delta e$ for some $\delta \neq 0$, and thus (4.3) can also be written as

$$\begin{aligned} \lambda &\geq 0, \\ \sum_{i \in I_+} \lambda_i - \sum_{i \in I_-} \lambda_i &= 0, \\ \lambda_i &= \delta \, \forall i \in I, \\ \sum_{i \in I_+} \lambda_i x_i - \sum_{i \in I_-} \lambda_i x_i &= 0, \end{aligned} \tag{4.4}$$

which is equivalent to the conditions in part 2. □

References

1. K. Bennett and E. Bredensteiner, *Duality and geometry in* SVM *classifiers*, Proc. 7th Internat. Conf. on Machine Learning (P. Langley, ed.), Morgan Kaufmann, San Francisco, 2000, pp. 57–64.
2. K. P. Bennett and C. Campbell, *Support vector machines: Hype or Hallelujah?*, SIGKDD Explorations **2** (2000), 1–13.
3. C. Burges, *A tutorial on support vector machines for pattern recognition*, Data Mining and Knowledge Discovery **2** (1998), 121–167.
4. E. Carrizosa, *Deriving weights in multiple-criteria decision making with support vector machines*, TOP **14** (2006), 399-424.
5. C. Cortes and V. Vapnik, *Support vector networks*, Machine Learning **20** (1995), 1–25.
6. N. Cristianini and J. Shawe-Taylor, *An introduction to support vector machines*, Cambridge University Press, Cambridge MA, 2000.
7. C. R. Johnson and P. Nylen, *Monotonicity properties of norms*, Linear Algebra Appl. **148** (1991), 43–58.
8. O. L. Mangasarian, *Arbitrary-norm separating plane*, Oper. Res. Lett. **24** (1999), 15–23.
9. ———, *Misclassification minimization*, J. Global Optim. **5** (1994), 309–323.
10. ———, *Generalized support vector machines*, Advances in Large Margin Classifiers (A. Smola, P. Barlett, B. Schölkopf, and D. Schuurmans, eds), MIT Press, Cambridge, MA, 2000, pp. 135–146.
11. J. M Moguerza and A. Muñoz, *Support vector machines with applications*, Statist. Sci. **21** (2006) 322–336.
12. R. Rifkin, M. Pontil, and A. Verri, *A note on support vector machine degeneracy*, AI Memo 1661, MIT, 1999; `http://five-percent-nation.mit.eduPersonalPages/rif/supvecdegen.ps`.
13. J. Zhu, S. Rosset, T. Hastie, and R. Tibshirani, L_1 *norm support vector machines*, Technical report, Standford University, 2003; `http://www-stat.stanford.edu/~hastie/pub.htm`.

FACULTAD DE MATEMÁTICAS, UNIVERSIDAD DE SEVILLA, TARFIA S/N, 41012 SEVILLA, SPAIN.
E-mail address: `ecarrizosa@us.es`

Centre de Recherches Mathématiques
CRM Proceedings and Lecture Notes
Volume **45**, 2008

0-1 Semidefinite Programming for Graph-Cut Clustering: Modelling and Approximation

Huarong Chen and Jiming Peng

Abstract. Graph cut optimization provides a plausible approach for the data clustering problem. In this paper, we present a unified framework for some graph-cut optimization problems arising from cluster analysis based on a new optimization model: 0-1 semidefinite programming (0-1 SDP). Secondly, we consider the issue of how to solve the underlying 0-1 SDP problem. We consider two approximation methods based on singular value decomposition (SVD) of the coefficient matrix and the SVD of the coefficient matrix in a projected subspace, respectively and prove that both algorithms can provide a 2-approximation to the original clustering problem. The complexity of these approximation algorithms are discussed and preliminary numerical results based on the new algorithm are reported.

1. Introduction

In general, clustering involves partitioning a given data set into subsets based on the closeness or similarity among the data. Clustering is one of major issues in data mining and machine learning with many applications arising from different disciplines including text retrieval, pattern recognition and web mining[15, 18]. There are many kinds of clustering problems and algorithms, resulting from various choices of measurements used in the model to measure the similarity/dissimilarity among entities in a data set. For a comprehensive introduction to the topic, we refer to the book [15, 18], and for more recent results, see survey papers [9] and [16].

In the present paper, we are mainly concerned about clustering algorithms based on graph model. Given a set of points $\mathcal{V} = \{v_i \in \Re^m : i = 1, \dots, n\}$, in clustering we first define a similarity matrix $W = [w_{ij}]$ where $w_{ij} = \phi(v_i, v_j)$ for some kernel function $\phi(\cdot)$, which can be further interpreted as the weight of the edge (v_i, v_j) in a graph with vertex set $\mathcal{V}$. We then solve a graph-cut optimization problem to cluster the data set.

2000 *Mathematics Subject Classification.* 90C11, 90C27.

This work was supported by the NSERC discovery grant # RPG 249635-02 of Canada and a PREA award in Ontario. It is also supported by the MITACS project "New Interior Point Methods and Software for Convex Conic-Linear Optimization and Their Application to Solve VLSI Circuit Layout Problems".

This is the final form of the paper.

There are many clustering algorithms depending on various choices of the similarity matrix W and the graph-cut model we solve [26, 1, 32, 21]. Among others, we shall focus on the following three specific clustering algorithms: (1) The so-called normalized cut for image segmentation introduced by Shi and Malik [26] and later investigated by Xing and Jordan [32]; (2) The classical K-means clustering based on minimum sum of squared errors [19]; (3) The kernel (weighted) K-means by Dhillon, Guan and Kulis [7]. We point out that the interrelations among these methods were first noted by Bach and Jordan and further explored in [7]. In [29], Verma and Meila compared several clustering algorithms that included the above mentioned methods and introduced the concept of the so-called perfect set for which the algorithm works well. The main purpose of this work is to present a unified framework for the above-mentioned methods and investigate how to solve the unified optimization model. However, as we shall see later, our unified framework covers not only these few algorithms, but also some other clustering scenarios as well.

This paper is inspired by our recent papers [24] where we reformulated the classical K-means clustering as the so-called 0-1 SDP and proposed a 2-approximation method based on SVD in a projected subspace to attack the underlying 0-1 SDP model. It has been observed that in addition to its excellent theoretical properties and high computational efficiency, the algorithm in [24] can be easily extended to deal with the so-called balanced clustering where the cardinality of the clusters is bounded.

The idea to reformulate the classical K-means clustering as a trace minimization problem of matrix argument can be dated back to [11] where the authors owed the idea to an anonymous referee. In the machine learning community, Zha *et al.* [33] first noted that the optimization problem in the classical K-means can be reformulated as a quadratic optimization problem subject to some nonlinear constraints. Further, the authors of [33] showed that a relaxed version of the quadratic optimization derived from K-means clustering can be solved by using singular value composition of the data matrix. Similar discussions for normalized cuts and kernel K-means can be found in [1, 32]. Since then, many results have been reported on how to use the eigenvectors of the affinity matrix to cluster the data set [6, 8, 33, 21], which is typically referred to as spectral clustering in the literature. A similar idea has also been adopted in the PCA-based approach for data analysis where we use the eigenvectors of the data matrix (or principal components) to reduce the dimension of the data, and then analyze the reduced data in lower dimension [17].

There are several issues need to be addressed in such an approach. First, how to design a rounding procedure to extract a feasible solution to the original clustering problem from the solution of the relaxed problem? Most rounding procedures in the literature apply K-means-type heuristics to a new data set derived from the solution of the relaxed problem. There are several ways to construct the new data set in the reduced space and all of them are based the SVD of the solution matrix of the relaxed problem. One way is to stack a certain number (typically k) of the unit eigenvectors corresponding to the largest eigenvalues to form a basis of the new data matrix. Every row of the basis matrix is then cast as a vector in a finite dimensional space. The algorithms in [1, 32] is based on such a strategy. Another way is to use various scaling technique to the basis matrix to get a new data set. For example, the algorithms in [12, 21] normalize every row of the basis matrix to have unit length. In [8, 24], the authors suggested to scale every column (which a unit eigenvector) of the basis matrix by the squared root of their corresponding eigenvalues. The last method can also be interpreted as an approximation to the original matrix in a lower dimension [8, 24].

The second issue in the SDP relaxation-based algorithms is how to estimate the quality of the solution obtained from the reduced problem. We note that most of the above-mentioned works did not provide a very satisfactory answer to the question except for [8], where Drineas *et al.* showed that their algorithm can give a 2-approximation to the original clustering problem. The algorithm in [24] follows a similar vein as the algorithm in [8] in the sense that both algorithms use the SVD of the coefficient matrix to reduce the dimension of the space and both algorithms can provide a 2-approximation solution to original problem. However, the motivation behind these algorithms are quite different. In [8], PCA is employed to reduce the dimension of data so that the reduced problem can be solved relatively easily, while the algorithm in [24] is based on the relaxation of the 0-1 SDP model. Note that there are several different ways to relax the 0-1 SDP model. On the other hand, the dimension of the working space in [24] is smaller than that in [8]. This simplifies the algorithm and improves its efficiency.

In this paper we first elaborate on how to characterize the graph-cut optimization problems precisely by means of matrix arguments. In particular, we show that all the three methods can be embedded into the so-called 0-1 semidefinite programming (SDP), which can be further relaxed to polynomially solvable linear programming (LP) and SDP. The unified framework not only opens novel avenues for these graph-cut optimization problems, but also provides an insightful analysis of spectral clustering. For example, the first algorithm we suggested in this paper extends the algorithm in [8] to the scenarios of the weighted (kernel) K-means clustering and the normalized cut problem. Our analysis shows that the algorithm can provide a 2-approximation to the original problem assuming that the global solution of the subproblem can be found. To the best of our knowledge, this is the first approximation with guaranteed quality to the normalized-cut problem reported in the literature. In the special scenario of the normalized 2-cut problem (which is still NP-hard), the proposed Algorithm 2 in the present paper can be cast as a refinement of the approach in [26], where Shi and Malik suggested to use the eigenvector corresponding to the second largest eigenvalue of the underlying coefficient matrix and then solve the normalized 2-cut problem in one dimensional space. However, in [26], the authors only employed some simple heuristics to solve the underlying optimization problem in one-dimensional space and thus could not guarantee that the obtained solution is the global solution of the underlying subproblem, while in the present work, by using the relation between the weighted K-means clustering and the normalized cut problem, we are able to develop a very simple algorithm that can find the global solution of the normalized 2-cut problem in one-dimensional space. This refinement further allows us to obtain a good approximation to the original graph-cut problem. Moreover, for the normalized 2-cut problem, our analysis shows that both Algorithms 1 and 2 in the present paper will provide identical partitioning for the original input data. This further implies that the algorithm in [1, 32] can also provide a 2-approximation to the underlying normalized 2-cut problem.

The paper is organized as follows. In Section 2, we show that all the three methods for spectral clustering can be modelled as 0-1 SDP, which allows convex relaxation such as SDP and LP. In Section 3, we consider two approximation algorithms for solving our 0-1 SDP model. The first algorithm follows a similar vein as the popular spectral clustering that uses the eigenvectors of the affinity matrix to cluster the data set. The second one is based on the eigenvectors of a projection of the affinity matrix, which can be viewed as a slight improvement of the first algorithm. Approximate ratios for both algorithms between the obtained solution and the global solution of the original spectral clustering are estimated. Preliminary experiments are reported in Section 4. Finally we close the paper by some remarks.

2. 0-1 SDP for Graph-cut Clustering

In this section, we present a unified framework for several clustering problems under the umbrella of 0-1 SDP model. The section has three parts. In the first part, we briefly describe SDP and 0-1 SDP. In the second part, we review the 0-1 SDP model for the classical K-means clustering and kernel K-means. In the last subsection, we establish the equivalence between 0-1 SDP and normalized k-cut problem.

2.1. 0-1 semidefinite programming. In general, SDP refers to the problem of minimizing (or maximizing) a linear function over the intersection of a polytope and the cone of symmetric and positive semidefinite matrices. The canonical SDP takes the following form

$$\text{(SDP)} \begin{cases} \min & \operatorname{Tr}(WZ) \\ \text{s.t.} & \operatorname{Tr}(B_i Z) = b_i \quad \text{for } i = 1, \ldots, m. \\ & Z \succeq 0 \end{cases} \tag{2.1}$$

Here Tr(.) denotes the trace of the matrix, and $Z \succeq 0$ means that Z is positive semidefinite. If we further require that the eigenvalues of the matrix argument in the above model take values 0 or 1, which implies Z is a projection matrix satisfying the relation $Z^2 = Z$[1], then we end up with the following problem

$$\text{(0-1 SDP)} \begin{cases} \min & \operatorname{Tr}(WZ) \\ \text{s.t.} & \operatorname{Tr}(B_i Z) = b_i \qquad \text{for } i = 1, \ldots, m. \\ & Z^2 = Z, \ Z = Z^T \end{cases} \tag{2.2}$$

We call it 0-1 SDP owing to the similarity of the constraint $Z^2 = Z$ to the classical 0-1 requirement in integer programming. It is easy to see that the SDP model is a relaxation of the 0-1 SDP model.

2.2. 0-1 SDP model for (weighted) K-means and Kernel K-means clustering. For a given data set $\mathcal{V} = \{v_i \in \Re^m : i = 1, \ldots, n\}$, the optimization problem involved in the weighted K-means clustering can be defined as follows

$$\min_{c_1,\ldots,c_k} \sum_{i=1}^{n} d_i \min\{\|v_i - c_1\|^2, \ldots, \|v_i - c_k\|^2\}$$

[1]Here Z^2 is the product of matrices, not componentwise.

where d_i is the weight associated with v_i. Let $X = [x_{ij}] \in \Re^{n\times k}$ be the assignment matrix defined by

$$x_{ij} = \begin{cases} 1 & \text{if } v_i \text{ is assigned to } C_j; \\ 0 & \text{otherwise.} \end{cases} \tag{2.3}$$

We can rewrite the optimization problem for weighted K-means as

$$\min_{x_{ij}} \sum_{j=1}^{k}\sum_{i=1}^{n} d_i x_{ij} \left\| v_i - \frac{\sum_{l=1}^{n} x_{lj} d_l v_l}{\sum_{l=1}^{n} d_l x_{lj}} \right\|^2 \tag{2.4}$$

$$\text{s.t.} \quad \sum_{j=1}^{k} x_{ij} = 1, \quad (i = 1, \ldots, n) \tag{2.5}$$

$$\sum_{i=1}^{n} x_{ij} \geq 1, \quad (j = 1, \ldots, k) \tag{2.6}$$

$$x_{ij} \in \{0, 1\}, \quad (i = 1, \ldots, n;\ j = 1, \ldots, k) \tag{2.7}$$

where the constraint (2.5) ensures that each point v_i is assigned to precisely one cluster, and (2.6) ensures that there are exactly k clusters.

Let us define

$$Z = X(X^T X)^{-1} X^T.$$

If all the weights are equal, i.e., $d_i = 1$ for $i \in \{1, \ldots, n\}$, then problem (2.4) can be stated as the following special form of 0-1 SDP

$$\begin{aligned} &\min \operatorname{Tr}\left(W(I - Z)\right) \\ &Ze = e, \quad \operatorname{Tr}(Z) = k, \\ &Z \geq 0, \quad Z = Z^T, \quad Z^2 = Z, \end{aligned} \tag{2.8}$$

as proved by the following result from [23, 24].

Proposition 2.1. *Finding a global solution of the classical* K*-means clustering equals to solving the* 0-1 SDP *problem* (2.8) *where* $w_{ij} = v_i^T v_j$.

We remark that the 0-1 SDP model (2.8) is essentially equivalent to the quadratic optimization model for K-means clustering in [33]. However, the new 0-1 SDP model can help us find various relaxations of the original problem and open new avenues for solving the problem.

Following a similar chain of arguments in the proof of [23, Theorem 2.1], we have

Proposition 2.2. [2] *Finding a global solution of the weighted* K*-means clustering equals to solving the following* 0-1 SDP *problem*

$$\begin{aligned} &\min \operatorname{Tr}\left(D^{\frac{1}{2}} W D^{\frac{1}{2}} (I - Z)\right) \\ &Z d^{\frac{1}{2}} = d^{\frac{1}{2}}, \quad \operatorname{Tr}(Z) = k, \\ &Z \geq 0, \quad Z = Z^T, \quad Z^2 = Z, \end{aligned} \tag{2.9}$$

where $d^{\frac{1}{2}}$ *is the vector whose ith element is the squared root of* d_i, $D^{\frac{1}{2}} = \operatorname{diag}(d^{\frac{1}{2}})$ *and* $w_{ij} = v_i^T v_j$.

---

[2]The proposition also follows from Theorem 2.1 by using the connection between the weighted K-means clustering and the normalized k-cut problem as shown in [32].

We mention that the 0-1 SDP model can also be extended to cover the scenario of the so-called kernel k-means clustering where the kernel matrix is defined via the kernel functions such as the Gaussian and polynomial kernel functions.

$$w_{ij} = \phi(v_i, v_j) = \exp^{-(\|v_i - v_j\|^2)/\sigma}, \quad \sigma > 0, \tag{2.10}$$

$$w_{ij} = (v_i^T v_j + c_0)^l. \tag{2.11}$$

The 0-1 SDP model can also be applied to the so-called balanced clustering [4, 24] where the number of points in every cluster is restricted and the so-called semi-supervised clustering [2] where prior knowledge needs to be incorporated into the model.

2.3. 0-1 SDP model for normalized K-cut problem. Recently, the normalized k-cut problem received much attention in the machine learning community, and many interesting results about such an approaches have been reported [7, 12, 20, 21, 26, 31, 32]. In particular, Xing and Jordan [32] first noted the connection between the weighted K-means and the normalized cut problem. They also discussed an SDP relaxation for the normalized k-cut problem. In [7], Dhillon *et al.* used the link between the weighted K-means and the normalized cut problem to develop a simple heuristics for the later one.

We next show that the normalized k-cut problem can be embedded into the 0-1 SDP model. Let us first recall the exact model for the normalized k-cut problem [32]. Let W be the affinity matrix defined by (2.10), $X = [x_{ij}] \in \Re^{n \times k}$ be the assignment matrix and e be the all 1 vector in suitable space. Let us define $\mathcal{F}_k$ be the set satisfying

$$\mathcal{F}_k = \{X : Xe^k = e^n, x_{ij} \in \{0, 1\}\}.$$

Let $d = We^n$ and $D = \text{diag}(d)$. The exact model for the normalized k-cut problem in [26, 32] can be rewritten as

$$\min_{X \in \mathcal{F}_k} \text{Tr}(D^{-1}W - X(X^T DX)^{-1}X^T W). \tag{2.12}$$

If we define[3]

$$Z = D^{\frac{1}{2}} X(X^T DX)^{-1} X^T D^{\frac{1}{2}}, \tag{2.13}$$

then we have

$$Z^2 = Z, \quad Z^T = Z, \quad Z \geq 0, \quad Zd^{\frac{1}{2}} = d^{\frac{1}{2}}, \quad \text{Tr}(Z) = k.$$

Therefore, we can rewrite the objective function in (2.12) as a linear function via using matrix argument. By adding the above conditions for the matrix Z, we obtain the following 0-1 SDP problem:

$$\min \text{Tr}\left(D^{-\frac{1}{2}} W D^{-\frac{1}{2}} (I - Z)\right) \tag{2.14}$$

$$Zd^{\frac{1}{2}} = d^{\frac{1}{2}}, \quad \text{Tr}(Z) = k, \tag{2.15}$$

$$Z \geq 0, \ Z^2 = Z, \ Z = Z^T. \tag{2.16}$$

We have

---

[3]We mention that in [32], the authors introduced the matrix $H = X(X^T DX)^{-\frac{1}{2}}$ and expressed the objective function in the normalized-cut as a quadratic function of H. It is easy to see that $Z = HH^T$.

Theorem 2.1. *Finding a global solution of problem* (2.12) *equals to solving the* 0-1 SDP *problem* (2.14).

PROOF. To prove the theorem, we first note that any feasible solution for problem (2.12) can be easily transferred into a feasible solution of problem (2.14). It remains to show that from any feasible solution of problem (2.14), we can construct a feasible assignment matrix X for problem (2.12).

Suppose that Z is a feasible solution of problem (2.14). We can define a matrix $\overline{Z} = D^{-\frac{1}{2}} Z D^{\frac{1}{2}}$. It follows from the constraints (2.15)–(2.16) that

$$\overline{Z}^2 = \overline{Z}, \quad \overline{Z}e = e, \quad \overline{Z} \geq 0. \tag{2.17}$$

Let

$$\bar{z}_{i_0 j_0} = \arg\max \bar{z}_{ij}, \quad \mathcal{J}_0 = \{i : \bar{z}_{i j_0} > 0\}.$$

Since $\overline{Z}^2 = \overline{Z}$ and $\sum_{j=1}^{n} \bar{z}_{i_0 j} = 1$, it must hold

$$\bar{z}_{i j_0} = \bar{z}_{i_0 j_0}, \quad \forall i \in \mathcal{J}_0,$$
$$\sum_{j \in \mathcal{J}_0} \bar{z}_{i_0 j} = 1, \quad \bar{z}_{i_0 j} = 0 \quad \forall j \notin \mathcal{J}_0.$$

Recall the definition of the matrix $\overline{Z}$, we can decompose the matrix $\overline{Z}$ into a bock matrix with the following structure

$$Z = \begin{pmatrix} \overline{Z}_{\mathcal{J}_0 \mathcal{J}_0} & 0 \\ 0 & \overline{Z}_{\overline{\mathcal{J}}_0 \overline{\mathcal{J}}_0} \end{pmatrix} \tag{2.18}$$

where $\overline{\mathcal{J}}_0 = \{i : i \notin \mathcal{J}_0\}$. Now we claim that $\overline{Z}_{\mathcal{J}_0 \mathcal{J}_0}$ is a submatrix with rank 1 for which all the elements in any column are equivalent. To see this, let us choose any column from the submatrix $\overline{Z}_{\mathcal{J}_0 \mathcal{J}_0}$ and consider the minimum element in that column, i.e., for a fixed $j \in \mathcal{J}_0$,

$$\bar{z}_{i_1 j} = \arg\min_{i \in \mathcal{J}_0} \bar{z}_{ij}.$$

From the relation (2.17) we have

$$\bar{z}_{i_1 j} = \sum_{k \in \mathcal{J}_0} \bar{z}_{i_1 k} \bar{z}_{kj} \geq \bar{z}_{i_1 j} \sum_{k \in \mathcal{J}_0} \bar{z}_{i_1 k} = \bar{z}_{i_1 j},$$

and the equality holds if and only if all the element in the column are equivalent. Since $\overline{Z}_{\mathcal{J}_0 \mathcal{J}_0}$ is a rank one matrix and the sum of each row equals 1, its trace also equals 1.

From the above discussion we can see that we can put all the points associated with the index set $\mathcal{J}_0$ into one cluster, and reduce the corresponding 0-1 SDP model (2.14) to a smaller problem as follows

$$\begin{aligned}
&\min \operatorname{Tr}\left(D_{\overline{\mathcal{J}}_0}^{-\frac{1}{2}} W_{\overline{\mathcal{J}}_0 \overline{\mathcal{J}}_0} D_{\overline{\mathcal{J}}_0}^{-\frac{1}{2}} (I - Z_{\overline{\mathcal{J}}_0 \overline{\mathcal{J}}_0})\right) \\
&Z_{\overline{\mathcal{J}}_0 \overline{\mathcal{J}}_0} d_{\overline{\mathcal{J}}_0}^{\frac{1}{2}} = d_{\overline{\mathcal{J}}_0}^{\frac{1}{2}}, \quad \operatorname{Tr}(Z_{\overline{\mathcal{J}}_0 \overline{\mathcal{J}}_0}) = k - 1, \\
&Z_{\overline{\mathcal{J}}_0 \overline{\mathcal{J}}_0} \geq 0, \quad Z^2_{\overline{\mathcal{J}}_0 \overline{\mathcal{J}}_0} = Z_{\overline{\mathcal{J}}_0 \overline{\mathcal{J}}_0}, \quad Z_{\overline{\mathcal{J}}_0 \overline{\mathcal{J}}_0} = Z^T_{\overline{\mathcal{J}}_0 \overline{\mathcal{J}}_0}.
\end{aligned}$$

Repeating the above process, we can reconstruct all the clusters from a solution of problem (2.14). This establishes the equivalence between the two models (2.12) and (2.14). □

It should be noted that the major difference among (2.8), (2.9) and (2.14) is the introduction of the scaling vector d in both the objective function and constraints. This coincides with the observation in [7, 32]. However, we would like to point out that it is not so easy to add extra constraints such as the number of points in each cluster to model (2.14), while adding balanced constraint to (2.8) is a trivial task.

Before closing this section, we give another equivalent 0-1 SDP model for the normalized k-cut problem. From Theorem 2.1, one can verify that solving the normalized k-cut problem is equivalent to finding the global solution of the following 0-1 SDP model

$$\begin{aligned} &\min \operatorname{Tr}\left((I + D^{-\frac{1}{2}} W D^{-\frac{1}{2}})(I - Z)\right) \\ &Zd^{\frac{1}{2}} = d^{\frac{1}{2}}, \quad \operatorname{Tr}(Z) = k, \\ &Z \geq 0, \quad Z^2 = Z, \quad Z = Z^T. \end{aligned} \tag{2.19}$$

It should be mentioned that if W is an affinity matrix of a graph, then $D + W$ is positive semidefinite, which in turn implies that the matrix $(I + D^{-\frac{1}{2}} W D^{-\frac{1}{2}}$ is positive semidefinite.

3. Algorithms for Solving 0-1 SDP

In this section we discuss how to solve the 0-1 SDP model for clustering. For simplification of our discussion, we restrict ourselves to the following unified model

$$\min \operatorname{Tr}\left(W(I - Z)\right) \tag{3.1}$$

$$Zs = s, \quad \operatorname{Tr}(Z) = k, \tag{3.2}$$

$$Z \geq 0, \quad Z^2 = Z, \quad Z = Z^T, \tag{3.3}$$

where s is a positive scalar vector satisfying norm$s = 1$. Throughout the paper, we further assume that the underlying matrix W is positive semidefinite. It is worthwhile mentioning that the affinity matrix W used in [26] is defined by (2.10), which implies W is positive semidefinite. For the normalized k-cut problem with a general graph, we can use the 0-1 SDP model (2.19), which provides a positive semidefinite coefficient matrix derived from the affinity matrix of the underlying graph. This justifies our assumption on the positive semidefiniteness of the coefficient matrix in problem (3.1).

The section consists of three parts. In the first subsection, we first elaborate on the interrelation between the 0-1 SDP model (3.1) and weighted k-means clustering. Then based on weighted k-means clustering, we propose an algorithm to find the global solution of problem (3.1) in the special case where W is of rank one, i.e., $W = vv^T$. In the second part, we first give a general introduction to algorithms for solving (3.1) based on relaxation and then describe an approximation algorithm based on the SVD of W. In the last subsection, we propose a new approximation method for (3.1) based on the SVD of a projection matrix of W.

3.1. Exact algorithms for 0-1 SDP with coefficient matrix in low rank. In this subsection we briefly discuss exact algorithms for the 0-1 SDP model (3.1). First we point out that in general, finding the global solution of problem (3.1) is NP-hard as proved in the appendix of the seminal paper [26]. However, as we can see from the following discussion, if the coefficient matrix W is of low rank, then we can still develop efficient algorithms for for the 0-1 SDP model. In what follows we first recall the connection between the 0-1 SDP model (3.1) and the so-called weighted k-means clustering, as observed in [32].

Theorem 3.1. *Suppose that W be a positive semidefinite coefficient matrix in the* 0-1 SDP *model* (3.1) *with $W = V^T V$, $V \in \Re^{d\times n}$ and s be a positive scaling vector. If we cast all the columns ($v_i, i = 1, \dots, n$) of the matrix V as vectors in $\Re^d$, then solving the* 0-1 SDP *model* (3.1) *equals to solving the following weighted k-means clustering problem*

$$\min_{c_1,\dots,c_k} \sum_{i=1}^{n} s_i^2 \min\left\{\left\|\frac{v_i}{s_i} - c_1\right\|^2, \dots, \left\|\frac{v_i}{s_i} - c_k\right\|^2\right\}. \tag{3.4}$$

Note that for given candidate centers $c_1, \dots, c_k$, the weighted K-means assigns the scaled data points precisely in the same manner as the classical K-means does. This allows us to use the so-called *Voronoi Partitions* in computational geometry [14] to find the global optimal solution of problem (3.1). Based on ref. [14, Theorem 3 and Theorem 5], the algorithm runs in $O(n^{dk+1})$ time. Since the focus of the present work is on approximation algorithm for problem (3.1), we leave the details for such an approach to the interested reader.

In what follows we discuss a special case of (3.1) when $W = vv^T$ with $v \in \Re^n$ and $k = 2$. In such a case, we can use the ordering of the scaled data points to design a more efficient algorithm for problem (3.1). From Theorem 3.1, we can use the following algorithm to solve problem (3.1).

Refined weighted K-means in one-dimensional space.

Step 0: Input the data set $\mathcal{V} = \{v_1, v_2, \dots, v_n\}$ and calculate the vector $\overline{v}$ by $\overline{v}_i = v_i/s_i$, $i = 1, \dots, n$;

Step 1: Sort the sequence $\overline{v}_i$ so that

$$\overline{v}_{i_1} \geq \overline{v}_{i_2} \cdots \geq \overline{v}_{i_n},$$

where $\{i_1, \dots, i_n\}$ is a permutation of the index set $\{1, \dots, n\}$;

Step 2: For $l = 1$ to n, set

$$C_1^l = \{v_{i_1}, \dots, v_{i_l}\}, \quad C_2^l = \{v_{i_{l+1}}, \dots, v_{i_n}\},$$

and calculate the objective function

$$f(C_1^l, C_2^l) = \sum_{v_i \in C_1^l}\left(v_i - \frac{\sum_{v_i\in C_1^l} s_i v_i}{\sum_{v_i \in C_1^l} s_i^2}\right)^2 + \sum_{v_i \in C_2^l}\left(v_i - \frac{\sum_{v_i\in C_2^l} s_i v_i}{\sum_{v_i \in C_2^l} s_i^2}\right)^2.$$

based on the partition (C_1^l, C_2^l);

Step 3: Find the optimal partition (C_1^*, C_2^*) such that

$$(C_1^*, C_2^*) = \arg\min_{l\in\{1,\dots,n\}} f(C_1^l, C_2^l),$$

and output it as the final solution.

We have

Theorem 3.2. *Suppose that $W = vv^T$ with $v \in \Re^n$ and $k = 2$. Then the* 0-1 SDP *model* (3.1) *can be solved by the refined weighted K-means in one-dimensional space. In particular, if $v = s$, then any feasible partitioning is optimal.*

PROOF. To prove the theorem, we first recall Theorem 3.1, which implies that there exist two distinct weighted centers corresponding to the optimal solution of problem (3.1), denoted by c_1^* and c_2^* respectively. Without loss of generality, we can assume that $c_1^* > c_2^*$. Based on Theorem 3.1, the optimal partition can be obtained by assigning the scaled data points to these two centers based on the distances from every point to the two centers. Let us denote the final clusters by (C_1, c_1^*), (C_2, c_2^*). Since the data set is in one-dimensional space, it is easy to see that

$$\overline{v}_1 \in C_1, \quad \overline{v}_2 \in C_2 \Longrightarrow \overline{v}_1 > \overline{v}_2.$$

This further implies the optimal partition can be obtained by making use of the ordering information of the scaled data points. It is straightforward to verify the second conclusion. This finishes the proof of the theorem. □

We remark that the above special case of the 0-1 SDP model had been mentioned in [28]. In [26], Shi and Malik suggested several simple heuristics to attack the underlying problem. However, both works [28, 26] did not address the issue of how to find the exact solution of the problem.

3.2. Approximation algorithms for 0-1 SDP based on SVD. We start with a description of the generic scheme of approximation algorithms for (3.1).

Approximation algorithm based on relaxation.

Step 1: Choose a relaxation model for (3.1),
Step 2: Solve the relaxed problem for an approximate solution,
Step 3: Use a rounding procedure to extract a feasible solution to (3.1) from the approximate solution.

The relaxation step has an important role in the whole algorithm. For example, if the approximation solution obtained from Step 2 is feasible for (3.1), then it is exactly the optimal solution of (3.1). On the other hand, when the approximation solution is not feasible regarding (3.1), we have to use a rounding procedure to extract a feasible solution.

Various relaxations and rounding procedures have been proposed for solving several special scenarios of (3.1) in the literature. For example, in [23], Peng and Xia considered a relaxation of the classical K-means clustering based on linear programming and a rounding procedure was also proposed in that work. Xing and Jordan [32] considered the SDP relaxation for normalized k-cuts and proposed a rounding procedure based on the singular value decomposition of the solution Z of the relaxed problem,i.e., $Z = U^T U$. In their approach, every row of U^T is cast as a point in the new space. Then the weighted K-means clustering is performed over the new data set in $\Re^k$. Similar works for spectral clustering can also be found in [12, 20, 21, 31, 33] where the singular value decomposition of the underlying matrix W is used and a K-means-type clustering based on the eigenvectors of W is performed. In the above-mentioned works, the solutions obtained from the weighted K-means algorithm for the original problem and that based on the eigenvectors of W has been compared, and simple theoretical bounds have been derived based on the eigenvalues of W.

The idea of using the singular value decomposition of the underlying matrix W is natural in the so-called principal component analysis (PCA) [17]. In [6], the link between PCA and K-means clustering was also explored and simple bounds were derived. In particular, Drineas *et al.* [8] proposed to use singular value decomposition to form a subspace, and then perform K-means clustering in the subspace $\Re^k$. They proved that the solution obtained by solving the K-means clustering in the reduced space can provide a 2-approximation to the solution of the original K-means clustering.

In what follows we consider approximation algorithms based on various relaxations of (3.1). First of all, the constraint $Z^2 = Z$ implies that Z must be a projection matrix. This implies $0 \preceq Z \preceq I$. Moreover, the nonnegativity of Z indicates that there exists at least one nonnegative eigenvector corresponding to its largest eigenvalue(see [3, Theorem 1.3.2]). On the other hand, recall that s is a positive eigenvector of Z with eigenvalue 1 and the fact that Z is real and symmetric. This implies that eigenvectors of Z corresponding to distinct eigenvalues are orthogonal to each other. Suppose that Z has an eigenvalue larger than 1 with nonnegative eigenvector v. It follows immediately $s^T v = 0$. This is impossible because s is positive. Therefore, the largest eigenvalue of Z must be 1. In such a case, the constraint $Z \preceq I$ becomes superfluous and can be waived without any influence. We thus consider only the following relaxed SDP problem

$$\begin{aligned} &\min \operatorname{Tr}\big(W(I - Z)\big) \\ &\operatorname{Tr}(Z) = k, \quad Z \succeq 0, \\ &Zs = s, \quad Z \geq 0. \end{aligned} \tag{3.5}$$

The above problem is essentially equivalent to the problem P1 in [32] where the constraint $I - Z \succeq 0$ is included. We can apply many existing optimization solvers such as interior-point methods to solve (3.5).

However, we would like to point out here that although there exist theoretically polynomial algorithms for solving (3.5), most of the present optimization solvers are unable to handle the problem in large size efficiently.

By further removing the nonnegative requirement and the scaling constraint $Zs = s$ in (3.5), we obtain the following simple optimization problem

$$\min \operatorname{Tr}\big(W(I - Z)\big) \tag{3.6}$$

$$\operatorname{Tr}(Z) = k, \quad I \succeq Z \succeq 0, \tag{3.7}$$

which can be solved by using the singular value decomposition of W [22], i.e.,

$$W = U\operatorname{diag}(\lambda_1, \ldots, \lambda_n)U^T,$$

where λ_i are the eigenvalues of W listed in the decreasing order, and U is an orthogonal matrix whose i-column is the unit eigenvector corresponding to λ_i. It follows immediately that

Theorem 3.3. *Suppose Z^* is a global solution to problem* (3.1), *we have*

$$\operatorname{Tr}\big(W(I - Z^*)\big) \geq \operatorname{Tr}(W) - \sum_{i=1}^{k} \lambda_i \geq 0.$$

To extract a feasible clustering for the original problem, we use the first k eigenvectors multiplied by the squared roots of their corresponding eigenvalues, i.e., $\lambda_1^{\frac{1}{2}} u_1, \ldots, \lambda_2^{\frac{1}{2}} u_k$, to construct a new matrix in $\Re^{n\times k}$. Then we can cast each row of the matrix as a point in $\Re^k$ and we thus obtain a new data set $\overline{V} \in \Re^k$. Finally we perform the weighted K-means clustering on $\overline{V}$. From Theorem 3.1, this amounts to solving problem (3.1) where the coefficient matrix W is replaced by its projection onto the subspace generated by the eigenvectors $u_1, \ldots, u_k$, i.e.,

$$W_k = \sum_{i=1}^{k} \lambda_i u_i u_i^T.$$

The algorithm scheme can be described as follows:

Algorithm 1: approximation based on the SVD of W.

S.1: Perform a singular value decomposition for the coefficient matrix W.
S.2: Calculate the matrix W_k by using the first k largest eigenvalues of W and their corresponding eigenvectors.
S.3: Solve the 0-1 SDP model (3.1) with the coefficient matrix W_k approximately by performing K-means-type clustering on the reduced data set $\overline{V}$ and cluster the data set based on the obtained assignment matrix.

We remark that Algorithm 1 can be viewed as an extension of the algorithm in [8] for the K-means clustering. Following a similar arguing chain as in [8], we can show that if the subproblem in Step 3 can be solved precisely, then we can obtain a 2-approximation solution to the original problem. For self-completeness, we also include a proof here.

Theorem 3.4. *Suppose Z^* is a global solution to the original problem* (3.1) *and Z_k^* is a global solution to the reduced problem* (3.1) *where W is replaced by W_k. Then we have*

$$\operatorname{Tr}\left(W(I - Z_k^*)\right) \leq 2 \operatorname{Tr}\left(W(I - Z^*)\right).$$

PROOF. Let us define

$${}_k = \sum_{i=1}^{k} u_i u_i^T.$$

From Theorem 3.3, we have

$$\operatorname{Tr}\left(W(I - U_k)\right) \leq \operatorname{Tr}\left(W(I - Z^*)\right).$$

Since

$$\operatorname{Tr}\left(W(I - Z_k^*)\right) = \operatorname{Tr}\left(W(I - U_k)\right) + \operatorname{Tr}\left(W(U_k - Z_k^*)\right),$$

to prove the theorem, it suffices to show that

$$\operatorname{Tr}\left(W(U_k - Z_k^*)\right) \leq \operatorname{Tr}\left(W(I - Z^*)\right),$$

which can be stated as

$$\operatorname{Tr}\left(W(I - U_k + Z_k^* - Z^*)\right) \geq 0. \tag{3.8}$$

From the choice of U_k, we have

$$W_k = WU_k = U_k W U_k \succeq 0, \quad W_k(I - U_k) = 0;$$
$$W - W_k = (I - U_k)W(I - U_k) \succeq 0, \quad (W - W_k)U_k = 0.$$

Since Z_k^* is a solution of problem (3.1) with a coefficient matrix W_k, we have

$$\operatorname{Tr}\left(W_k(I - Z_k^*)\right) \leq \operatorname{Tr}\left(W_k(I - Z^*)\right). \tag{3.9}$$

It follows

$$\begin{aligned}\operatorname{Tr}\left(W(I-U_k+Z_k^*-Z^*)\right) &= \operatorname{Tr}\left(W_k(I-U_k+Z_k^*-Z^*)\right) \\ &\quad + \operatorname{Tr}\left((W-W_k)(I-U_k+Z_k^*-Z^*)\right) \\ &= \operatorname{Tr}\left(W_k(Z_k^*-Z^*)\right) \\ &\quad + \operatorname{Tr}\left((W-W_k)(I+Z_k^*-Z^*)\right) \\ &\geq \operatorname{Tr}\left(W_k(Z_k^*-Z^*)\right) + \operatorname{Tr}\left((W-W_k)(I-Z^*)\right) \\ &\geq \operatorname{Tr}\left((W-W_k)(I-Z^*)\right) \geq 0,\end{aligned}$$

where the first inequality is given by the fact that both the matrices Z_k^* and $W-W_k$ are positive semidefinite, the second inequality by (3.9), and the last one by $I-Z^* \succeq 0$. This proves the relation (3.8), which further yields the theorem. □

It should be pointed out that although we have proved that the algorithm based on the SVD of the coefficient matrix W can provide a 2-approximation solution, it indeed requires to find the global solution of the reduced problem. In general, this is still a nontrivial task. For example, if we use the *Voronoi Partitions* to solve the subproblem [14], then it takes $O(n^{k^2+1})$ time to solve the subproblem. Even for the special case $k=2$, such an algorithm is clearly impractical for large data set.

3.3. An approximation algorithm based on the SVD of the projected coefficient matrix. In this subsection, we propose a new approximation algorithm based on another relaxation form of the model (3.1). Let us recall that in the relaxed model (3.5), we stipulate that s is an eigenvector of the final solution matrix Z. Since we already know this fact in advance, we can keep such a simple constraint in our relaxed problem. We therefore obtain another form of relaxation For example, if we remove only the nonnegative requirement in the relaxation form (3.5), then we obtain the following SDP problem:

$$\begin{aligned}&\min \operatorname{Tr}\left(W(I-Z)\right) \\ &Zs = s, \quad \operatorname{Tr}(Z) = k, \\ &I \succeq Z \succeq 0.\end{aligned} \tag{3.10}$$

We next discuss how to solve the above problem. First, we note that for any feasible solution of (3.10), let us define

$$\overline{Z} = Z - ss^T.$$

It is easy to see that

$$\overline{Z} = (I-ss^T)Z = (I-ss^T)Z(I-ss^T), \tag{3.11}$$

i.e., $\overline{Z}$ represents the projection of the matrix Z onto the null subspace of s. Moreover, since norm$s=1$, it is easy to verify that

$$\operatorname{Tr}(\overline{Z}) = \operatorname{Tr}(Z) - 1 = k-1.$$

Let $\overline{W}$ denote the projection of the matrix W onto the null space of s, i.e.,

$$\overline{W} = (I-ss^T)W(I-ss^T). \tag{3.12}$$

Then, we can reduce (3.10) to

$$\begin{aligned} \min \operatorname{Tr}\left(\overline{W}(I-\overline{Z})\right) & \\ \operatorname{Tr}(\overline{Z}) = k-1, & \\ \overline{Z} \succeq 0. & \end{aligned} \tag{3.13}$$

Let $\lambda_1, \ldots, \lambda_{n-1}$ be the eigenvalues of the matrix $\overline{W}$ listed in the order of decreasing values. The optimal solution of (3.13) can be achieved if and only if

$$\operatorname{Tr}(\overline{W}\overline{Z}) = \sum_{i=1}^{k-1} \lambda_i.$$

This gives us an easy way to solve (3.13) and correspondingly (3.10). The algorithmic scheme for solving (3.10) can be described as follows:

SVD of the projected coefficient matrix.

Step 1: Calculate the projection $\overline{W}$ via (3.12);

Step 2: Use singular value decomposition method to compute the first $k-1$ largest eigenvalues of the matrix $\overline{W}$ and their corresponding eigenvectors $u_1, \ldots, u_{k-1}$,

Step 3: Set

$$Z = ss^T + \sum_{i=1}^{k-1} u_i {u_i}^T.$$

Let Z^* be the global optimal solution of (3.1), and $\lambda_1, \ldots, \lambda_{k-1}$ be the first $k-1$ largest eigenvalues of the matrix $\overline{W}$. From our previous discussion, we immediately have

$$\operatorname{Tr}\left(W(I-Z^*)\right) \geq \operatorname{Tr}(W) - s^T W s - \sum_{i=1}^{k-1} \lambda_i. \tag{3.14}$$

In the sequel we propose a rounding procedure to extract a feasible solution for (3.1) from a solution of the relaxed problem (3.10) provided by the SVD of the projected coefficient matrix. Our rounding procedure follows a similar vein as the rounding procedure in the previous subsection. In case no confusion occurs, we use the notation introduce in the previous subsection. Let

$$U_{k-1} = \sum_{i=1}^{k-1} u_i u_i^T$$

be the solution matrix obtained from the projected coefficient matrix, and

$$U_k = ss^T + U_{k-1}, \quad \overline{W}_{k-1} = \overline{W} U_{k-1}.$$

We can formulate a matrix in $\Re^{n\times(k-1)}$ whose ith column is $\lambda_i^{\frac{1}{2}} u_i$. Then we cast each row in such a matrix as a point in $\Re^{k-1}$. We thus obtain a data set of n points in $\Re^{k-1}$. Then we perform the clustering task for the new data set. In other words, we need to solve the 0-1 SDP model (3.1) with a new coefficient matrix $\overline{W}_{k-1}$. Finally, we partition all the points in the original space based on the obtained clusters for the new data set.

The whole algorithm can be described as follows.

Algorithm 2: approximation method based on the SVD of the projected coefficient matrix.

Step 1: Calculate the projected matrix $\overline{W}$ of the matrix W onto the null space of s;

Step 2: Use singular value decomposition to compute the first $k-1$ largest eigenvalues of the matrix $\overline{W}$ and their corresponding eigenvectors $u_1, \ldots, u_{k-1}$. Formulate a new data set $\overline{V}$ and compute the matrix $\overline{W}_{k-1}$;

Step 3: Solve problem (3.1) with the coefficient matrix $\overline{W}_{k-1}$ approximately by performing K-means-type clustering on the data set $\overline{V}$ and assign all the points in the original space based on the obtained assignment.

The above algorithm can be viewed as an improved version of the algorithm based on the SVD of the coefficient matrix. In particular, in case of binary clustering, the subproblem involved in the above algorithm needs only to cluster a data set in one dimension, which can be done in $O(n \log n)$ time as shown in the refined weighted K-means in one dimension. This improves the efficiency of the algorithm substantially and allows us to deal with large-scale data set.

We also point out that a similar idea had been employed by Shi and Malik [26] in their seminal paper on normalized k-cut for image segmentation. In that case, the 0-1 SDP model takes the form as in (2.14) with $k = 2$. Since $d^{\frac{1}{2}}$ is the eigenvector corresponding to the largest eigenvalue of the underlying coefficient matrix, Shi and Malik proposed to use the eigenvector corresponding to the second largest eigenvalue of the coefficient matrix to separate the data set. Shi and Malik also proposed several simple heuristics to solve the underlying 0-1 SDP model in one-dimensional space. However, these heuristics in [26] can not find the global solution of the subproblem in general. On the other hand, in our algorithm we need to perform a sorting first and then find the best breaking point in term of the objective function. As we shall see from our later discussion, the extra effort taken in our algorithm allows us to obtain an approximation with guaranteed quality.

We next progress to estimate the solution obtained from Algorithm 2. We have

Theorem 3.5. *Suppose that Z^* is a global solution to problem* (3.1) *and Z_k^* is a global solution to the subproblem in Step* 3 *of by Algorithm* 2. *Then, we have*

$$\operatorname{Tr}\left(W(I - Z_k^*)\right) \le 2 \operatorname{Tr}\left(W(I - Z^*)\right).$$

PROOF. Let Z^* be a global solution to (3.1) and Z_k^* is the solution provided by Algorithm 2. From the choices of U_{k-1} and U_k it follows

$$\operatorname{Tr}\left((I - U_k)\overline{W}_{k-1}\right) = 0; \tag{3.15}$$

$$\operatorname{Tr}\left(U_k(\overline{W} - \overline{W}_{k-1})\right) = 0. \tag{3.16}$$

From inequality (3.14), we have

$$\operatorname{Tr} W(I - Z^*) \ge \operatorname{Tr} W(I - U_k). \tag{3.17}$$

It follows

$$\begin{aligned}&\operatorname{Tr}\left(W(I - Z_k^*)\right)\\ &\quad = \operatorname{Tr}\left(W(I - U_k + U_k - Z_k^*)\right) \le \operatorname{Tr}\left(W(I - Z^*)\right) + \operatorname{Tr}\left(W(U_k - Z_k^*)\right).\end{aligned}$$

To prove the conclusion in the theorem, it suffices to show

$$\operatorname{Tr}\left(W(U_k - Z_k^*)\right) \le \operatorname{Tr}\left(W(I - Z^*)\right), \tag{3.18}$$

or equivalently

$$\mathrm{Tr}\left(W(I - Z^* + Z_k^* - U)\right) \geq 0. \tag{3.19}$$

By the choices of Z^*, Z_k^* and U_k, it is easy to verify

$$(I - Z^* + Z_k^* - U)s = 0, \tag{3.20}$$

$$(I - ss^T)(I - Z^* + Z_k^* - U)(I - ss^T) = I - Z^* + Z_k^* - U. \tag{3.21}$$

It follows immediately that

$$\begin{aligned}
\mathrm{Tr}\, W(I - Z^* + Z_k^* - U) &= \mathrm{Tr}\left(\overline{W}(I - Z^* + Z_k^* - U_k)\right) \\
&= \mathrm{Tr}\left(\overline{W}_{k-1}(I - Z^* + Z_k^* - U_k)\right) \\
&\quad + \mathrm{Tr}\left((I - Z^* + Z_k^* - U_k)(\overline{W} - \overline{W}_{k-1})\right) \\
&= \mathrm{Tr}\left((Z_k^* - Z^*)\overline{W}_{k-1}\right) + \mathrm{Tr}\left((I - Z^* + Z_k^*)(\overline{W} - \overline{W}_{k-1})\right) \\
&\geq \mathrm{Tr}\left(\overline{W}_{k-1}(Z_k^* - Z^*)\right),
\end{aligned}$$

where the last equality is given by (3.15) and (3.16), and the last inequality is implied by the fact that $I - Z^* + Z_k^* \succeq 0, \overline{W} - \overline{W}_{k-1} \succeq 0$. Recall that Z_k^* is the global solution of problem (3.1) with the coefficient matrix $\overline{W}_{k-1}$ while Z^* is only a feasible solution of the same problem, we therefore have

$$\mathrm{Tr}\left(\overline{W}_{k-1}(Z_k^* - Z^*)\right) \geq 0,$$

which further implies (3.18). This finishes the proof of the theorem. □

We remark that for K-means clustering, Algorithm 2 reduces to the algorithm in [24]. For the normalized cut problem, since $s = d^{\frac{1}{2}}$ is also an eigenvector of the coefficient matrix in the 0-1 SDP model corresponding to its largest eigenvalue, from Theorems 3.1 and 3.2 we can conclude that the vector $s = d^{\frac{1}{2}}$ does not play any role in the partitioning process of the weighted K-means clustering. Therefore, if we employ the weighted K-means to the subproblems in Algorithms 1 and 2, then we will end up with the same partition. The same conclusion holds for the Algorithm 1 in [1] and one variant of the algorithms in [32] based on spectral relaxation, both algorithms provide a 2-approximation to the normalized 2-cut problem.

Our above results can be extended to scenarios of constrained clustering where the number of points in every cluster is bounded and semi-supervised clustering. In such cases, we need to add some extra constraints in the subproblem involved in Algorithm 2. In [24], the authors considered the case of balanced clustering. Since the discussions for these scenarios follow a similar chain of reasoning as in the proofs of Theorem 3.5, we leave the details to interested readers.

In the sequel we discuss briefly the complexity of Algorithm 2. In the second step of the algorithm, we need to perform the singular value decomposition for the matrix $\overline{W}$. In general, this takes $O(n^3)$ time. If we use the power method [10] to calculate the first $k-1$ largest eigenvalues and their corresponding eigenvectors, then the complexity can be reduced to $O(kn^2)$. If the matrix W has a certain sparsity, then we can use its sparse structure to speedup the process. In the context of the classical K-means clustering, we can use the structure of the underlying matrix W to improve the process. Recall that for K-means clustering, we have $W = W_x W_x^T$ where $W_x \in \Re^{n \times m}$ is a matrix such that every row represents a point in $\Re^m$. In such a case, it is not necessary to calculate the matrix W to estimate its eigenvalues and eigenvectors exactly. Note that we can perform a singular value decomposition on the matrix W_v directly, which can proceed in the following way. We first compute a matrix $\widetilde{W} = W_v^T W_v \in \Re^{m\times m}$ which takes $O(nm^2)$ time. It is easy to see that the matrix $\widetilde{W}$ has the same spectrum as that of the matrix W. Therefore, we can perform a singular value decomposition on $\widetilde{W}$ directly such that

$$\widetilde{W} = V\mathrm{diag}(\lambda_1, \dots, \lambda_m)V^T,$$

where $V \in \Re^{m\times m}$ is an orthogonal matrix such that every column is an eigenvector of $\widetilde{W}$. This takes $O(m^3)$ time. We then calculate the matrix $\widetilde{U} = W_x V$ in $O(nm^2)$ time. One can easily verify that the ith column of the matrix $\widetilde{U}$ is an eigenvector of W corresponding to eigenvalue λ_i. Therefore, the total computational cost to calculate the eigenvalues and their corresponding eigenvectors is $O(nm^2 + m^3)$. If m is not very large, say $m < 1000$ (which is true for most data sets in practice), then we can obtain the eigenvalues and eigenvectors of W very quickly. However, this is not true if we use some other kernel matrices such as in normalized cut.

Next we discuss the complexity in solving the subproblem in Algorithm 2. There are several different ways to solve the subproblem in Algorithm 2. For example, we can use the so-called *Voronoi Partitions* to the scaled data set in the subproblem [14], which takes $O(n^{(k-1)^2+1})$ time to find the global solution of the subproblem in Step 3 of Algorithm 2.

Another way to solve the subproblem in Algorithm 2 is to use the hierarchical approach as suggested in [26] and perform a binary clustering task subsequently. By using the refined weighted K-means in one dimension, the subproblem in Algorithm 2 for binary clustering can be done in $O(n \log n)$ time. Therefore, the total complexity of the algorithm will be $O(kn \log n + kn^2)$. This allows us to cope with relatively large data set in high dimension.

4. Numerical Experiments

In this section, we report some numerical experiments based on our algorithms. The section consists of two parts. In the first subsection, we compare our algorithms with some general partitioning methods in the literature. In the second subsection, we compare our hierarchical approach with the hierarchical approach in [26].

4.1. Test results based on partitioning algorithms. We have implemented our Algorithms 1 and 2 in `matlab 6.5` combined with `C` language. The tests are done on a PC with `AMD Athlon 1.24G CPU` and `512M RAM`. In our implementation, the largest eigenvalue and its corresponding eigenvector of the coefficient matrix are computed by Lanczos tridiagonalization with the modified partial orthogonalization described in [25].

When $k > 2$, we also develop an algorithm based on the hierarchical approach to partition the set into k clusters as follows. We first cut the data set into two parts by Algorithm 2. Then we cut each subset into two parts based on a similar strategy, and select the best cut in term of the reduction of the objective function values. We repeat such a process until we obtain k clusters. We point out that to get the ith cut, we only need to perform the binary separation for the two subsets obtained from the $(i-1)$th cut.

We have tested our algorithms including Algorithms 1, 2 and the hierarchical approach on several examples in the literature. For comparison, we have also tested the algorithm proposed in [32] and the standard weighted K-means rounding procedure, denoted by Xing-Jordan and Rank-N respectively.

Except for Hierarchical Approach, all other algorithms are run 100 times with random initial points and the lowest objective values are recorded.

The test data sets are listed as follows.

- The Soybean dataset(small) is from from the UCI machine Learning Repository. The data set consists of 47 instances and each instance has 35 normalized features with four groups.
- The human fibroblast gene expression is from `http://genome-www.stanford.edu/serum/`, see also [7]. The data set has 517 instances and each instance has 18 features. It is known that the data set has 5 clusters.
- The protein dataset is from `http://www.nersc.gov/ cding/protein`, see also [32]. It is known as that the data set has 27 clusters. We collected the instances within first six largest clusters as the testing data.
- The Pendigits data set is also from the UCI machine Learning Repository. The data set contains 7494 digits and each digit is represented as a vector in 16-dimensional space. It is known that the data set has 10 clusters. Due to the computational facility, ten percent data was randomly subsampled as the testing data.

Note that since it has been observed that the classical K-means performs well for the Soybean data set. Therefore, in our experiments, we compare the performance of various algorithms for the Soybean data set based on K-means clustering. For the other three data sets, we use the same objective as in the normalized cut [32]. In the following tables, Rank-N means that we use all the eigenvalues and their corresponding eigenvectors to cluster the data set.

The results for the other three data sets are summarized in Tables 3 and 4.

From the above tables it is easy to see that Rank-N provides worst objective values although it spent less CPU time than others. For the Soybean data, all the algorithms achieve same objective value except Rank-N. For the gene expression data, the objective values provided by Algorithms 1 and 2 are slightly better than what obtained in Xing-Jordan's algorithm. For the protein data and the pendigits data, Algorithm 1, Algorithm 2, and the Xing-Jordan algorithm report similar results. In the contrast, our hierarchical approach presents much better objective

TABLE 1. Objective values of algorithms on Soybean data.

K	Algorithm I	Algorithm II	Rank-N	Hier. Appr.
2	404.4593	404.4593	404.4593	404.4593
3	246.4593	246.4593	251.2196	246.4593
4	208.5167	208.5167	207.2167	205.9637

TABLE 2. CPU time(s) of algorithms on Soybean data.

K	Algorithm I	Algorithm II	Rank-N	Hier. Appr.
2	0.031	0.047	0.031	0.031
3	0.031	0.046	0.032	0.046
4	0.031	0.047	0.032	0.063

TABLE 3. Objective values of algorithms.

	gene	protein	pendigits
Alg. 1	1.6148	$3.5987*10^{-1}$	$1.3544*10^{-5}$
Alg. 2 II	1.6204	$3.6031*10^{-1}$	$1.4799*10^{-5}$
Rank-N	2.2398	1.0608	$4.1616*10^{-2}$
Xing-Jordan	2.5153	$3.6035*10^{-1}$	$1.5912*10^{-5}$
Hier. Appr.	$3.0115*10^{-3}$	$2.1847*10^{-1}$	$3.8424*10^{-4}$

TABLE 4. CPU time(s) of algorithms.

	gene	protein	pendigits
Alg. 1	52.93	3.00	119.53
Alg. 2	53.98	3.31	120.70
Rank-N	17.73	1.94	44.15
Xing-Jordan	53.68	3.12	119.46
Hier. Appr.	26.80	3.88	59.22

values than the other algorithms with less computation time for the gene expression data set and the protein data set.

From the above experiments, we can conclude that if the number of clusters is not too large, our hierarchical approach is a good method to partition data sets by considering the trade-off between the quality of clustering and the CPU time. Moreover, it is also independent of initial starting points.

4.2. Test results on the hierarchical approach. To further test the performance of our algorithms, we also compare our hierarchical approach with the algorithm in [26] where the authors used a simple heuristics to attack the subproblem of Algorithm 2 in one-dimensional space. The numerical results are summarized in the following tables and figures. For convenience, we use SDPCut and Ncut to denote the algorithm in the present paper and the algorithm in [26].

We have tested our algorithm on several examples in the literature. These include the data point set from [26], an image from the Berkeley segmentation data set and benchmark[4], and two images from Adobe Photoshop for image segmentation. Due to the limit of the computational facility in our experiment, we resized all the images proportional to their original sizes. To compare the performance of our algorithm on the test problems, we use both tables and figures. In the tables we list the values of the objective function obtained by our algorithm (SDPCut) and the algorithm (Ncut) in [26], respectively. The CPU time used to obtain the partitioning is also reported. The figures visualize the final clusters that can help us to better understand the practical performance of these two algorithms.

4.2.1. *Point data set.* The first test problem is the data point set in [26], which consists of 120 points in two-dimensional space. Table 5 records the values of the objective function and CPU time for both algorithms for k from 2 to 4. Although both algorithms have exactly the same partition for the data point set when $k = 4$, the values of the objective function provided by SDPCut is better than what provided by Ncut when $k = 2, 3$. The numerical improvement has been

[4] `http://www.eecs.berkeley.edu/Research/Projects/CS/vision/grouping/segbench/`

TABLE 5. Numerical test on circle data point set.

	Objective Value		CPU Time(s)	
k	SDPCut	Ncut	SDPCut	Ncut
2	0.000000000	0.134910574	0.44	0.27
3	0.000666557	0.002806832	0.53	0.23
4	0.003473389	0.003473389	0.63	0.27

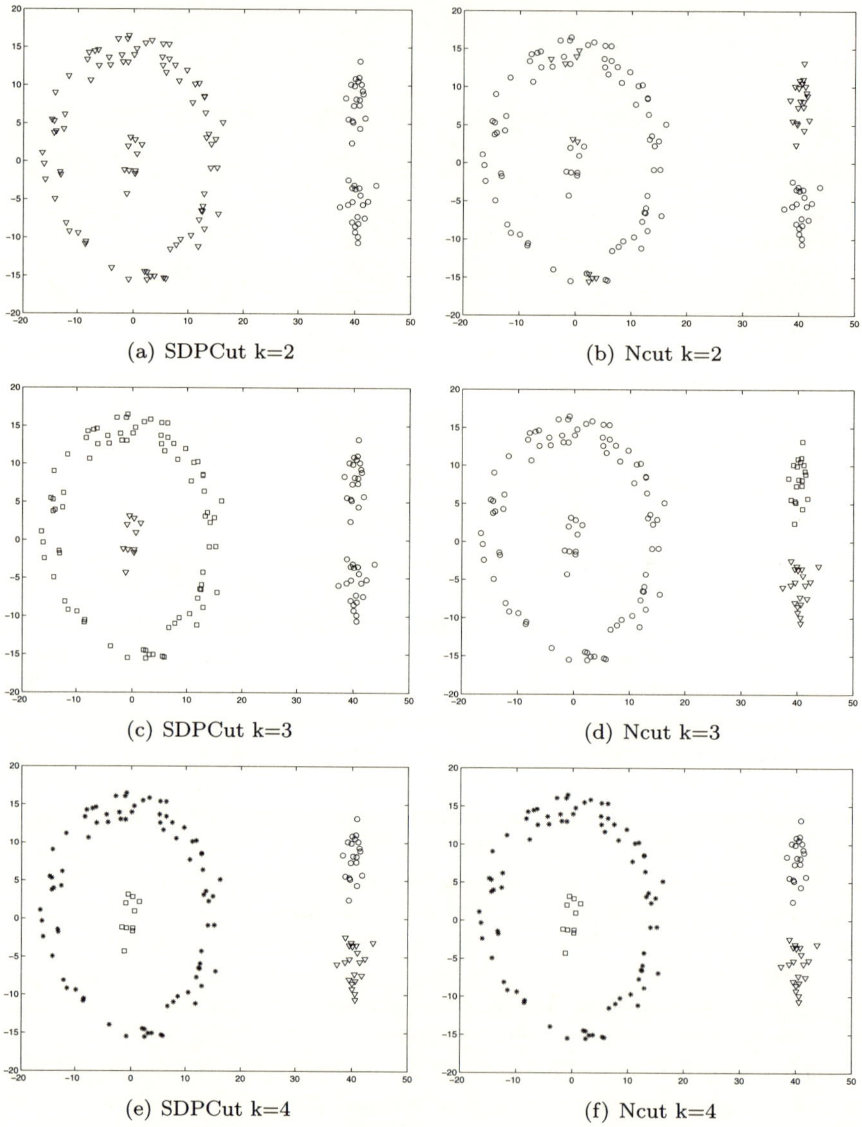

(a) SDPCut k=2

(b) Ncut k=2

(c) SDPCut k=3

(d) Ncut k=3

(e) SDPCut k=4

(f) Ncut k=4

FIGURE 1. Partition of the circle data point set.

confirmed by Fig. 1, which shows when $k = 2$, the cut provided by our algorithm separate the data set into two parts clearly, while the separation provided by Ncut

TABLE 6. Woman face image.

	Objective Value		CPU Time(s)	
k	SDPCut	Ncut	SDPCut	Ncut
2	0.000373422	0.00267505	27.22	36.97
3	0.000951509	0.01056890	44.45	35.81
4	0.003592392	0.01115780	57.90	35.36
5	0.012211270	0.05035214	74.00	33.09
6	0.043300375	0.05025529	88.13	35.58
7	0.070311355	0.16297105	95.49	34.44

(a) SDPCut (b) Ncut

FIGURE 2. Segmentations of the woman face for $k = 7$.

is unclear. When $k = 3$, though both methods managed to separate the data set, the final clusters are different.

4.2.2. *Woman face.* The second test problem is the image of a woman's face with two hands on her cheek from the Berkeley segmentation data set and benchmark. The original size is 481×321 pixels and it was resized as 137×91 pixels in the experiment. Table 6 records the performance of segmentations of the woman face for K from 2 to 7. Both algorithms are able to find the boundary of the face and two eyeballs. However, SDPCut did a better job on extracting the shape of fingers of both hands than Ncut. Figure 2 shows two different segmentations when $k = 7$.

4.2.3. *Ducky image.* The third test problem is a picture of a toy ducky with bright background from the sample pictures of Adobe Photoshop. The original size

TABLE 7. Ducky image.

	Objective Value		CPU Time(s)	
k	SDPCut	Ncut	SDPCut	Ncut
2	0.006035151	0.006252636	44.80	94.03
3	0.010534337	0.012925502	77.80	86.23
4	0.022542201	0.027505889	102.47	86.11
5	0.033214296	0.042012681	121.35	82.22
6	0.052768125	0.094367725	137.57	94.39
7	0.108721513	0.199546366	147.52	83.69

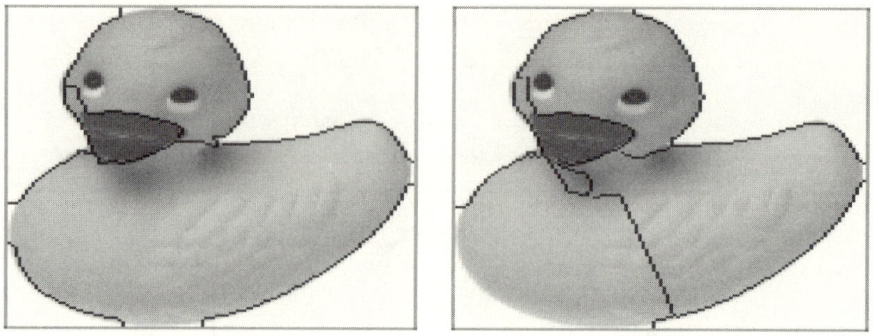

(a) SDPCut (b) Ncut

FIGURE 3. Segmentations of the ducky for $k = 7$.

(a) SDPCut (b) Ncut

FIGURE 4. Segmentations of the ranch house for $k = 8$

is 546×500 pixels and was resized as 130×119 pixels in our experiment. Table 7 summarizes the performance of the two algorithms for the ducky image when k from 2 to 7. It can be seen that the values of the objective function provided by SDPCut is always better than what obtained by Ncut, and the difference becomes prominent when $k = 7$. This can also be verified via Fig. 3, which shows that SDPCut clearly outlined the shape of the ducky mouth, while Ncut did not.

4.2.4. *Ranch house.* The last test problem is the scene of a ranch house from the sample pictures of Adobe Photoshop whose original size is 692×589 pixels. It was resized as 140×119 pixels in our experiment. Table 8 gives the performance of segmentations of the Ranch House image for k from 2 to 8. It indicates that SDPCut has better computational results compared to Ncut. For example, when $k = 6$, SDPCut was able to outline the contour of the shoes while Ncut only drew out the shadow of the door. In addition, SDPCut also gave a clearer shape of the shadow of the cloth than Ncut when $k = 8$. Figure 4 illustrates the segmentations of the image for $k = 8$.

5. Conclusions

In the present work, we presented a novel unified framework for various clustering scenarios and proposed two different approximation algorithms for solving the unified 0-1 SDP model based on the SVD of the coefficient matrix and the

TABLE 8. Ranch House Image

	Objective Value		CPU Time(s)	
k	SDPCut	Ncut	SDPCut	Ncut
2	0.000618508	0.000830058	50.69	94.34
3	0.014763275	0.014937043	88.19	94.55
4	0.044782877	0.067792577	112.19	90.03
5	0.078499524	0.108438306	126.78	84.45
6	0.086836495	0.173986842	138.06	97.83
7	0.095264966	0.182797317	149.09	95.44
8	0.117966244	0.235106454	161.67	99.13

SVD of a projection matrix of the coefficient matrix, respectively. We have shown that both algorithms can provide a 2-approximation to the original clustering problem, while the algorithm based on the SVD of the projected coefficient matrix is computationally more attractive. A hierarchical approach is also developed based our new model. Our results not only open new avenues for solving these clustering problems, but also provide insightful analysis for several existing algorithms in the literature. Preliminary experiments illustrate that our new algorithm not only enjoys theoretical efficiency, but also performs very well in practice.

There are several open questions regarding the new 0-1 SDP model. First, we note that there are several different ways to relax the 0-1 SDP model that have not been investigated. For example, as proposed in [32], we can solve the relaxed model (3.5) to find an approximation solution, which will give us a tighter bound than the relaxation based on SVD. However, it is unclear how to design a rounding procedure that can extract a good approximation to the solution of the original problem. Secondly, both algorithms in the present paper require to solve the subproblems exactly, which turns out still to be a challenge when $k \geq 3$. More study is necessary to address these questions.

References

[1] F. R. Bach, and M. I. Jordan, *Learning spectral clustering*, 2003 Conference on Adv. in Neural Inform. Processing Systems, (S. Thrun, L. K. Saul and B. Schölkopf, eds.), vol. 16, 2004, MIT Press, Cambridge, pp. 305–312.

[2] S. Basu, M. Bilenko, and R. J. Mooney, *A probabilistic framework for semi-supervised clustering*, KDD04, Seattle, WA, 2004, pp. 59–68.

[3] A. Berman and R. J. Plemmins, *Nonnegative matrices in the mathematical sciences*, Classics in Appl. Math., vol. 9, Society for Industrial and Applied Mathematics (SIAM), Philadelphia, PA, 1994.

[4] P. Bradley, K. Bennet, and A. Demiriz, *Constrained K-means clustering*, MSR-TR-2000-65, Microsoft Research, 2000.

[5] I. Davidson and S. Ravi, *Clustering with constraints: feasibility issues and the k-means algorithm*, SIAM Internat. Conf. on Data Mining, 2005.

[6] C. Ding and X. He, *K-means clustering via principal component analysis*, 21st Internat. Conf. on Machine Learning (Banff, 2004).

[7] I. S. Dhillon, Y. Guan, and B. Kulis, *Kernel k-means, spectral clustering and normalized cuts*, 10th ACM SIGKDD, 2004, pp. 551-556.

[8] P. Drineas, A. Frieze, R. Kannan, R. Vempala, and V. Vinay, *Clustering large graphs via singular value decomposition*, Machine Learning **56** (2004), 9–33.

[9] J. Ghosh, *Scalable clustering methods for data mining*, Handbook of Data Mining (Nong Ye, ed.), Chapter 10, Lawrence Ealbaum Assoc., Mahwah, NJ, 2003, pp. 247–277

[10] G. Golub and C. V. Loan, *Matrix computation*, 3rd ed., John Hopkins University Press, Baltimore, MD, USA, 1996.

[11] A. D. Gordon and J. T. Henderson, *An algorithm for Euclidean sum of squares classification*, Biometrics **33** (1977), 355–362.

[12] M. Gu, H. Zha, C. Ding, X. He, and H. Simon, *Spectral relaxation models and structure analysis for k-way graph clustering and bi-clustering*, Technical report, Pennsylvania State University, PA, 2001.

[13] P. Hansen, B. Jaumard, and N. Mladenović, *Minumum sum of squares clustering in a low dimensional space*, J. Classification **15** (1998), 37–55.

[14] M. Inaba, N. Katoh, and H. Imai, *Applications of weighted voroni diagrams and randomization to variance-based k-clustering: (extended abstract)*, 10th Annual Symposium on Computational Geometry, ACM Press, New York, 1994, pp. 332–339.

[15] A. K. Jain and R. C. Dubes, *Algorithms for clustering data*, Englewood Cliffs, NJ, Prentice Hall, 1988.

[16] A. K. Jain, M. N. Murty, and P. J. Flynn, *Data clustering: a review*, ACM Computing Surveys **31** (1999), 264–323.

[17] I. Jolliffe, *Principal component analysis*, 2nd ed., Springer, New York, 2002.

[18] L. Kaufman and P. Rousseeuw, *Finding groups in data: an introduction to cluster analysis*, Wiley Ser. Probab. Math. Statist., A Wiley-Interscience Publication, John Wiley & Sons, Inc., New York, 1990

[19] J. McQueen, *Some methods for classification and analysis of multivariate observations*, 5th Berkeley Symposium on Mathematical Statistics and Probability (L. Le Cam and J. Neyman, eds.), University of California Press, Berkeley, CA, 1967, pp. 281297.

[20] M. Meila, and J. Shi, *A random walks view of spectral segmentation*, AI & Statistics, 2001.

[21] A. Y. Ng, M. I. Jordan, and Y. Weiss, *On spectral clustering: analysis and an algorithm*, Proc. of Neural Info. Processing Systems, NIPS **14** (2001).

[22] M. L. Overton, and R. S. Womersley, *Optimality conditions and duality theory for minimizing sums of the largest eigenvalues of symmetric matrices*, Math. Programming **62** (1993), 321–357.

[23] J. M. Peng and Y. Xia, *A new theoretical framework for K-means-type clustering*, Studies in Fuzziness and Soft Computing, vol. 180, Springer Berlin/Heidelberg, 2005, pp. 79–96.

[24] J. Peng and Y. Wei, *Approximating K-means-type clustering via semidefinite programming*, Technical report, McMaster University, 2005.

[25] S. Qiao, *Orthogonalization techniques for the Lanczos tridiagonalization of complex symmetric matrices*, Adv. Signal Processing Algorithms, Architectures, and Implementations XIV (T. L. Franklin, ed.), Proc. of SPIE, vol. 5559, 2004, pp. 423–434.

[26] J. Shi and J. Malik, *Normalized cuts and image segmentation*, IEEE Trans. on Pattern Analysis and Machine Intelligence **22** (2000), no. 8, pp. 888–905.

[27] H. Späth, *Algorithms for data reduction and classification of objects*, John Wiley & Sons, Ellis Horwood Ltd., 1980.

[28] D. A. Spielman, and Sh. Teng, *Spectral partitioning works: planar graphs and finite element meshes*, 37th Annual Symposium on Foundations of Computer Science (Burlington, 1996), IEEE Comput. Soc. Press, Los Alamitos, CA, 1996, pp. 96–105.

[29] D. Verma and M. Meila, *Comparison of spectral clustering methods*, Technical report, University of Washington, 2005.

[30] J. H. Ward, *Hierarchical grouping to optimize an objective function*, J. Amer. Statist. Assoc. **58** (1963), 236–244.

[31] Y. Weiss, *Segmentation using eigenvectors: a unifying view*, 7th IEEE Internat. Conf. (Greece, 1999) Computer Vision, vol. 2, 1999, pp. 975–982.

[32] E. P. Xing, and M. I. Jordan, *On semidefinite relaxation for normalized k-cut and connections to spectral clustering*, Technical report CSD-03-1265, University of California, Berkeley, 2003.

[33] H. Zha, C. Ding, M. Gu, X. He, and H. Simon, *Spectral relaxation for K-means clustering*, Neural Information Processing Systems (NIPS 2001) (Vancouver 2001), vol. 14, 2001, pp. 1057–1064.

ADVANCED OPTIMIZATION LAB., DEPARTMENT OF COMPUTING AND SOFTWARE, MCMASTER UNIVERSITY, HAMILTON (ONTARIO), L8S 4K1, CANADA.

E-mail address, H. Chen: chenh4@mcmaster.ca

E-mail address, J. Peng: pengj@mcmaster.ca

Centre de Recherches Mathématiques
CRM Proceedings and Lecture Notes
Volume **45**, 2008

Artificial Attributes in Analyzing Biomedical Databases

Zsolt Csizmadia, Peter L. Hammer, and Béla Vizvári

ABSTRACT. This paper consists of two parts. The first one is devoted to some theoretical and algorithmic problems of generating artificial attributes. After providing a mathematical proof of the importance of introducing linear combinations of subsets of attributes as new artificial attributes in data analysis problems, efficient methods are proposed for the identification of significant artificial attributes.

In the second part, by applying a variety of classification methods to a series of publicly available biomedical datasets it is shown that accuracy increases substantially when artificial attributes are introduced.

1. Introduction

The importance of analyzing information contained in data was realized for a long time and led to the development of a large variety of analytic tools, including those of statistics, machine learning, data mining, artificial intelligence etc. One of the tools developed for this purpose is the Logical Analysis of Data (LAD) ([13, 15, 16]), a methodology based on combinatorics, optimization and logic. Many of the analytic tools mentioned above have already been applied successfully to the solution of numerous real life problems arising in biology, medicine, finance, marketing, text mining etc.

One of the most important problems faced by all data analysis methods consists in the selection and identification of the most significant attributes describing a phenomenon. In numerous cases the original description of the problem contains large (sometimes very large) numbers of redundant attributes, and their identification and elimination from further analysis can have significant impact on the accuracy of the resulting models. On the other hand, it is frequently possible to identify new, artificial attributes, synthesized from the given ones, and carrying increased amounts of information. The study of a class of such artificial attributes is the object of this paper.

2000 *Mathematics Subject Classification.* Primary: 68T27; Secondary: 68R99.

Peter Hammer died after he submitted this article.

The authors acknowledge the partial support of NSF and NIH under grant numbers NSF-IIS-0312953, and NIH- HL-072771-01 and NIH-DK-067468-01. The research of the first and third authors was carried out during their visit to RUTCOR in 2004 and 2005.

This is the final form of the paper.

We shall evaluate the usefulness of artificial attributes by including them along with the original ones in several real life datasets and measuring the impact of their presence on the accuracy of various data analysis methods. A prominent role in this analysis will be played by LAD, whose basic elements will be discussed in Section 5.6.

In this paper we shall consider a special class of artificially synthesized attributes, obtained by an arithmetic transformation of a subset $x, y, \ldots, z$ of original attributes, and associating to them a new attribute $ax + by + \cdots + cz$, where $a, b, \ldots, c$ are real numbers. Clearly, the significance of such a new attribute depends on a concept of "degree", representing the number of attributes $x, y, \ldots, z$ used for the transformation. The lower this number, the more easy it is to identify the new attribute, and the higher its significance. In this paper we shall put heavy emphasis on the case of artificial attributes obtained by using only two of the original attributes. A second important characteristic of "valuable" artificial attributes is their ability to separate the positive observations from the negative ones. We shall be primarily interested in the synthesis of those artificial attributes which define separating (or almost separating) hyperplanes, i.e. such that all (or almost all) the positive observations are on one side of these hyperplanes, and all (or almost all) the negative observations are on the other side of them. It will be seen in the paper, that the inclusion of these artificial attributes together with the original ones allows the identification of numerous other, highly significant patterns of very low degrees (usually not more than 1, 2 or 3).

2. Importance of Artificial Attributes in Data Analysis

The aim of this section is to provide a clear mathematical justification of the usefulness of artificial attributes for the analysis of data. We shall show below that the use of artificial attributes allows a substantially more compact representation of information contained in a dataset, than that which uses only the original attributes.

We shall consider now the following problem. Given a finite set Ω^+ of *positive points* in the plane, and a finite set Ω^- of *negative points* in the plane, we shall consider families of "boxes" (i.e. intervals), the union of which covers all the points in Ω^+, and is disjoint from Ω^-. The family $\mathcal{F}$ consist of *boxes* whose sides are of the form $x = c_1$, $x = c_2$, $y = c_3$ and $y = c_4$.

Let us consider now a linear transformation of $\mathbb{R}^2$ which maps the original attributes x and y into the *artificial attributes* $\widehat{x}$ and $\widehat{y}$, respectively. Let $\widehat{\mathcal{F}}$ be a family of boxes in this new space, the union of which covers the points in Ω^+, and is disjoint from Ω^-.

It is natural to ask whether the linear transformation described above could allow us to replace in the transformed space the family $\mathcal{F}$ covering the points in Ω^+, and disjoint from the set Ω^-, by a smaller family $\widehat{\mathcal{F}}$ with similar properties. Let $\mathcal{F}^*$ (respectively $\widehat{\mathcal{F}}^*$) be a minimal family of boxes covering all the points in Ω^+, and disjoint form Ω^-. The aim of this section is to prove the following result.

Lemma 2.1. *The ratio*

$$\frac{|\mathcal{F}^*|}{\left|\widehat{\mathcal{F}}^*\right|}$$

can be arbitrarily large.

PROOF. Let k be an arbitrary positive integer, and let

$$\Omega^+ = \{(i,i) \mid i = 1, \ldots, k\},$$

and

$$\Omega^- = \{(i+1,i) \mid i = 1, \ldots, k\}.$$

It is obvious that if $1 \le i < j \le k$ then if a box contains the positive points (i,i) and (j,j) (the negative points $(i+1,i)$ and $(j+1,j)$) then it contains all the positive points (l,l) (all the negative points $(l+1,l)$), with $i \le l \le j$. If a box contains the positive points (i,i), and $(i+1,i+1)$ (the negative points $(i+1,i)$, and $(i+2,i+1)$) then it contains the negative point $(i+1,i)$ (the positive point $(i+1,i+1)$). Therefore for each $\omega^+ \in \Omega^+$ there must exists a box R_{ω^+} which does not cover any other point of Ω^+. Hence,

$$|F^*| \ge k$$

and therefore,

$$|F^*| = k.$$

On the other hand let us consider the box, whose vertices are the points

$$\left(\frac{3\sqrt{2}}{4}, \frac{-\sqrt{2}}{4}\right), \left(\frac{3\sqrt{2}}{4}, \frac{\sqrt{2}}{4}\right), \left(k\sqrt{2} + \frac{\sqrt{2}}{4}, \frac{\sqrt{2}}{4}\right), \left(k\sqrt{2} + \frac{\sqrt{2}}{4}, \frac{-\sqrt{2}}{4}\right).$$

This box is obviously not a member of $\mathcal{F}^*$, since it does not contain any point of Ω^+. However if we rotate $\mathbb{R}^2$ by 45^0 we get the box with vertices $(\frac{1}{2}, 1), (1, \frac{1}{2}), (k+\frac{1}{2}, k), (k, k+\frac{1}{2})$, which contains every positive point, and none of the negative points. Thus

$$|\hat{F^*}| = 1,$$

proving the statement. □

It follows from this lemma, that there is a possibility of simplifying various models (including LAD) by replacing the original concepts (e.g. patterns) by their generalized forms, in which the edges are not necessarily parallel to the axes. The simplification which is achieved in this way has the potential to reduce substantially the number of patterns used in these classification models.

3. Generation of Artificial Attributes

In this section we shall describe some heuristic and an exact method to determine efficient linear transformations for the generation of artificial attributes. The presentation below will involve small subsets (in most cases only two) of the original attributes.

In the algorithms to be proposed we shall first select a pair of attributes x, y, associate to it an "*artificial attribute*" $ax + by + c$, and then project the points of $\Omega = \Omega^+ \cup \Omega^-$ on the plane determined by x and y. The aim of the algorithms is to find pairs of attributes for which there exists a line $ax + by + c = 0$ in this plane, separating "approximately well" the projections of Ω^+ and Ω^-.

The half-plane $(x,y) : ax+by+c \ge 0$ will be called *positive*, and its complement *negative*. The quality of a separation is measured by the number of points in Ω^-

lying in the positive half-plane, plus the number of points in Ω^+ lying in the negative half-plane.

We shall present below 6 methods of iteratively generating and evaluating artificial attributes. The first two methods frequently provide artificial attributes of relatively low quality, which however are very easy to compute. These two methods are proposed for the preprocessing of large datasets, i.e. for identifying those pairs of attributes which are "promising" to generate high quality attributes using the last four, more sophisticated, methods in this section.

The generation process will be repeated for a large set of attribute pairs (in certain cases for every attribute pair, and in other cases only for certain subsets of attribute pairs); the choice of the family of pairs to be considered will be discussed in Section 5.

3.1. The method of estimated direction. In this section we shall describe a simple and fast heuristic algorithm aimed at determining all those subsets S of attributes, which have sufficiently low cardinalities, and on which one can define a linear discriminant which "approximately" separates the projections of Ω^+ and Ω^- on S. By "approximate" separation we mean that the total number of observation whose projections on S lying on the "wrong" side of the separating hyperplane is below a prescribed threshold value. The algorithm is described in this section for the special case of $|S| = 2$.

Let us first consider two attributes x_i and x_j (denoted for simplicity as i and j), and let us consider an arbitrary observation $\omega \in \Omega^+ \cup \Omega^-$. Let us denote by ω_{ij} the projection of ω on the two-dimensional Euclidean space defined by i and j.

Let us consider now a positive observation $\pi \in \Omega^+$, and a negative observation $\nu \in \Omega^-$. We shall denote by $\nu_{ij}(\pi_{ij})$ the negative observation whose projection on $\mathbb{R}^2$ is nearest in Euclidean distance to π_{ij}. Let us further denote by

$$\mathbf{s}(\pi) = \begin{cases} \dfrac{\pi_{ij} - \nu_{ij}(\pi_{ij})}{\|\pi_{ij} - \nu_{ij}(\pi_{ij})\|} & \text{if } \|\pi_{ij} - \nu_{ij}(\pi_{ij})\| \neq 0, \\ 0 & \text{otherwise.} \end{cases} \tag{3.1}$$

We shall consider the family of those lines whose normal vector is

$$\mathbf{s} := \sum_{\pi \in \Omega^+} \mathbf{s}(\pi), \tag{3.2}$$

and shall select from them one, which minimizes the number of errors, i.e. one for which the number of negative points on the positive side plus the number of positive points on the negative side is minimized. It is easy to remark, that in order to calculate this minimum it is sufficient to restrict our attention to the family of those lines which pass through points of Ω. An additional simplification which is easy to notice is that this number does not change if we further restrict our attention only to lines which pass through at least one of the points in either Ω^+ or Ω^-. In the algorithm described below, we shall only consider those lines which pass through at least one of the points in Ω^-. The smallest number of errors detected in this way will be denoted by μ_{ij}.

We shall assume that a numerical threshold τ is given, and we shall call an attribute pair (i, j) *admissible* if $\mu_{ij} \leq \tau$. The purpose of the algorithm described below is the enumeration of all admissible pairs (i, j).

We shall also assume that we are given a *numerical gap* ε; a pair consisting of a positive and a negative a observation will be called *distinguished* if their Euclidean distance is at least ε. In order to avoid numerical instability we shall replace in the algorithm below the definition of $s(\pi)$ by

$$\mathbf{s}(\pi) = \begin{cases} \dfrac{\pi_{ij} - \nu_{ij}(\pi_{ij})}{\pi_{ij} - \nu_{ij}(\pi_{ij})} & \text{if} \quad \|\pi_{ij} - \nu_{ij}(\pi_{ij})\| \geq \varepsilon, \\ 0 & \text{otherwise.} \end{cases} \tag{3.3}$$

Algorithm MED

Input data:
  $\Omega^+ \cup \Omega^-$, τ, ε
Begin
  Forall attribute pairs (i, j), $i \neq j$
    $\mathbf{s} := 0 \in \mathbb{R}^2$
    Forall positive observation $\pi_{ij} \in \mathbb{R}^2$
      If $(\|\pi_{ij} - \nu_{ij}(\pi_{ij})\| > \varepsilon)$ then
        $\mathbf{s} := \mathbf{s} + \frac{\pi_{ji} - \nu_{ji}(\pi_{ij})}{\|\pi_{ji} - \nu_{ji}(\pi_{ij})\|}$
      Endif
      value $:= \infty$
      Forall negative observation $\nu_{ij} \in \mathbb{R}^2$
        Let $\mathbf{s}^T\mathbf{x} := \mathbf{s}^T\nu_{ij}$ be a candidate for the "separating" line
        value := min(number of observations on wrong side,*value*)
      Endforall
      If (*value* $< \tau$) then
        save rotated attribute
      Endif
    Endforall
  Endforall
End

The results of a series of computational experiments with the above algorithm applied to frequently used benchmark problems will be presented in Section 5, showing its efficiency in identifying large numbers of accurate linear separators, each one of which involves only two attributes.

Example. Consider the dataset

index	attributes			class
1	0	3	0	+
2	2	3	1	+
3	0	1	0	+
4	2	1	1	+
5	1	2	3	-
6	3	2	4	-
7	1	0	2	-
8	3	0	3	-

involving 8 observations and 3 attributes. Similarly to the notation used above e.g. 1_{23} denotes the first observation projected to the plane of the second and third attributes. If the observations are projected into the plane $(1,2)$ then the following equations are obtained:

$$\frac{1}{4}1_{12} + \frac{1}{4}2_{12} + \frac{1}{4}3_{12} + \frac{1}{4}4_{12} = 5_{12}$$

and

$$\frac{1}{4}5_{12} + \frac{1}{4}6_{12} + \frac{1}{4}7_{12} + \frac{1}{4}8_{12} = 4_{12}.$$

Since the centers of gravity of Ω^+ and Ω^- coincide, no separation of Ω^+ and Ω^- is possible, see, e.g. [21]. On the other hand the separation is possible on both of the planes $(1,3)$, and $(2,3)$. In the case of plane $(2,3)$ the nearest negative neighbour of the positive observations are the followings:

observation		vector
positive	negative	
1_{23}	5_{23}	$\frac{1}{\sqrt{10}}(1,-3)$
2_{23}	5_{23}	$\frac{1}{\sqrt{5}}(1,-2)$
3_{23}	7_{23}	$\frac{1}{\sqrt{5}}(1,-2)$
4_{23}	7_{23}	$\frac{1}{\sqrt{2}}(1,-1)$

Thus the normal vector of the separating line suggested by MED is

$$\left(\frac{1}{\sqrt{10}} + \frac{2}{\sqrt{5}} + \frac{1}{\sqrt{2}}, -\left(\frac{3}{\sqrt{10}} + \frac{4}{\sqrt{5}} + \frac{1}{\sqrt{2}}\right)\right) \approx (1.918, -3.445).$$

The cut point $c = 0$ separates the sets Ω^+, and Ω^- perfectly. In Fig. 1, the value of c is -1.527 as it is allowed that some observations may lie on the separating line. The equations of the separating lines on planes $(1,2)$ and $(1,3)$ are

$$-2.823w_1 + 2.828w_2 \approx 2.828,$$

and

$$0.391w_1 - 3.445w_3 \approx -2.663,$$

respectively. The second one separates perfectly. The left-hand sides of the equations, i.e. the homogeneous linear terms, are the new high quality artificial attributes even here and in all further continuation of the example.

3.2. Regression. The role of MED is to select those pairs of attributes, which can jointly generate a separating hyperlane. The same role can be played by regression on two independent attributes.

Let ω_i be an observation, ω_{ip} its pth component, and let w_p be the vector of components $\omega_{1p}, \ldots, \omega_{|\Omega|p}$. Let χ be the characteristic vector of the positive observations, i.e.

$$\chi(\omega_i) = \begin{cases} 1 & \text{if} \quad \omega_i \in \Omega^+, \\ 0 & \text{if} \quad \omega_i \in \Omega^-. \end{cases}$$

If p and q are the indices of two components of the observations ω_i ($i = 1, \ldots, |\Omega|$), then a regression on the independent variables w_p, w_q can provide that linear combination of w_p, w_q, which approximates optimally χ in the norm L_2. Let that linear combination be

$$\chi \approx aw_p + bw_q + c.$$

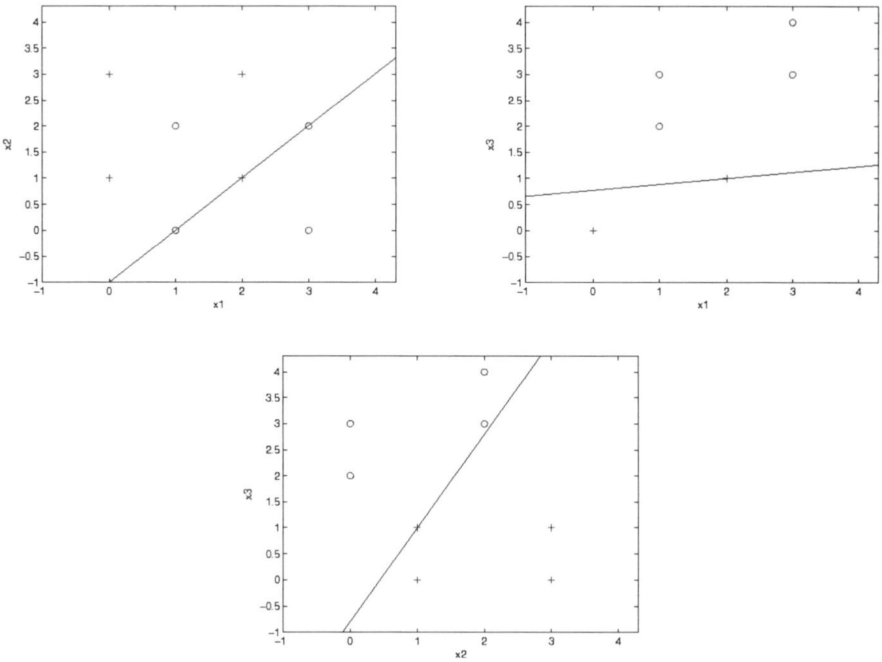

FIGURE 1. Separation lines generated by MED.

Then the expression

$$aw_p + bw_q$$

can be viewed as an artificial attribute, along with its naturally associated cut point

$$\tfrac{1}{2} - c.$$

For a given pair of attributes, regression—similarly to MED—can be viewed as an attribute generation procedure, which is successful if the number of observations assigned to the wrong side of the above cut point is sufficiently small.

Example (cont.). The linear regression formulae obtained in the above mentioned example are as follows:

$$\begin{aligned}
\chi_{12} &\approx -0.1616w_1 + 0.1616w_2,\\
\chi_{13} &\approx +0.1892w_1 - 0.4324w_3 + 0.4724,\\
\chi_{23} &\approx +0.1096w_2 - 0.3014w_3 + 0.3630.
\end{aligned}$$

One obtains a separating line by requiring the right-hand side to be equal to 0. Both the second and the third lines separate perfectly.

3.3. Separating buffer zones. We shall discuss in this subsection three closely related optimization models for artificial attribute construction. The first and the third models were initially introduced by Mangasarian in [19], see also [14], who proposed to use as classification tools the hyperplanes associated to their optimal solutions. As opposed to their original role, these hyperplanes are used in

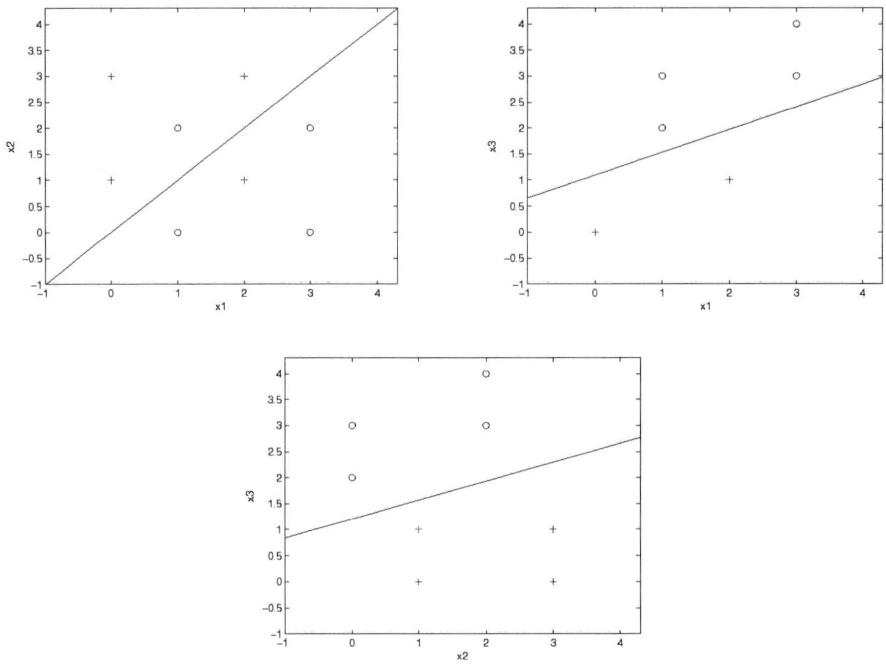

FIGURE 2. Separation lines generated by regression.

this paper to define new, high quality artificial attributes. The second model is a particular case of the well-known support vector machine [14].

In these models the separating hyperplane are not defined in the complete attribute space, but in the subspace of attributes belonging to a specific subset S. It cannot be expected in general that for every particular subset of attributes S, a separating hyperplane would exists. Therefore each of the proposed methods aims at keeping as low as possible the number of those observations which are on the wrong side of the proposed hyperplane. Technically, we replace the search for a separating hyperplane by the search for a separating buffer zone, contained between two parallel hyperplanes; we shall aim at constructing the separating buffer zone in such a way that the half-space lying on one side of it should be disjoint (or almost disjoint) from Ω^+, and the half-space lying on the other side of the zone should be disjoint (or almost disjoint) from Ω^-. The observations falling inside the zone will be considered as classification errors.

We shall need the following notations. For a set S of attributes and for a positive (negative) observation $\omega^+ \in \Omega^+$ ($\omega^- \in \Omega^-$), let ω_S^+ (ω_S^-) denote the projection of the vector ω^+ (ω^-) on the components belonging to S.

In order to formulate a mathematical model describing separating buffer zones, let us introduce the following sets of variables:

(i) the components of the normal vector $\mathbf{a}$ describing the hyperplane,
(ii) the constant terms $b + \varepsilon$ and $b - \varepsilon$ of the right hand sides of the two parallel hyperplanes defining the buffer zone, ε being a fixed (strictly) positive constant, and

(iii) nonnegative correction variables u_ω associated to every $\omega \in \Omega$, introduced to assure the feasibility of the model.

The two essential inequalities describing the requirements that (i) all the observations on the positive side of one of the hyperplanes belong to Ω^+, and (ii) all the observations on the negative side of the other hyperplane belong to Ω^-, can be written as:

$$\forall \omega^+ \in \Omega^+ \mathbf{a}^T \omega_S^+ + u_{\omega^+} \geq b + \varepsilon, \tag{3.4}$$

$$\forall \omega^- \in \Omega^- : \mathbf{a}^T \omega_S^- - u_{\omega^-} \leq b - \varepsilon, \tag{3.5}$$

$$\forall \omega \in \Omega : u_\omega \geq 0. \tag{3.6}$$

The constraints (3.4))–(3.6) are common to each of the three models described in this subsection; these models differ only in their objective functions.

If for a given dataset we construct several different separating hyperplanes based on a common set S of attributes, than in order to maintain the comparability of the models, the constant ε should take the same value in the corresponding inequalities (3.4)–(3.5).

Notice that the inequality system (3.4)–(3.6) has always a solution: $\mathbf{a} = \mathbf{0}$, $b = 0$, and $\forall \omega \in \Omega : u_\omega = \varepsilon$. This will be called their *trivial solution.*

3.3.1. *Linear models.* In the first model our goal is to minimize the sum of all corrections, i.e. our problem reduces to

$$\min \sum_{\omega \in \Omega} u_\omega. \tag{3.7}$$

under the constraints (3.4), (3.5), (3.6).

The objective function (3.7) minimizes the correcting variables in the norm $\|\cdot\|_1$. It is natural to consider also the minimization of the same variables in the norm $\|\cdot\|_\infty$. In order to do this we introduce a new variable, denoted by μ, defined by the constraint

$$\forall \omega \in \Omega : \mu \geq u_\omega. \tag{3.8}$$

The problem becomes then to find

$$\min \mu \tag{3.9}$$

under the constraints (3.4), (3.5), (3.6), (3.8). If the trivial solution is optimal then the optimal value is ε and vice versa if the optimal value is ε then the trivial solution is optimal.

Example (cont.). If the norm $\|\cdot\|_1$ of the vector of the correcting variables is minimized then the following separating lines are obtained:

$$w_1 - 0.3333w_2 \approx 2 \quad \text{and} \quad -1.6667w_1 + 1.3333w_3 \approx 0 \quad \text{and} \quad -w_2 \approx -2.$$

In the case when the norm $\|\cdot\|_\infty$ is minimized the trivial solution is optimal on the plane $(1, 2)$. The two further separating lines are the following:

$$-1.6667w_1 + 0.6667w_3 \approx -1.6667 \quad \text{and} \quad -1.6667w_1 + 0.6667w_3 \approx -1$$

Notice that—as mentioned above—the required high quality artificial attributes are the left-hand sides of the equations.

It should be noted, that if a separating hyperplane exists, then both linear models presented above will find it.

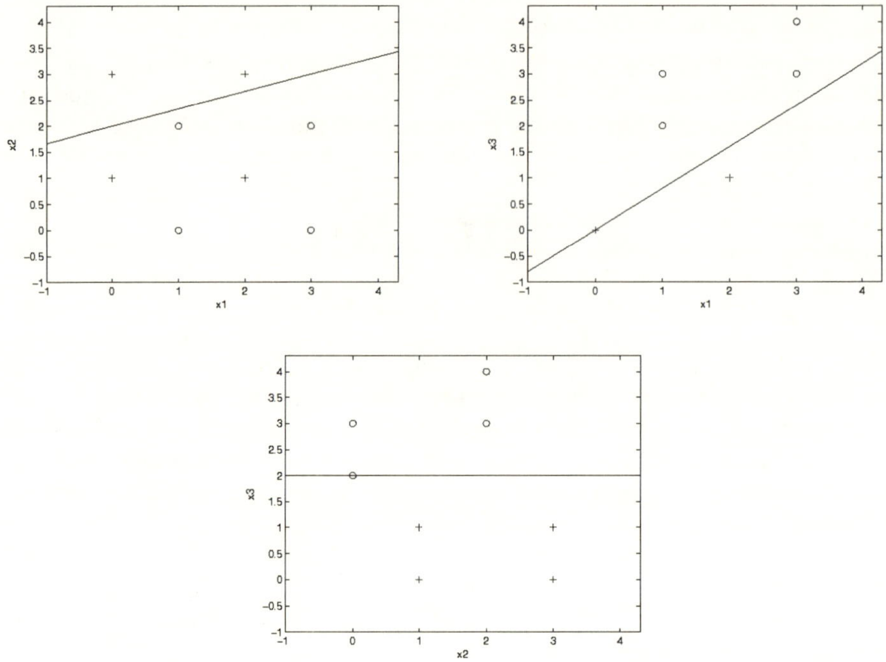

FIGURE 3. Separation lines generated by the linear programming model in $\|\cdot\|_1$ norm.

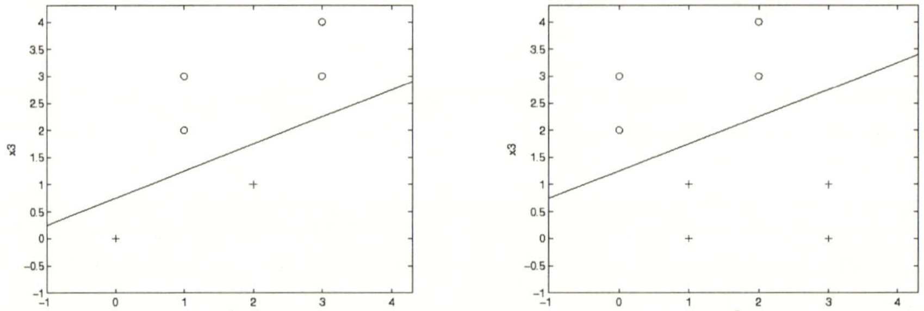

FIGURE 4. Separation lines generated by the linear programming model with $\|\cdot\|_\infty$ norm.

3.3.2. *A quadratic model.* Let us assume that the value of the constant ε is 1. In an ideal case the value of every correcting variable should be zero, i.e. $\mathbf{u} = \mathbf{0}$. In this case the points on the boundary of the two half-spaces determined by the inequalities (3.4) and (3.5) will satisfy the relations

$$\mathbf{a}^T \omega = b + 1 \quad \text{and} \quad \mathbf{a}^T \omega = b - 1, \tag{3.10}$$

and the Euclidean distance of these two hyperplanes is

$$\frac{2}{\|\mathbf{a}\|_2}. \tag{3.11}$$

The robustness of the separation of the positive and negative observations increases with the distance between the two parallel hyperplanes. Therefore, both the minimization of the norm of the vector **u** and the maximization of the distance between the two hyperplanes appear as reasonable objectives.

In order to ensure that the direction of optimization in the two problems (concerning the sum of correcting variables, respectively the distance between the hyperplanes) coincide, we shall not maximize (3.11), but minimize its reciprocal.

For every fixed $0 \leq \lambda < 1$ we shall consider a compromise objective of the form

$$(1-\lambda)\sum_{\omega\in\Omega} u_\omega + \frac{\lambda}{2}\|\mathbf{a}\|_2^2. \tag{3.12}$$

and optimize it under the constraints (3.4)–(3.6).

It should be remarked that the constant λ expressing the relative importance of the two goals can not take the value 1, since in that case the correction variables do not play a role and therefore the optimal solution would automatically be given by $\mathbf{a} = \mathbf{0}$, $b = 0$.

Example (cont.). If $\lambda = 0.5$ then the three separating lines are the following:

$$0.9999w_1 - 0.9907w_2 \approx 0, \ -1.3333w_1 + 0.6667w_3 \approx -1$$

and

$$-1.2w_2 + 0.6w_3 \approx -1.4.$$

3.4. The mixed integer linear programming model. In view of the fact, that datasets admitting separating hyperplanes are rare, we have described in the previous subsection an "approximately separating model" consisting of two parallel hyperplanes such that one of the two "outside" regions is positive, the other is negative, and the buffer zone contained between the two hyperplanes is mixed, containing a (hopefully) small number of positive and negative observations.

The model to be described in this section presents another idea of approximating separating hyperplanes. We shall attempt now to identify a hyperplane such that one of the two half-spaces determined by it is "mostly" positive, the other being "mostly" negative, while allowing for a (hopefully) small number of negative observations on the positive side, and a (hopefully) small number of observations on the negative side. In order to achieve this goal, we shall present a discrete optimization model, to identify a hyperplane which minimizes the number of points on the "wrong" side.

The variables in this model are the components of the normal vector **a** of the hyperplane, the right hand side b of the hyperplane's equation, and zero-one variables u_ω associated to every observation $\omega \in \Omega^+ \cup \Omega^-$, indicating the "position" of the point which represents the observation ω relative to the hyperplane. The variables u_ω will take the value 0 in those observations which are on the correct side of the hyperplane, and the value 1 for those which are on the wrong side.

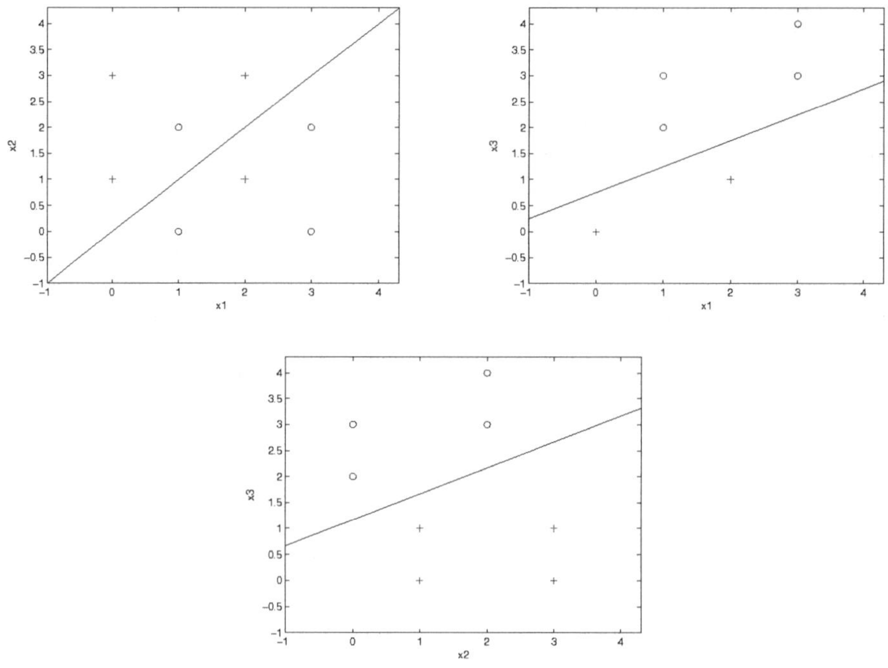

FIGURE 5. Separation lines generated by the quadratic programming model.

Similarly to the previous models the following discrete optimization problem is obtained:

$$\min \sum_{\omega \in \Omega^+ \cup \Omega^-} u_\omega. \tag{3.13}$$

subject to

$$\forall \omega^+ \in \Omega^+ : \mathbf{a}^T \omega_S^+ + M u_{\omega^+} \geq b + \varepsilon, \tag{3.14}$$

$$\forall \omega^- \in \Omega^- : \mathbf{a}^T \omega_S^- - M u_{\omega^-} \leq b - \varepsilon, \tag{3.15}$$

$$u_\omega \in \Omega^+ \cup \Omega^- : u_\omega \in \{0, 1\}, \tag{3.16}$$

where ε is a small fixed positive number having the same role as in (3.4) and (3.5).

Example (cont.). The separating line generated on the planes $(1,3)$ and $(2,3)$ are similar to the lines obtained from other models, but on the plane $(1,2)$ one gets a very different (and better) line:

$$50w_1 + 16w_2 \approx 149, -1.3333w_1 + 0.6667w_3 \approx -1$$

and

$$-1.3333w_2 + 0.6667w_3 \approx -1.6667.$$

The main advantage of this model is that it finds a separating hyperplane if one exists, and produces an approximation of it with a minimum number of errors in the other case. Its disadvantage is a consequence of the computational difficulties

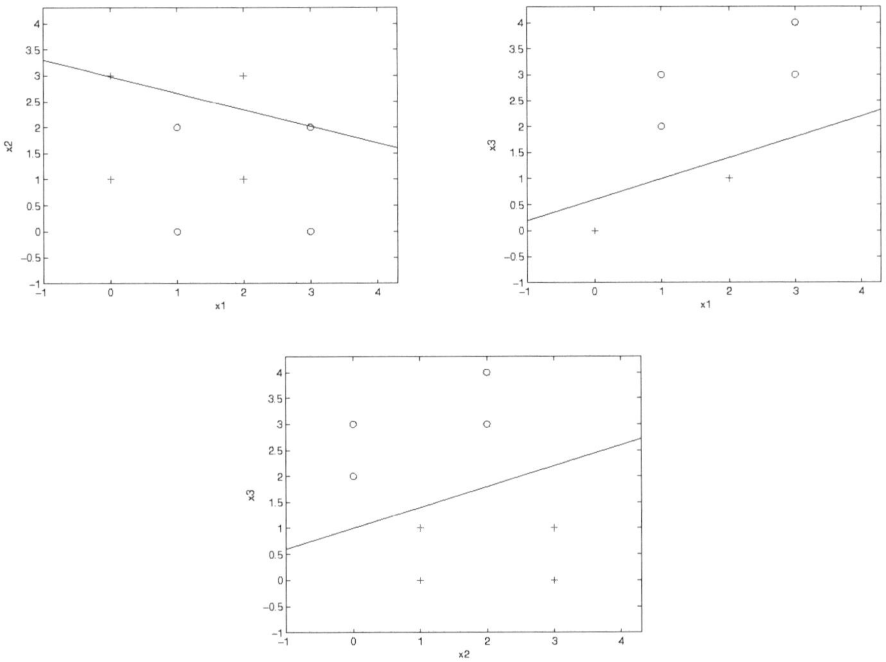

FIGURE 6. Separation lines generated by the mixed integer programming model.

of solving discrete optimization problems with $|\Omega^+ \cup \Omega^-|$ attributes, regardless of the cardinality of S.

3.5. Combined algorithm. For large datasets where the application of the MILP model for every pair of attributes is computationally infeasible, we shall proceed as follows.

We propose a combined algorithm which makes use of the specific advantages of the three algorithms described in the previous subsections. In the first stage of the combined algorithm we apply the first two heuristic methods described is Subsections 3.1 and 3.2 to identify "promising" pairs of attributes. In a second stage we solve the integer programming algorithm proposed in Subsection 3.4 for each one of the so identified promising pairs of attributes.

Although the algorithms described in the first two subsections of this section are very efficient and allow to investigate many attribute pairs, the number of attributes in certain databases (i.e. the genomic and proteomic ones) is too high to allow the checking of all attribute pairs, and additional heuristics are needed to reduce the number of those pairs which are to be examined. As an example, one could use a straightforward preprocessing technique to create a subset $\mathcal{A}$ of attributes whose Pearson correlations with the outcome exceed a prescribed threshold, and examine only those pairs of attributes which have at least one of their components in $\mathcal{A}$. This is the heuristic used in our computational experiments described in Subsection 5.3.

3.6. Iterative application of the models. The generation and evaluation of artificial attributes is carried out within an iterative process. In the first iteration we only consider pairs of original attributes. When executing iteration $k+1$, the attributes created at iteration k will be called the "new attributes", while all the attributes given originally, or created in iterations $1, 2, \ldots k-1$, will be called "old attributes". At iteration $k+1$ we shall examine pairs including at least one of the new attributes.

After creating a set of new attributes in iteration k, the quality of each of them is evaluated. Those new attributes which taken individually can distinguish the sets of positive and negative attributes with a (dataset dependent) sufficiently high accuracy will be termed "efficient", and retained for the next iteration. The procedure stops when no new attributes of acceptable quality are found. The set of all efficient attributes found at different iterative steps will be called the "efficient set of attributes" of the problem.

4. Datasets Used for Computational Experiments

We have carried out a series of computational experiments for creating artificial attributes for seven publicly available datasets. The objective of these experiments was to evaluate the number and the separating quality of the hyperplanes defining the artificial attributes, obtained by using the models described in Section 3.

The seven datasets used in this study are the following:

- Balance Scale Weight & Distance Database[1] (*balance*).
- Breast Cancer Wisconsin Database (*bcw*) [20].
- BUPA Liver Disorders (*bupa*).
- Lymphoma Database (*lymphoma*).
- Ovarian Cancer Database (*ovarian*).
- Breast Cancer Database (*breast*).

The first three datasets were taken from the Machine Learning Repository of the University of California at Irwine (`http://www.ics.uci.edu/~mlearn/MLRepository.html`). The *lymphoma*, the *breast* cancer and the *ovarian* datasets were taken respectively from the following websites:

▷ `http://www.genome.wi.mit.edu/MPR/lymphoma`.

▷ `http://www.rii.com/publications/2002/v\_antveer.htm`.

▷ `http://clinicalproteomics.steem.com`.

The first three datasets contain clinical information on the patients, while the last three datasets contain proteomic or genomic data.

The datasets were slightly modified by the elimination of those observations and attributes which contained large numbers of missing values. Some basic information on these reduced datasets is presented in Table 4.

5. Use of Artificial Attributes for Classification

We have carried out a series of computational experiments in order to determine the efficiency of the proposed algorithms for determining artificial attribute-defining hyperplanes which involve only a small number of original attributes.

---

[1]The original dataset contains 3 classes: "left", "right" and "middle"; in this study we have deleted the observations in class "middle" (representing 7.84% of the total number of observations).

TABLE 1. Database sizes.

Database	Number of Attributes Numerical	Categorical	Number of Observations Positive	Negative
balance	4	0	288	288
bcw	0	9	241	458
bupa	6	0	200	145
breast	23972	0	33	44
lymphoma	7129	0	58	19
ovarian	15155	0	162	91

The computations were carried out on a 2400MHz processor. The running time of the proposed heuristics turned out to be far less then 1 second for every pair of attributes, making it possible to complete the algorithms for each one of the three small datasets, and for a large number of selected pairs even for some large proteomic and genomic datasets, as will be explained in Section 5.3. The exact MILPM method was applied only for those pairs, for which the heuristic methods indicated the existence of promising solutions. The running time of MILPM was about 5 seconds for most pairs of attributes in the proteomic and genomic datasets, and in the case of the other datasets (which involve less attributes, but substantially larger numbers of observations) it was limited to at most 2 minutes. It is also interesting to note, that the most time consuming attribute pairs turned out to be those whose quality was low, i.e. those which produced only poorly separating hyperplanes.

5.1. Separation by original and by artificial attributes. A linear combination of two or more attributes defines—as remarked in Section 2—a new artificial attribute, which—if added to the set of original attributes— may be successfully used for separating the positive observations from the negative ones.

In this section we shall explore ways of using artificial attributes for separating the sets of positive and of negative observations in the dataset. We shall first explore the possibility of using a single artificial attribute for this purpose, and shall extend afterwards this idea to that of combining the separating powers of several artificial attributes.

5.2. Separation quality with and without use of artificial attributes. Let us consider a dataset with binary outcomes, and let x be one of the numerical attributes in it. It is natural to try to find a value x_0 of x with the property that the values of x in all the positive observations should exceed (or be below) x_0, while its values in all the negative observation should be below (or exceed) x_0. Obviously it is not realistic to expect such values x_0 to exist in many real life datasets. However, it is true that one can identify in many such datasets good artificial separating attributes, along with associated cut points, separating reasonably correctly the sets of positive and negative observations.

We have applied this idea to the seven datasets discussed in this paper. We have first analyzed the possibility of separation using only the set of original attributes. Afterwards, we have reconsidered the problem using only artificial attributes associated to pairs of original attributes.

TABLE 2. Best separation obtained using one original or one artificial attribute.

Database	Original Best Separation	Artificial Best Separation
balance	68.92%	76.04%
bcw	92.70%	96.38%
bupa	63.18%	70.72%
breast	79.22%	89.61%
lymphoma	93.50%	100%
ovarian	98.02%	100%

TABLE 3. Description of attributes used in Table 5.2

Database	Original variable Name	Cut Point	Artificial Variable $\alpha x + \beta y$	Cut Point
balance	$(Left - Weight)$	2.5	$(LeftWeight)+$ $(-1)(RightDistance)$	0
bcw	$(Uniformity$ $of\ Cell\ Size)$	3.5	$0.6(UniformityCellSize)+$ $0.4\text{“}BareNuclei''$	2.7
bupa	$(gamma - glutamyl$ $transpeptidase)$	20.5	$-11.8(sgpt)+$ $10.8(gammagt)$	−86
breast	$(Contig31312\_RC)$	0.025	$-0.48(NM\_015849)+$ $1.48(NM\_006544)$	0.052
lymphoma	$(D55716\_at)$	844	$2.3(D55716\_at)-$ $0.7(D78134\_at)$	0
ovarian	$(M/Z\ 244.95245)$	52.8015	$-1.467(M/Z\ 435.46452)+$ $0.466(M/Z\ 682.35609)$	−23.03

The results of these experiments are shown in Tables 5.2 and 5.2. The numbers appearing in the two columns indicate the percentages of positive and negative observations in the datasets which can be separated using a single original or a single artificial attribute, along with a single cut-point associated to it.

It can be seen, that for every single dataset the use of artificial attributes makes it possible to increase the percentage of observations which can be separated in the proposed way, and that in some of the datasets this increase is very substantial (for instance, in the breast cancer dataset the increase is more than 10%, and in the lymphoma dataset it is more than 5%), even using exclusively artificial attributes depending on only two original attributes. The effect of more complex artificial attributes is explored in more details in the next section.

It should also be remarked that for some of the datasets (e.g. ovarian cancer) we identified several artificial attributes and associated cut points which provide a *perfect* (100%) *separation* of the positive observations from the negative ones. For illustration, we present in Fig. 7 two such perfect separations, obtained by using artificial attributes in the case of the *lymphoma* and *ovarian* datasets. Several such hyperplanes (artificial attributes) constructed as linear combinations of two original attributes, with the property that they provide a perfect separation of positive and

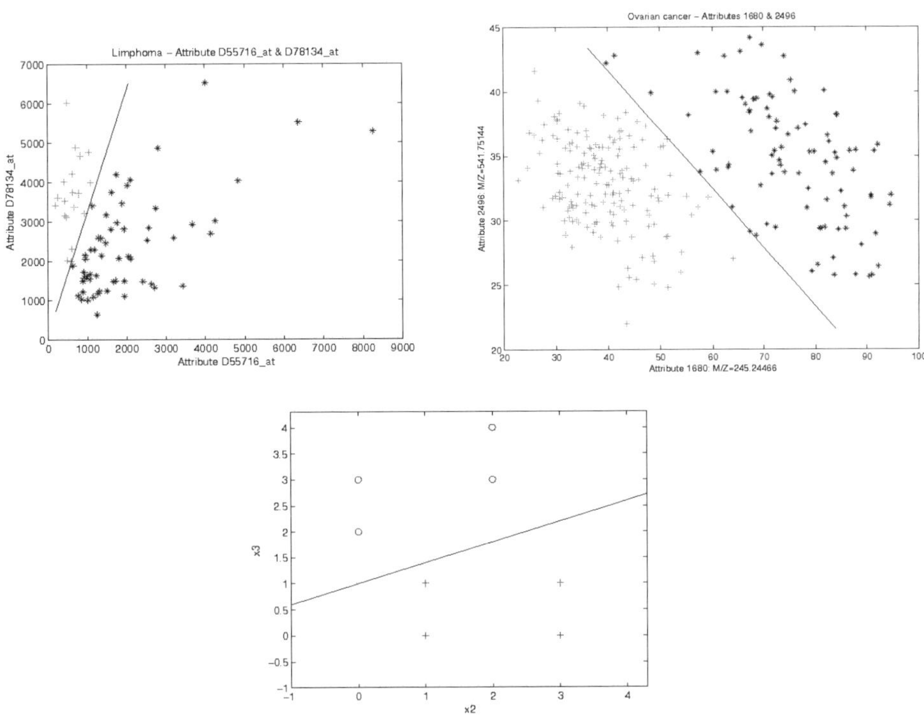

FIGURE 7. Perfect separation of the *lymphoma* and *ovarian* cancer datasets.

negative observations appearing in the ovarian cancer dataset have been previously identified by G. Alexe [1].

5.3. Separation using several artificial attributes. In view of the demonstrated separating power of even the simplest individual artificial attributes, it is to be expected that the combined use of several artificial attributes could further increase the proportion of those positive and negative observations which can be separated. We have experimentally applied this idea to the seven datasets.

Separation on "small" datasets. We present first the results obtained by the iterative method suggested in Section 3.6. The results were cross-validated by 10 5-folding[2] experiments.

The procedure works on each fold independently. Therefore different pairs may prove to be of high quality in the different folds, and even the number of iterations may vary from fold to fold.

We report in the Table 4 the separation accuracy achieved when using one single attribute as classifier. The table shows the relative accuracy obtained by the use of an original attribute, versus that of using an artificial attribute.

[2] k-folding is a well known statistical technique for evaluating the accuracy of a classification procedure, by (1) randomly partitioning the dataset into k (more or less) equally sized subsets of observations, (2) using k-1 of the subsets as a "training set", and the remaining one as the "test set", (3) rotating the selection of that subset which plays the role of "test set", and (4) reporting the average accuracies obtained in classifying the observations in the k "test sets".

TABLE 4. Separation by one attribute.

Database:		Sum of infeasibilities	Quadratic programming	Minimize maximal dist.	Regression
Balance	Original best	72.21%	72.21%	72.21%	72.21%
	LB	89.56%	91.11%	90.04%	79.34%
	AVG	90.51%	91.88%	90.22%	82.56%
	UB	95.13%	94.27%	90.39%	95.31%
Bcw	Original best	93.56%	93.56%	93.56%	93.56%
	LB	95.70%	95.56%	95.55%	94.27%
	AVG	95.99%	95.95%	95.99%	94.72%
	UB	97.71%	97.00%	96.71%	96.28%
Bupa	Original best	61.74%	61.74%	61.74%	61.74%
	LB	62.89%	64.03%	63.23%	63.62%
	AVG	62.97%	64.34%	64.66%	63.76%
	UB	65.21%	66.08%	65.21%	65.50%

The artificial attributes are obtained by one of four methods proposed in Section 3: sum of infeasibilities, quadratic programming, maximal distance minimization, regression. We did not consider artificial attributes produced by MED, which is viewed only as a preprocessing method. We have also eliminated from this table, the artificial attributes obtained by MILP, since the repeated application of this method (within the iterative framework) for creating large numbers of artificial attributes, was considered to be too time consuming. The statistics concerning the artificial attributes obtained by the four methods are represented in the four columns of the table.

As the 5-folding cross validation is repeated 10 times, there are altogether 50 folds generated by the procedure. It follows from the definition of the iterative method, that the separating power of the artificial attributes generated in the very last iteration is higher than the separating power of any attribute generated in any earlier iteration. Therefore Table 4 reports the separation quality of the family of artificial attributes obtained in the last iterations of the 50 folds. By UB (respectively LB) we mean the average of the 50 best (respectively worst) accuracies. By AVG we mean the average of the 50 average separation accuracies corresponding to the individual artificial attributes appearing in the last iterations.

An efficient attribute is a linear combination of original and artificial attributes generated along the iterative process. If for a given efficient attribute x we express each of the artificial attributes as a function depending only on the original attributes used in the process of its generation, then x will appear as a linear combination of a set $S(x)$ of original attributes. The set $S(x)$ will be called an "efficient set".

An efficient attribute will be called "robust" if it occurs several times during the k-folding experiments. If an efficient attribute x is robust, then the set $S(x)$ will be called a "robust set". The "robustness indicator" of an artificial attribute is the percentage of the ten 5-folding (i.e. 50) experiments in which that attribute is present.

Table 5, concerning the same set of ten 5-folding experiments discussed before, presents information about the number of efficient sets of attributes, and the presence and frequency of robust sets among them. For each of the datasets, we

TABLE 5. Efficiency and robustness statistics.

Database:		Sum of infeasibilities	Quadratic programming	Minimize maximal dist.	Regression
Balance	Avg. # iter.	2.6	3	3.4	2
	# of efficient sets	10.5	11	9.5	11
	Max. rob. indicator	100%	100%	100%	100%
	% of most robust sets	100%	100%	22.22%	100%
Bcw	Avg. # iter.	2.2	2.6	2.6	2.4
	# of efficient sets	107.3	109.7	35.2	29
	Max. rob. indicator	100%	100%	100%	100%
	% of most robust sets	18.69%	19.26%	8.57%	20.68%
Bupa	Avg. # iter.	3.2	3.6	1	4.4
	# of efficient sets	11	10.2	4.7	15.7
	Max. rob. indicator	90%	84%	82%	92%
	% of most robust sets	18.18%	20.00%	33.33%	13.33%

present in the first row the average number of iterations used for generating artificial attributes. The second row indicates simply the average number of efficient sets discovered along the process. The third row presents the maximum robustness indicator associated to an efficient set. The fourth row of this table shows the percentage of robustness indicators of maximum value.

Separation on "large" datasets. We have applied to the three larger datasets (genomic and proteomic) the proposed procedure based on the application of the "combined algorithm" of Subsection 3.5. The defining parameters of the algorithm consist of (i) the maximum number of admissible errors in the definition of the "promising" pairs of attributes, i.e. the number of observations lying on the wrong side of the heuristically determined hyperplanes (MED), and (ii) the maximum number of errors allowed for an artificial attribute generated by the selected method from a promising pair suggested by MED, and (iii) in case of the MILP model a time limit on each individual MIP optimization. An artificial attribute provided either by the heuristics or by the other methods is accepted if the number of errors is less than the bound of (ii). In case of the large datasets it is impossible to investigate all pairs because of the combinatorial explosion. In order to reduce the number of pairs, the heuristic search for promising pairs was restricted to pairs having one member from the first 200 attributes best correlated to the classification vector and the other member from the first 200 attributes of the same order (and the first 100 with the first 200 for the MILP).

As shown in Section 5.2 several perfectly separating artificial attributes exist, both for the *ovarian* and the *lymphoma* datasets. Clearly, the number of perfectly separating artificial attributes may increase in a k-folding experiment, since the training set in each experiment is a subset of the original dataset. Based on this observation, the values of the thresholds used in the heuristic algorithm in step (i) were chosen as 4 and 8, and in step (ii) as 0 and 1 for the *ovarian* and *lymphoma* datasets, respectively. The results of the ten 5-folding experiments are reported in Tables 6 and 7.

In the case of these datasets there exist pairs of original attributes yielding nearly perfect separation on the training set, and therefore the iterative method—not being able to identify pairs of attributes exceeding this accuracy—stops after one iteration. Unfortunately the results of applying the same idea on the breast cancer dataset did not produce comparably good results.

TABLE 6. Separation by one attribute on large datasets.

Database:		Sum of infeasibilities	Quadratic programming	Minimize maximal dist.	Regression	MILP
ovarian	Original best	98.03%	98.03%	98.03%	98.03%	98.03%
	LB	97.25%	97.81%	98.17%	97.95%	98.11%
	AVG	97.79%	98.71%	98.41%	98.80%	98.37%
	UB	100%	100%	100%	100%	100%
lymphoma	Original best	93.83%	93.83%	93.83%	93.83%	93.83%
	LB	83.94%	83.93%	85.78%	82.95%	86.91%
	AVG	86.88%	86.88%	89.47%	84.32%	90.77%
	UB	100%	100%	100%	95.78%	100%

TABLE 7. The robustness of separation by one attribute on large datasets.

Database:		Sum of infeasibilities	Quadratic programming	Minimize maximal dist.	Regression	MILP
ovarian	.Avg. # iter.	1	1	1	1	1
	# of efficient sets	702.2	583	231.4	456.5	802.2
	Max. rob. indicator	100%	100%	100%	100%	100%
	% of most robust sets	16.96%	10.70%	18.18%	15.73%	12.45%
lymphoma	Avg. # iter.	1	1	1	1	1
	# of efficient sets	804.2	724.6	362.5	207.2	914.7
	Max. rob. indicator	100%	100%	100%	100%	100%
	% of most robust sets	1.21%	1.21%	4.32%	1.44%	3.56%

5.4. Application to classification. The most natural way of making use of the information contained in artificial attributes is to simply expand the set of original attributes of a dataset by including the artificial ones, and use this extended representation of the observations for their classification. We shall present in this section strong evidence for the validity and usefulness of this approach in supplementing the classification capabilities of various methods.

It is clear that the mathematical ideas underlining classification methods range from statistical to combinatorial. An interesting conclusion of the experimental studies to be presented below is that the classification methods which benefit most from the inclusion of artificial attributes in the datasets are those in which the combinatorial elements prevail.

In view of the above observation we have subdivided this section into two subsections. In the first one we examine the effects of including artificial attributes when using a variety of classification methods. For this purpose we have used six well-known algorithms taken from the publicly available WEKA package [23]. In the second subsection we have examined separately the advantages of including artificial attributes in the dataset, when classification is carried out by the quintessentially combinatorial LAD (Logical Analysis of Data) methodology.

The efficiency of many classification methods examined (including many in the WEKA package, as well as LAD) depends on the choice of some control parameters. In order to exclude any ambiguity resulting from the better or worse choices of parameter values in various methods, we have adopted the uniform policy of using for each method the proposed default values.

We have not included in the study the genomic and proteomic datasets, since their examination would require a preprocessing step (reduction of the original set of attributes to a minimal support set), which can be carried out in a variety of ways, and would therefore not allow the drawing of conclusions with the same degree of clarity as for other datasets.

TABLE 8. Abbreviations of WEKA methods used.

Abbreviation	WEKA method
J48	trees.J48
SL	functions.simpleLogistic
KS	lazy.Kstar
SMO	functions.SMO
VP	functions.voted perceptron
JR	rules.jrip

TABLE 9. Average accuracies of classification based only on original attributes.

Original	J48	SL	KS	SMO	VP	JR
balance	86.77%	94.83%	96.25%	95.00%	94.75%	86.91%
bcw	94.73%	96.25%	95.25%	96.67%	89.85%	95.38%
bupa	65.30%	68.41%	65.97%	57.97%	63.24%	64.70%

5.5. Inclusion of artificial attributes in classification by WEKA algorithms. For evaluating the effects of expanding the set of attributes in a dataset by the inclusion of both original and artificial attributes as described in the previous sections, we have applied 6 classification methods taken from the publicly available WEKA package to the three smaller datasets, systematically used in this study[3].

For each dataset we have applied the 5-folding validation procedure, and repeated it 10 times. The artificial attributes were always extracted using only information taken from the training sets. In each experiment the average classification accuracies using original attributes only are reported in Table 9. The changes in accuracies resulting from the inclusion of artificial attributes, generated using one of the six proposed methods, are shown in the six subtables of Table10.

It should be noted that since the extended sets of attributes cannot be considered to be statistically independent, the Bayesian methods are not applicable, and have not been included in this study.

Table 10 shows clearly that in about $\frac{2}{3}$ of the cases, the introduction of artificial attributes along with the original ones, increases accuracy, and in many cases this increase is significant. It must also be noted, that in about $\frac{1}{3}$ of cases, the accuracy decreases. We believe that the possible decreases, if any, are due to accumulated noise, and in practical applications we suggest to first test the possible changes of accuracy for the specific dataset and the specific method to be used. It should be noted that the possible gain of accuracy in a favourable case is significant, and strictly dominates the possible loss in the worst case; indeed, the average best increase of accuracy is of 5.73%, while in the worst case the average decrease of accuracy is 1.89%. The positive effect of the use of artificial attributes is also supported by the fact that the average of all the numbers (i.e. classification accuracies) in Table 10 is positive.

In an iterative procedure, two strategies may be applied. The first one is that in every iteration all of the generated artificial attributes are kept and taken into

[3]Special thanks are due to Ms. Anupama Reddy for having compared our results on several datasets with those obtained by applying to them algorithms contained in the WEKA package.

TABLE 10. Changes in average classification accuracies with inclusion of artificial attributes.

Original + MIP	J48	SL	KS	SMO	VP	JR
balance	+8.54%	+1.33%	−0.13%	+0.43%	+0.21%	+6.35%
bcw	+2.61%	+0.46%	+0.61%	+0.04%	+7.29%	−0.86%
bupa	−0.08%	−1.19%	+0.99%	+6.96%	+1.98%	+0.81%
Original + Sum. Infeas.	J48	SL	KS	SMO	VP	JR
balance	+9.92%	+0.47%	+1.14%	−0.05%	+0.37%	+10.30%
bcw	+0.55%	+0.32%	+0.04%	−0.24%	+6.73%	+0.19%
bupa	−2.11%	−0.58%	−0.17%	+6.38%	+1.11%	−0.64%
Original + QP	J48	SL	KS	SMO	VP	JR
balance	+9.75%	−0.40%	+1.31%	+0.13%	+0.55%	+9.78%
bcw	+0.55%	+0.18%	−0.11%	−0.24%	+6.58%	+1.05%
bupa	−0.96%	−0.87%	−3.36%	+6.67%	+1.40%	−1.22%
Original + MinMax	J48	SL	KS	SMO	VP	JR
balance	+0.88%	−0.40%	+0.96%	−0.74%	+0.72%	+8.91%
bcw	−0.31%	+0.61%	+0.32%	+0.05%	+0.15%	+0.34%
bupa	−1.83%	+0.00%	+1.86%	+0.87%	+0.24%	−2.09%
Original + Regression	J48	SL	KS	SMO	VP	JR
balance	+8.88%	−0.05%	+2.18%	−0.57%	+0.72%	+10.13%
bcw	+0.84%	+0.03%	+0.18%	+0.05%	+5.30%	+0.77%
bupa	+2.52%	−0.87%	−4.52%	+7.54%	+0.53%	+2.55%

consideration in the next iteration. The second strategy is to keep only artificial attributes having good separation quality. It is natural to ask whether it is preferable to generate at each step of the iterative procedure artificial variables associated to every pair of attributes, or to just generate those artificial attributes which provide a high quality separation of the positive observations from the negative ones. We have examined this question for each of the methods, with the exception of MIP, where the iterative examination of all artificial attributes associated to every pair of present attributes would be computationally infeasible. For all of the datasets (*balance*, *bcw*, *bupa*) the proposed selective generation of attributes gives better results than the association of a new attribute to every existing pair.

5.6. Inclusion of artificial attributes in classification by LAD. The Logical Analysis of Data (LAD) is a prominently combinatorial technique, whose essence consists in identifying large collections of "patterns". The patterns in LAD's terminology represent boxes in the Euclidean space (in the sense explained in Section 2), defined by conditions of the type $x_i \geq a_i$ or $x_i \leq b_i$ $(i \in I)$, where x_i are attributes, a_i and b_i are real numbers (called cut points), and the sets I have low cardinality (usually $|I| \leq 2, 3$, or at most 4). LAD has been first proposed in [15] and [16], and a description of its basic ideas can be found in [13] and [17].

The basic concept of LAD is that of *patterns.* The definition of a pattern uses that of an elementary conjunction. An *elementary conjunction* is defined by a set of conditions requiring that a subset of attributes should take sufficiently high values $(x_i \geq u_i, \forall i \in I)$, while another subset of attributes should take sufficiently low values $(x_j < v_j, \forall j \in J)$. An elementary conjunction defines a *positive* (respectively *negative*) *pattern* if (i) its defining conditions are satisfied by a nonempty subset of the given positive (respectively negative) observations in the dataset, and (ii) its defining conditions are not satisfied by any negative (respectively positive) observation in the dataset.

The most significant patterns related to a dataset are those which have two important properties: they have low *degrees* and high *prevalences.* The degree of

TABLE 11. Classification improvement using LAD on the *balance* and *bcw* datasets.

Database	Iterative method:	Accuracy on train set original	extended	Accuracy on test set original	extended	Accuracy on all original	extended
balance							
	Sum. infeas.	91.81%	100%	86.42%	95.05%	87.29%	99.02%
	QP	91.81%	100%	86.42%	96.16%	87.29%	99.24%
	Minmax	91.81%	99.26%	86.42%	94.43%	87.29%	98.30%
	Regression	91.81%	100%	86.42%	95.99%	87.29%	99.20%
bcw							
	Sum. infeas.	97.50%	98.17%	95.42%	95.16%	97.08%	97.57%
	QP	97.50%	98.25%	95.42%	95.00%	97.08%	97.60%
	Minmax	97.50%	98.32%	95.42%	94.99%	97.08%	97.65%
	Regression	97.50%	98.32%	95.42%	94.71%	97.08%	97.60%

a pattern is simply the number $|I| + |J|$ of conditions describing the elementary conjunction which defines the pattern. The prevalence of a, say positive, pattern is the proportion of the positive observations in the dataset "covered" by the pattern, i.e. satisfying all the conditions which describe the elementary conjunction defining the pattern. Efficient algorithms are described in [4, 7, 10, 12] for identifying large families of positive and negative patterns.

Many important interpretations of patterns are at hand. First, in frequent applications the original attributes describe symptoms, and the patterns can be naturally interpreted as logical or synthetic syndromes (small, significant collections of related symptoms). Second, a pattern can be viewed as a function which takes the value 1 in observations covered by it, and 0 in other observations. Third, a pattern can be viewed as a *logically synthesized artificial attribute* or *variable* or *feature*. The information captured in these artificial attributes is highly valuable; e.g. it has been shown in [3] that clustering in the space of these artificial attributes can provide substantial information which is not captured by the original (natural) attributes.

It is intuitively clear that since the patterns of LAD are described by conditions requiring that a few attributes are restricted only to small or only to large values, the number of points "covered" by a pattern can be substantially increased by a "rotation of the axes", e.g. by the type of operations which are inherent in the idea of artificial attributes. A mathematical reflection of this basic idea appears in Lemma 2.1.

The LAD software used was S. Alexe's Datascope [8].

In computational experiments concerning the use of artificial variables in LAD, we have obtained promising results for all the small datasets. In case of the *balance* dataset, the selective generation of only high quality artificial attributes turned out to be more successful than the one step (non iterative) generation of all the artificial attributes associated to every pair of original attributes. On the other hand, for the *bupa* dataset, the addition of all the artificial attributes deduced from all pairs of original attributes proved to give superior results compared to the selective iterative method.

The results presented in Tables 11 and 12 represent average accuracies of our usual 10 times repeated 5-folding experiments, in which LAD has been used to classify the three small datasets, using first only the original attributes, and afterwards the expanded set consisting the original and artificial attributes obtained by the MILP method. Since in the case of LAD, the increase of accuracy was very substantial, the training set statistics are also included in the tables.

TABLE 12. Average classification accuracy using LAD and MILP-generated artificial attributes.

	Accuracy on train set		Accuracy on test set		Accuracy on all	
Database	original	ext.	original	ext.	original	ext.
balance	91.81%	99.83%	86.42%	98.93%	87.29%	99.44%
bcw	97.50%	98.50%	95.02%	96.00%	97.08%	98.00%
bupa	66.24%	81.59%	61.70%	69.57%	65.33%	79.19%

TABLE 13. Average classification increase using LAD and MILP-generated artificial attributes.

	Change		
Database	on train set	on test set	on all
balance	+8.02%	+12.51%	+12.15%
bcw	+1.00%	+0.98%	+0.92%
bupa	+15.35%	+7.87%	+13.86%

Tables 11 and 12 show that the reduction of error rates in LAD, using MILP generated artificial attributes is very substantial, averaging 7.21%. Moreover, the presence of such reductions both in the training and in the test set, shows the robustness of the augmented datasets.

6. Conclusions

The study has clearly demonstrated that

- in all the tested publicly available datasets, there exist extremely informative artificial attributes, capable of separating (alone or in combination with other attributes) the positive observations from the negative ones;
- artificial attributes with the above property can be efficiently identified using the algorithms proposed in this paper;
- the addition of artificial attributes to the original ones leads generally to improved accuracies of most classification methods, with particularly strong enhancements of the combinatorial type classification techniques;
- the accuracy of the Logical Analysis of Data allows extremely strong improvements by the inclusion of artificial attributes in the dataset along with the original ones.

References

1. G. Alexe, *Separation*, Private communication, 2003.
2. G. Alexe, S. Alexe, D. E. Axelrod, P. L. Hammer, and D. Weissmann, *Logical analysis of diffuse large B-cell lymphomas*, Artificial Intelligence in Medicine **34** (2005), 235–267.
3. G. Alexe, S. Alexe, and P. L. Hammer, *Pattern-based clustering and attribute analysis*, Soft Computing **10** (2006), no. 5, 442–452.
4. G. Alexe, S. Alexe, P. L. Hammer, and A. Kogan, *Comprehensive vs. comprehensible classifiers in logical analysis of data*, Discrete Appl. Math., 2007 (in press).
5. G. Alexe, S. Alexe, P. L. Hammer, and B. Vizvari, *Pattern-based feature selection in genomics and proteomics*, Ann. Oper. Res. **148** (2006), 189–201.
6. G. Alexe, S. Alexe, L. Liotta, E. Petricoin, M. Reiss, and P. L. Hammer, *Ovarian cancer detection by logical analysis of proteomic data*, Proteomics **4** (2004), no. 3, 766–783.

7. G. Alexe, and P. L. Hammer, *Patterns in the logical analysis of data*, Discrete Appl. Math. **154** (2006), no. 7, 1039–1049.
8. S. Alexe, *DataScope*, `http://rutcor.rutgers.edu/~salexe/LAD_kit/SETUP-LAD-DS-SE20.zip`.
9. S. Alexe, E. Blackstone, P. L. Hammer, H. Ishwaran, M. Lauer, and C. P. Snader, *Coronary risk prediction by logical analysis of data*, Ann. Oper. Res. **119** (2003), 15–42.
10. S. Alexe and P. L. Hammer, *Accelerated algorithm for pattern detection in logical analysis of data*, Discrete Appl. Math. **154** (2006), no. 7, 1050–1063.
11. E. Balas, C. Ceria, and G. Cornuéjols, *A lift-and-project cutting plane algorithm for mixed* $0-1$ *programs*, Math. Programming **58** (1993), 295–324.
12. T. Bonates, P. L. Hammer, and A. Kogan, *Maximum patterns in datasets*, RUTCOR Research report 9-2006, 2006.
13. E. Boros, P. L. Hammer, T. Ibaraki, A. Kogan, E. Mayoraz, and I. Muchnik, *An implementation of logical analysis of data*, IEEE Trans. Knowledge and Data Engineering **12** (2000), no. 2, 292–306.
14. P. Bradley, *Mathematical programming approaches to machine learing and data mining*, Ph.D Theses, University of Wisconsin–Madison, 1998.
15. Y. Crama, P. L. Hammer, and T. Ibaraki, *Cause-effect relationships and partially defined Boolean functions*, Ann. Opera. Res. **16** (1988), 299–326.
16. P. L. Hammer, *Partially defined Boolean functions and cause-effect relationships*, Internat. Conf. on Multi-Attribute Decision Making Via OR-Based Expert Systems, University of Passau, Germany, 1986
17. P. L. Hammer and T. Bonates, *Logical analysis of data: from combinatorial optimization to medical applications*, Ann. Oper. Res. **148** (2006), 203–225.
18. M. S. Lauer, S. Alexe, E. H. Blackstone, H. Ishwaran, C. P. Snader, and P. L. Hammer, *Use of the logical analysis of data method for assessing long-term mortality risk after exercise electrocardiography*, Circulation **106** (2002), 685–690.
19. O. L. Mangasarian, *Mathematical programming in data mining*, Data Mining and Knowledge Discovery **1** (1997), 183–201.
20. O. L. Mangasarian and W. H. Wolberg, *Cancer diagnosis via linear programming*, SIAM News **23** (1990), no. 5, 1–18.
21. S. Muroga, *Threshold logic and its applications*, John Wiley & Sons, New York, 1971.
22. R. S. Siegler, *Three aspects of cognitive development*, Cognitive Psychology **8** (1976), 481–520.
23. I. H. Witten, E. Frank, and M. Kaufmann, *Data mining: practical machine learning tools with java implementations*, Morgan Kaufmann Publishers, 2000.

MATHEMATICAL DOCTORAL SCHOOL, EÖTVÖS LÓRÁND UNIVERSITY OF BUDAPEST, PÁZMÁNY PÉTER SÉTÁNY 1/C, H-1117 HUNGARY.
E-mail address: `csisza@cs.elte.hu`

RUTCOR, RUTGERS UNIVERSITY, 640 BARTHOLOMEW ROAD, PISCATAWAY, NJ 08854-8003, USA.
Current address: Department of Industrial Engineering, Eastern Mediterranean University, Gazi Magusa.

RUTCOR, RUTGERS UNIVERSITY AND DEPARTMENT OF OPERATIONS RESEARCH, EÖTVÖS LÓRÁND UNIVERSITY OF BUDAPEST, H-1088 BUDAPEST, PÁZMÁNY PÉTER SÉTÁNY 1/C, H-1117 HUNGARY.
E-mail address: `vizvari@cs.elte.hu`

Centre de Recherches Mathématiques
CRM Proceedings and Lecture Notes
Volume **45**, 2008

Recent Advances in Mathematical Programming for Classification and Cluster Analysis

Ya-Ju Fan, Cem Iyigun, and W. Art Chaovalitwongse

ABSTRACT. This chapter is focused on recent advances in mathematical programming methodologies in data mining research, which is a rapidly emerging interdisciplinary research area. The main focus of this review chapter lies on classification (supervised learning) and clustering (unsupervised learning), which are among the most studied data mining tasks. We give a thorough discussion on the mathematical modeling aspect of classification and clustering problems.

1. Introduction

Recent advances in digital data collection and storage technology over the past decade have resulted in the growth of massive databases or data avalanches. Not only have these rapidly growing data avalanches occurred in everyday life, but they have also occurred in a variety of scientific and engineering research applications [33] as well as medical applications [55]. Research in finding meaningful patterns to make sense out of these massive data sets is often known as *Data Mining*. Data mining can be broadly defined as the analysis of large observational data sets to find hidden and unsuspected relationships and to summarize the data in novel ways that are both understandable and useful to the data owner [36]. The term data mining is also often known as knowledge discovery, which is referred to the process of extracting useful information from databases. Since data mining has to deal with data of massive size (curse dimensionality), most data mining tasks can naturally be optimized. Classification and clustering are the most critical tasks in data mining and are fundamentally optimization problems. Mathematical programming methodologies formalize the problem definition and make use of recent advances in optimization theory and applications for the efficient solution of the corresponding formulations. One of the major advantages of a formulation based on mathematical programming is the ease in incorporating explicit problem specific constraints. Mathematical programming approaches, particularly linear programming, have long been used in data mining tasks. The pioneering works in

2000 *Mathematics Subject Classification.* Primary: 62B0; Secondary:90C90.
This material is based upon work supported by the National Science Foundation under Grant No. 0546574.
This is the final form of the paper.

classification began in the 60's [52, 53] and the pioneering works in clustering began in the 70's [38].

In this chapter, without loss of generality we shall assume that the input data are expressed in the form of n-dimensional feature vectors $x \in X \subseteq \mathbb{R}^n$. A general goal of data mining is to estimate an explicit, or implicit, function that maps points of the feature vector from the input space, $X \subseteq \mathbb{R}^n$, to an output space, C. The data mining tasks can thus be defined based on the nature of the mapping C. One can, in turn, categorize the data mining tasks into (1) supervised and (2) unsupervised learning. For the supervised learning, we are usually given a finite representative subset of the original data with the mapping C, known as the training set, and we wish to make predictions on new unlabeled elements of another data set (testing set). In other words, if the predicted quantity is a categorical value and we know the value that corresponds to each element of the training set then the supervised learning is to identify the mapping that connects the feature vector and the corresponding categorical value (class). This problem is also known as a classification problem. Another related problem in supervised learning is the feature selection problem, which is associated with superfluous information in the feature vector. In data mining, it is believed that having a minimal number of features leads to simpler models, better generalization and easier interpretation. One of the fundamental issues in supervised learning is therefore to identify the least number of features, subset of the original set of features, that best address the two issues previously defined. On the other hand, for unsupervised learning, the class assignment is not known and we seek to identify whether a small, yet unknown, number of classes exist and then define the mapping assigning the features to classes. This problem is also known as a clustering problem.

Nice reviews of classification and clustering were recently presented in [33, 36]. In this chapter, we will concentrate on mathematical programming methodologies based on reformulating the classification, feature selection, and clustering questions as optimization problems. This chapter is organized as follows. In Section 2, we give a brief discussion on classification problems and their connections to discriminant analysis models. In Section 3, we provide a thorough discussion on well studied mathematical programs for classification problems. Later in Section 4, we give an overview of current research on clustering and how they are categorized based on methodologies used. In Section 5, we provide comprehensive mathematical programming models for various clustering approaches.

2. Classification

In general, a classification problem has to deal with categorizing a data sample or an entity into one of G ($G \geq 2$) mutually exclusive groups based upon n (positive integer) specific measurable features of the entity. A classification rule is typically formed from a sample of data points (training data) where the group categorizations are known. Then the rule is used to categorize new data points (test data). In other words, the main goal in classification is to predict a categorical variable (class) based on the values of the feature vector. The classification problems can be grouped based on the number of classes: two-class classification and multi-class classification. The two-class classification problem can be formulated as the search of a function that assigns a given input vector x into two disjoint point sets A and B. The data are represented in the form of matrices. Assuming that the set

A has m elements and the set B has k elements, then $A \in \mathbb{R}^{m \times n}$, $B \in \mathbb{R}^{k \times n}$, describe the two sets respectively. The multi-class classification problem is aimed at finding a partition $\{R_0, R_1, \ldots, R_G\}$ of $\mathbb{R}^m$ for m number of features. Two-group classification methods can be adapted to accommodate more than two group classifications.

The general families of methods for addressing this classification problem include [36]:

- Estimation of the conditional probability of observing class C given the feature vector x (also known as Baysian-based methods).
- Analysis of various proximity metrics and based the decision of class assignment based on proximity (also known as distance-based methods).
- Recursive input space partitioning to maximize a score of class purity (also known as tree-based methods).

One major group of classification methods that is related to the distance-based and tree-based methods is the hyperplane-based method, which is aimed at finding a hyperplane to classify two groups, such as in the form of $x^T\omega = \gamma$, where x denotes points on the hyperplane. The decision rule is then made by the weighting vector ω, and the scalar γ, which are best selected by solving a mathematical programming model. The goal is to have data points of one group lie on one side of the hyperplane and points of the other class lie on the other side of the hyperplane. Different objective functions form different hyperplanes.

The pioneering work in hyperplane-based classification is presented in [52, 53], where it demonstrated how to formulate the problem of constructing planes to separate linearly separable sets of points. This method is known as Support Vector Machines (SVM). This method generally tries to construct a hyperplane that minimizes the upper bound on the out-of-sample error. There have been many variations of mathematical programming models for SVM. One of the most successful models uses the idea that once a data set is transformed to points in a high dimensional space, which is called kernel transformation, every point can be classified by the separating plane [15]. On the other hand, the hybrid model proposed in [24, 25, 32] does not depend on data transformation. The objective is to find a plane that minimizes violations and maximizes satisfactions of the classified groups. Instead of linear programming formulations, there are mixed integer programming (MIP) formulations, such as Glover [31, 1983] that extends the hybrid model by adding binary variables indicating misclassified data points. Other MIP formulations include Abad and Banks [1, 1993], Glen [27, 1999], and Glen [28, 2003]. On the other hand, as opposed to separating two groups by "one" hyperplane, Glen [29, 2005] has piecewise-linear models that gives convex separating planes. In addition, multi-hyperplane formulations that generate multiple linear hyperplanes simultaneously with the consequence of forming a binary decision tree is proposed in [12]. SVM have been generalized to apply to multi-class classification as in the decision tree models [9]. Each decision in a tree involves a support vector machine or a hyperplane. The results usually rely on kernel transformations that project data into a high-dimensional space so that the data becomes almost separable.

Later the feature selection problem is introduced to improve the performance of classification. Mathematical programming methods can be formulated for selecting well represented features/attributes. Mangasarian [51, 1997] gives a feature selection formulation such that the objective not only separates data points into two groups, but also suppresses as many of nonsignificant features as possible. Chaovalitwongse *et al.* [16, 2006] proposed Support Feature Machine (SFM) formulations that will find a set of features that gives the highest performance.

Asparouhov and Danchev [5, 1997] proposed a mathematical programming model that does not need the construction of separating hyperplanes. They formulated a mixed integer programming model with binary variables that are conformed with the Bayesian decision theoretic approach. That study is along the same line with Anderson [3, 1969], where a formulation that incorporates the population densities and prior probabilities is developed.

3. Mathematical Programming for Classification

3.1. Support vector machines: a hyperplane classification model. Support Vector Machines (SVM) is aimed at finding an optimal hyperplane that separates data into two groups, say A and B. The optimal plane is first obtained from training data that has been labeled, which means we know which group each entity comes from. Then the plane can be used for classifying new observations. All data points from A and B will be separated by the plane under the assumption that A and B are separable. This assumption can be achieved if there exits a proper kernel function that projects all data points from A and B into a high dimensional space. The projection into sufficiently high dimensional space will lead to a separable data set. A set of data of two groups may have many possible separating planes. However, there may be only one optimal SVM hyperplane for a data set. Define a margin as the minimum distances from the plane to elements in a group, A or B. Then the objective of SVM is to find a separating hyperplane with the largest margin. One can express a plane as $x^T\omega = \gamma$, where $\omega \in \mathbb{R}^n$ is an n-dimensional vector of real numbers, n is total number of features used to represent a data point, and $\gamma \in \mathbb{R}$ is a scalar. An example of an SVM hyperplane is shown in Fig. 1.

We then represent the set A by the matrix $A \in \mathbb{R}^{m\times n}$ and the set B by the matrix $B \in \mathbb{R}^{k\times n}$, where m and k are number of data points in group A and B,

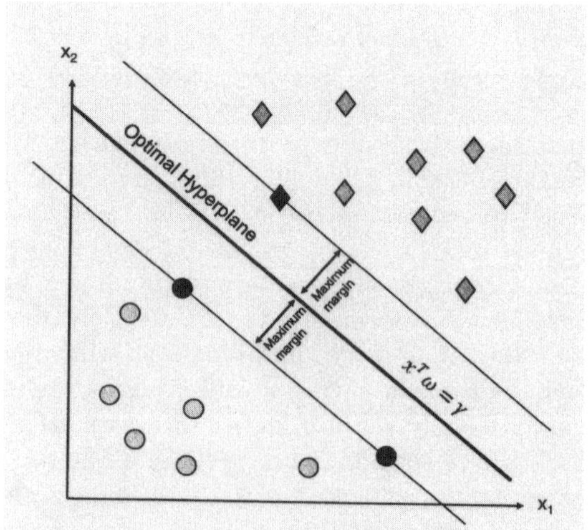

FIGURE 1. Flow chart of methods.

respectively. The plane $x^T\omega = \gamma$ divides the space into two open half spaces. The set A must lie in the open half space $\{x \mid x \in \mathbb{R}^n, x^T\omega > \gamma\}$ and the set B in the open half space $\{x \mid x \in \mathbb{R}^n, x^T\omega < \gamma\}$. Let e denote a vector of ones. Then the following constraints must be satisfied:

$$A\omega > e\gamma, \quad B\omega < e\gamma. \tag{3.1}$$

A linear programming (LP) problem can be formulated to determine vectors ω and γ. However, the constraints in eq. (3.1) are strict inequalities, which can not be used directly in an LP formulation. Equation (3.1) can be reformulated as:

$$A\omega \geq e\gamma + e, \quad B\omega \leq e\gamma - e. \tag{3.2}$$

The main advantage of SVM is that it can incorporate a suitable nonlinear mapping $\phi(\cdot)$, known as kernel function, used to project data points into a sufficiently high dimension so that the data from two classes can always be separated by a hyperplane [15, 20]. The performances of SVM, therefore, rely heavily on the data transformation to represent patterns from the data in a high dimension. Examples of kernel functions include linear, polynomial, radial basis function (RBF) and sigmoid kernel. Shimodaira *et al.* [60, 2001] gives a Dynamic Time-Alignment Kernel that can deal with time series data.

3.2. Robust support vector machines. In practice, it is almost impossible that we can satisfy the assumption that A and B are perfectly separable. In other words, it is possible that the inequalities in eq. (3.1) have no solution, since the data may not be perfectly separable. For this reason, one should try to construct a separating plane that minimizes misclassifications. This leads to the development of robust SVM, which is an improved linear programming model minimizing an average sum of violations as shown in Bennett *et al.* (1992) [10]. The model is given by

$$\begin{aligned} &\min_{\omega,\gamma,y,z} \frac{e^Ty}{m} + \frac{e^Tz}{k} \\ &\text{s.t. } A\omega - e\gamma - e \geq y, \\ &\quad - B\omega + e\gamma - e \geq z, \\ &\quad y \geq 0, \ z \geq 0. \end{aligned} \tag{3.3}$$

The variables y and z in the constraints of this problem satisfy the conditions: $y \geq \max\{0, -A\omega + e\gamma + e\}$ and $z \geq \max\{0, B\omega - e\gamma + e\}$. Hence, y and z are vectors containing violations of (3.2). Minimizing the objective function $e^Ty/m + e^Tz/k$ leads to the minimum average misclassifications. The minimization is equivalent to

$$\frac{1}{m}\| \max\{0, -A\omega + e\gamma + e\}\|_1 + \frac{1}{k}\| \max\{0, B\omega - e\gamma + e\}\|_1 \geq 0.$$

The objective value will be zero if and only if the constraints in eq. (3.2) are satisfied. An extension of this model can be used as a feature selection model as shown in Mangasarian (1997) [51]. A new term is added to the objective function to suppress as many of the components of ω as possible, which gives less emphasis on unnecessary features. This model will be discussed in greater detail later in this section.

3.3. Hybrid discriminant models. A hybrid discriminant model proposed in Glover *el al.* [32, 1988] also provides a hyperplane of the form $x^T\omega = \gamma$. Again, there may be many possible hyperplanes. The goal of the hybrid model is to ensure that data points from one class lie on one side of the hyperplane and the data points from other class lie on the other side. More specifically, the objective is to construct a hyperplane that minimizes misclassifications, and at the same time, maximizes the satisfaction of the classified groups. As a result, the solution is invariant to any data transformations. It is guaranteed to give an optimal solution regardless of the nature of the data. This model is improved to avoid many shortcomings in the previous formulations, which are reviewed and discussed in [45].

Suppose that there are m attributes. Denote each data point by a row vector A_i with length of m. The membership in Group 1 or Group 2 is indicated by $i \in G_1$ or $i \in G_2$, respectively, $(G = G_1 \cup G_2)$. Let A denotes all the row vector A_i's. The hyperplane can be denoted as $x^T\omega = \gamma$, where x^T represents a point on the hyperplane like A_i for some i. The solution is seeking a weighting vector $\omega \in \mathbb{R}^m$, and a scalar $\gamma \in \mathbb{R}$. The optimal plane is to assure that $A_i\omega < \gamma$, for $i \in G_1$ and $A_i\omega > \gamma$, for $i \in G_2$ to the highest possibility. The hybrid LP model is expressed as follows:

$$
\begin{aligned}
&\min h_0 y_0 + \sum_{i\in G} h_i y_i - k_0 z_0 - \sum_{i \in G} k_i z_i \\
&\text{s.t. } \mathbf{A}_i\omega - y_0 - y_i + z_0 + z_i = \gamma, && i \in G_1,\\
&\qquad \mathbf{A}_i\omega + y_0 + y_i - z_0 - z_i = \gamma, && i \in G_2,\\
&\qquad z_0 + \sum_i z_i = 1, && i \in G,\\
&\qquad y_0, z_0 \geq 0, \quad y_i, z_i \geq 0, && i \in G,\ \omega, \gamma \text{ unrestricted.}
\end{aligned}
\tag{3.4}
$$

The variables y_i and z_i (for $i = 0$ and $i \in G$) represent external and internal deviation variables, which are referred to the point violations and satisfactions of classification decision. Particularly, they are the magnitudes how far the points lie outside or inside their correct half spaces. If a point lies outside its targeted half space, there will be a violation, otherwise if a point lies inside, there will be a satisfaction. Especially, the variable y_0 weights the maximum external deviation. On the other hand, the variable z_0 weights the minimum internal deviation. The objective is to minimize violations and maximize satisfactions. In order to achieve this goal, two coefficients are introduced in the objective. The coefficient h_i's (for $i = 0$ and $i \in G$) discourage external deviations and the coefficient k_i's (for $i = 0$ and $i \in G$) encourage internal deviations. They should satisfy $h_i \leq k_i$ for $i = 0$ and $i \in G$. Equation 3.4 is used for normalization which is necessary in order to avoid a trivial solution where all $\omega_j = 0$ and $\gamma = 0$. In addition, Glover [30, 1990] identifies more normalization methods to conquer the null weighting problem.

The MIP model proposed in Glover [31, 1993] is formulated as a simple variant of the hybrid LP model, which gives a hyperplane that minimizes the number of misclassified points. The binary variables z_i ($i \in G$) are used to indicate the number of misclassified points. The new formulation is given by

$$\begin{aligned}
&\min \sum_{i \in G} z_i \\
&\text{s.t. } A_i\omega - Mz_i + \beta_i = \gamma, \quad i \in G_1 \\
&\qquad A_i\omega + Mz_i - \beta_i = \gamma, \quad i \in G_2 \\
&\qquad \beta_i \geq 0,\ z_i \in \{0,1\}, \qquad i \in G,\ \omega, \gamma \text{ unrestricted.}
\end{aligned} \tag{3.5}$$

The classification plane is in the form of $x^T\omega = \gamma$, where x^T is any point on the plane. The β_i ($i \in G$) variables can be viewed as slack variables. The constant M should be chosen large enough so that when a point i from G_1 is misclassified ($z_i = 1$ for some $i \in G_1$), the constraint $A_i\omega \leq \gamma + Mz_i$ will be redundant and when a point i from G_2 is misclassified ($z_i = 1$ for some $i \in G_2$), the constraint $A_i\omega \geq \gamma - Mz_i$ will also be redundant. To encourage a nontrivial solution, it is suggested to include a normalization constraints. The suggested normalization is given by $(-n_2 \sum_{i \in G_1} A_i + n_1 \sum_{i \in G_2} A_i)\omega = 1$, where n_1 and n_2 are the number of data points in G_1 and G_2, respectively. Glover [31, 1993] also provides a theoretical result that this normalization is equivalent to requiring a meaningful separation and eliminates the null weighting ($\omega = 0$) as a feasible solution. Similar to the LP hybrid model, the optimal values of z and β are unchanged for all transformations of the problem data A_i's ($i \in G$).

3.4. Feature selection linear program. An extension of the Robust Linear Programming formulation proposed in Bennett *et al.* (1992) [10] and Mangasarian (1997) [51] can be used for feature selection. A new term is added in the objective function in (3.3) to suppress as many of the components of ω as possible, which gives less emphasis on unnecessary features. Let v denote the absolute value of the weight vector ω, and log is the base of the natural logarithm. $\lambda \in (0,1)$. The mathematical program with a concave objective function and linear constraints for feature selection is as follows

$$\begin{aligned}
&\min_{\omega,\gamma,y,z,v} (1-\lambda)\left(\frac{e^T y}{m} + \frac{e^T z}{k}\right) + \lambda e^T(e - \log^{-\alpha v}) \\
&\text{s.t. } A\omega - e\gamma - e \geq y, \\
&\qquad -B\omega + e\gamma - e \geq z, \\
&\qquad y \geq 0,\ z \geq 0,\ -v \leq \omega \leq v.
\end{aligned} \tag{3.6}$$

When $\lambda = 0$, the problem gives a plane that separates A and B without considering feature suppression. On the other hand, when $\lambda > 0$, the objective function not only tries to separate A and B, but also tries to suppress as many of w components as possible. For each v_i, $i = 1, \ldots, m$, an exponential smoothing of the step function $(1 - \log^{-\alpha v_i})$ is minimized. This step function enables the suppression of irrelevant components of ω. There is also a finitely-terminating algorithm that solves this problem using successive linear programming [14].

3.5. Support feature machines. Suppose there are n samples and m features. Support Feature Machine (SFM) proposed in [16] is aimed at finding a set of features that gives the highest classification performance. Decision variables are $x_j \in \{0,1\}$ indicate if feature j is selected by SFM for $j = 1, \ldots, m$ and $y_i \in \{0,1\}$ indicate if sample i can be correctly classified by SFM for $i = 1, \ldots, n$. SFM has two models, voting and averaging, each with different weighting matrices. Those matrices should be provided by user's classification rule.

The objective function of voting SFM is to maximize the total number of correctly classified data points as in eq. (3.7). There are two sets of constraints used to ensure that the training samples are classified based on the voting nearest neighbour rule as in eqs. (3.8)–(3.9). There is a set of logical constraints in eq. (3.10) used to ensure that at least one feature is used in the voting nearest neighbour rule. The mixed-integer program for voting SFM is given by:

$$\max \sum_{i=1}^{n} y_i \tag{3.7}$$

$$\text{s.t.} \quad \sum_{j=1}^{m} a_{ij} x_j - \sum_{j=1}^{m} \frac{x_j}{2} \leq M y_i, \qquad \text{for } i = 1, \ldots, n \tag{3.8}$$

$$\sum_{j=1}^{m} \frac{x_j}{2} - \sum_{j=1}^{m} a_{ij} x_j + \varepsilon \leq M(1 - y_i), \quad \text{for } i = 1, \ldots, n \tag{3.9}$$

$$\sum_{j=1}^{m} x_j \geq 1 \tag{3.10}$$

$$x \in \{0,1\}^m, \; y \in \{0,1\}^n, \tag{3.11}$$

where $a_{ij} = 1$ if the nearest neighbour rule correctly classified sample i at electrode j, 0 otherwise, n is total number of training samples, m is total number of features, $M = m/2$, and $0 < \varepsilon < \frac{1}{2}$.

The objective function of averaging SFM is to maximize the total correct classification as in eq. (3.12). There are two sets of constraints used to ensure that the training samples are classified based on the distance averaging nearest neighbour rule as in eqs. (3.13)–(3.14). There is a set of logical constraints in eq. (3.15) used to ensure that at least one feature is used in the distance averaging nearest neighbour rule. The mixed-integer program for averaging SFM is given by:

$$\max \sum_{i=1}^{n} y_i \tag{3.12}$$

$$\text{s.t.} \quad \sum_{j=1}^{m} \bar{d}_{ij} x_j - \sum_{j=1}^{m} d_{ij} x_j \leq M_{1i} y_i, \qquad \text{for } i = 1, \ldots, n \tag{3.13}$$

$$\sum_{j=1}^{m} d_{ij} x_j - \sum_{j=1}^{m} \bar{d}_{ij} x_j \leq M_{2i}(1 - y_i) \qquad \text{for } i = 1, \ldots, n \tag{3.14}$$

$$\sum_{j=1}^{m} x_j \geq 1 \tag{3.15}$$

$$x \in \{0,1\}^m, \; y \in \{0,1\}^n, \tag{3.16}$$

where d_{ij} is the average statistical distance between sample i and all other samples from the same class at feature j (*intra-class distance*), $\bar{d}_{ij}$ is the average statistical distance between sample i and all other samples from different class at feature j (*inter-class distance*), $M_{1i} = \sum_{j=1}^{m} \overline{d_{ij}}$, and $M_{2i} = \sum_{j=1}^{m} d_{ij}$.

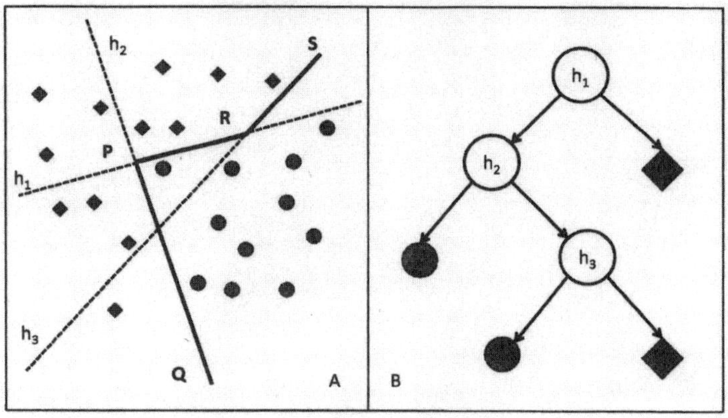

FIGURE 2. Example of SPS tree when $D = 3$.

3.6. Multi-hyperplane classification. Multi-hyperplane formulations [12] generate multiple linear hyperplanes simultaneously with the consequence of forming a decision tree. Instead of using kernel transformations that projects data into high dimensional space for better performance of SVM methods, the multi-hyperplane approach approximates a nonlinear discriminant function that aims to separate the data directly. Denote $d = 0$, which means that we are at a root node of a binary tree where none of the classifications have been done. At $d = D$ the tree has two leaf nodes corresponding to the final separation step. We consider D = 3 for the initial multi-hyperplane model. Figure 2 is an example of one particular type of SPS (Successive Perfect Separation) tree for $D = 3$.

In Fig. 2.(A) separating points of squares from points of circles is expected. There are three hyperplanes h_1, h_2 and h_3 that separates them. Hyperplane h_1 and h_2 construct boundary PQ and PR; hyperplane h_1 and h_3 build the boundary PR and RS. The combined boundary $PQRS$ formed by PQ, PR and RS of the three hyperplanes separates the two groups, squares and circles. Note that it is not necessary for a boundary to make a convex region. Figure 2(B) depicts the binary tree built according to the classifications made by the three hyperplanes, h_1, h_2 and h_3. At the tree node h_1, a *square* will be correctly classified by the hyperplane h_1 if it lies on the correct side of the hyperplane h_1; **or** by both hyperplane h_2 **and** h_3. Similarly, a *circle* will be correctly classified by both hyperplane h_1 **and** h_2, **or** by both h_1 **and** h_3. In order to explain the model, two definitions are as follows. *Successive Perfect Separation (SPS)* is defined a procedure that all elements of either group 1 (G_1) or group 2 (G_2) are compelled to lie on one side of the hyperplane at each node for any depth $d \in \{0, \ldots, D-1\}$. SPS is a special use of a variant based on a proposal of Glover [30, 1990]. A SPS *Decision Tree* is defined as a tree that results from the two-group classification applying the SPS procedure. At $d = 0$ the root node contains all data points in the data set, and at $d = D$ the two leaf nodes corresponds to the final separation step.

For the maximum depth D that is selected, an initial multi-hyperplane model considers each possible SPS tree type of depth d, for $d = 0, \ldots, D$. A root node is taken as a "problem" node where points from both groups need to be separated.

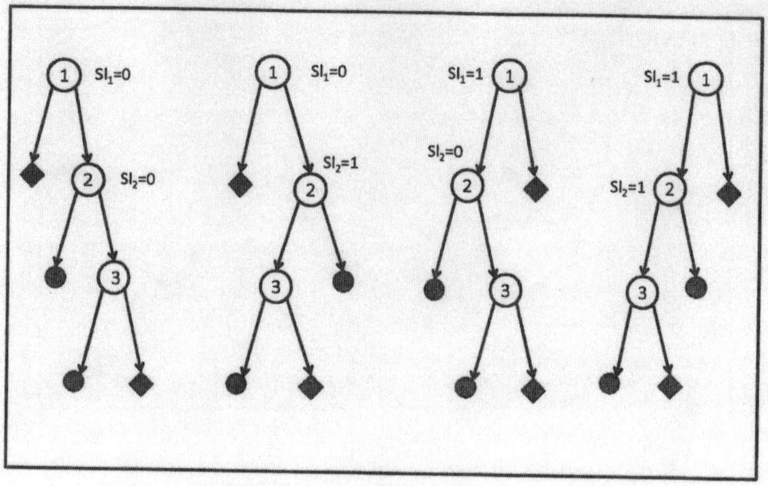

FIGURE 3. All possible SPS tree types for $D = 3$.

A leaf node, on the other hand, is considered a "decision" node where points are classified into two groups. Define slicing variables sl_i for $i \in \{1, \ldots, D-1\}$. There are total $D-1$ slicing variables needed for a tree having maximum depth D. Specifically, at depth $d = 1$, $\mathrm{sl}_1 = 0$ if the "left" node constitutes a leaf node while the "right" node constitutes a root (or problem) node. Also, $\mathrm{sl}_1 = 1$ implies that the "left" node constitutes a problem node and another hyperplane is required to classify the residual objects. Similarly, if $\mathrm{sl}_2 = 1$, and current depth $d = 2 < D$, the tree will slide to the "left" node while the "right" node becomes a leaf node. Figure 3 exhibits all possible SPS tree type for $D = 3$ with their corresponding values of slicing variables at each depth d, $d = 1, 2$.

The mathematical programming formulation is as follows. Let constants M and ε denote large and small positive numbers, respectively. Also, let G denote a set of the union of data points in G_1 and G_2. Two kinds of binary variables are defined as:

$$z_i^* = \begin{cases} 0 & \text{if object } i \text{ is correctly classified by the tree;} \\ 1 & \text{otherwise.} \end{cases}$$

$$z_{hi}^* = \begin{cases} 0 & \text{if object } i \text{ is correctly classified by hyperplane } h; \\ 1 & \text{otherwise.} \end{cases}$$

Traditional hyperplane constraints for each depth d of the tree as well as the normalization methods, which are similar to the mixed integer programming model in [30] are used in the formulation. Explicitly ε is added to prevent points from ending up lying on the hyperplane. Tree type constraints will identify the optimal tree structure for the data set, which will be in part of the optimal classification rule. Binary variables y_i are used for tree types (0,1) and (1,0) to activate or deactivate either-or constraints. This formulation only performs well for small depths and has computational limitations. Details for an improved and generalized structure model for all SPS trees can be found in [12]. The specific structure SPS decision tree formulation for the depth $D = 3$ is as follows:

$$\min \sum_{i=1}^{n} z_i^*$$

s.t.

$$A_i x_d - M z_{di} + \beta_i = b_d - \varepsilon, \quad i \in G_1, d = 1,2,3 \tag{3.17}$$

$$A_i x_d + M z_{di} - \beta_i = b_d + \varepsilon, \quad i \in G_2, d = 1,2,3 \tag{3.18}$$

$$M(\text{sl}_1 + \text{sl}_2) + z_i^* \geq z_{1i} + z_{2i} + z_{3i} - 2, \quad i \in G_1 \tag{3.19}$$

$$M(\text{sl}_1 + \text{sl}_2) + M z_i^* \geq z_{1i} + z_{2i} + z_{3i}, \quad i \in G_2 \tag{3.20}$$

$$M(2 - \text{sl}_1 - \text{sl}_2) + M z_i^* \geq z_{1i} + z_{2i} + z_{3i}, \quad i \in G_1 \tag{3.21}$$

$$M(2 - \text{sl}_1 - \text{sl}_2) + z_i^* \geq z_{1i} + z_{2i} + z_{3i} - 2, \quad i \in G_2 \tag{3.22}$$

$$M(1 + \text{sl}_1 - \text{sl}_2) + z_i^* \geq z_{1i} - M y_i, \quad i \in G_1 \tag{3.23}$$

$$M(1 + \text{sl}_1 - \text{sl}_2) + M z_i^* \geq z_{2i} + z_{3i} - M[1 - y_i], \quad i \in G_1 \tag{3.24}$$

$$M(1 + \text{sl}_1 - \text{sl}_2) + z_i^* \geq z_{1i}, \quad i \in G_2 \tag{3.25}$$

$$M(1 + \text{sl}_1 - \text{sl}_2) + z_i^* \geq z_{2i} + z_{3i} - 1, \quad i \in G_2 \tag{3.26}$$

$$M(1 + \text{sl}_1 - \text{sl}_2) + z_i^* \geq z_{1i}, \quad i \in G_1 \tag{3.27}$$

$$M(1 + \text{sl}_1 - \text{sl}_2) + z_i^* \geq z_{2i} + z_{3i} - 1, \quad i \in G_1 \tag{3.28}$$

$$M(1 + \text{sl}_1 - \text{sl}_2) + z_i^* \geq z_{1i} - M y_i, \quad i \in G_2 \tag{3.29}$$

$$M(1 + \text{sl}_1 - \text{sl}_2) + M z_i^* \geq z_{2i} + z_{3i} - M[1 - y_i], \quad i \in G_2 \tag{3.30}$$

$$\sum_{j=1}^{n} \sum_{d=1}^{3} x_{jd} = 1 \tag{3.31}$$

$$z_i^* \in \{0,1\}, \; z_{di} \in \{0,1\}, \; y_i \in \{0,1\}, \quad i \in G, d = 1,2,3$$

$$\text{sl}_k \in \{0,1\} \; k = 1,2, \; x, b \text{ unrestricted},$$

where the constraints in eqs. (3.17)–(3.18) are the hyperplane constraints, eqs. (3.19)–(3.20) are the constraints for tree type (0,0), eqs. (3.21)-(3.22) are the constraints for tree type (1,1), eqs. (3.23)–(3.26) are the constraints for tree type (0,1), eqs. (3.27)–(3.30) are the constraints for tree type (1,0), and eq. (3.31) is the normalization constraint.

3.7. Separation probabilistic model by Anderson. The above-mentioned deterministic models for classification assume that the data are separable. In the case that the data may not be well separated, using the deterministic models may lead to high misclassification rate. The classification models that incorporate probabilities may be a better option for those kind of data. When the population densities and prior probabilities are known, there is a formulation that considers constrained rules with a reject option [3]. The objective is to maximize the probability of correct allocation subject to constraints on the misclassification probabilities. The mathematical model can be formally defined as follows. Let f_h, $h = 1, \ldots, G$, denote the group conditional density functions. Let π_g denote the prior probability that a randomly selected entity is from group g, $g = 1, \ldots, G$, and α_{hg}, $h \neq g$, are constants between 0 and 1. The probabilistic classification model is then given by

$$\min \sum_{g=1}^{G} \pi_g \int_{R_g} f_g(w)\,dw \tag{3.32}$$

$$\text{s.t.} \int_{R_g} f_h(w)\,dw \le \alpha_{hg}, \quad h, g = 1, \ldots, G,\ h \neq g. \tag{3.33}$$

The optimal rule that can be used as a classification method is given by

$$R_g = \{x \in \mathbb{R}^k : L_g(x) = \max_{h \in 0,1,\ldots,G} L_h(x)\}, \quad g = 0, \ldots, G, \tag{3.34}$$

where $L_0(x) = 0$, and $L_h(x) = \pi_h f_h(x) - \sum_{i=1, i\neq h}^{G} \lambda_{ih} f_i(x)$, for $h = 1, \ldots, G$. In general, there exist nonnegative constants $\lambda_{ih}, i, h \in 1, \ldots, G,\ i \neq h$, such that this optimal rule holds. The procedure for deriving a discriminant rule is composed of two stages. The first stage is to compute $\hat{f}_h$, which are estimated density functions f_h and $\hat{\pi}_h$, which are estimated prior probabilities π_h, for $h = 1, \ldots, G$. There are many methods proposed for density estimation. The second stage is to estimate the optimal $\lambda'_{jh}s$, given the estimates $\hat{f}'_h s$ and $\hat{\pi}'_h s$. For estimating the $\lambda'_{jh}s$, there is a MIP approach proposed in [26], and a LP approach proposed in [50].

The MIP approach uses binary variables to record whether each entity was allocated to each region. This approach measures the probabilities of correct classification and misclassification for any candidate set of $\lambda'_{ih}s$, which are calculated as the proportion of training samples that fall into each of the regions. The objective is to maximize a linear combination of variables representing correct allocation. The proportions of training samples misclassified were incorporated in constraints on misclassification probabilities. On the other hand, the LP approach does not have binary variables to incorporate proportions of misclassified training data points, and to provide a mechanism for modeling a priori bounds on misclassification probabilities. Instead, the LP approach provides a mechanism for estimating $\lambda'_{ih}s$ that balances the minimization of misclassifications and the maximization of correct classifications. This can be demonstrated as follows. Redefine the function L_h, $h = 1, \ldots, G$, as

$$L_h(x) = \pi_h p_h(x) - \sum_{i=1, i\neq h}^{G} \lambda_{ih} p_i(x), \tag{3.35}$$

where $p_i(x) = f_i(x)/\sum_{t=1}^{G} f_t(x)$. This is analogous to the definition of original p_i in eq. (3.34) since R_g can be expressed as

$$R_g = \{x \in \mathbb{R}^k : L_g(x) \le L_h(x) \text{ for all } h = 0, \ldots, G\},$$

if and only if $\big(1/(\sum_{t=1}^{G}) f_t(x)\big) L_g(x) \le \big(1/(\sum_{t=1}^{G}) f_t(x)\big) L_h(x)$. Note that this new definition of L_h is just an assumption. In addition, we also assume that we have a training sample of n data points whose group classifications are known. There are n_g data points in group g and $\sum_{g=1}^{G} n_g = n$. For notational convenience, let $\gamma = 1, \ldots, G$ and $N_g = 1, \ldots, n_g$. Each data point x has k attributes, denoted as $x^{gj} \in \mathbb{R}^k$ for $g = 1, \ldots, G$ and $j = 1, \ldots, n_g$.

3.7.1. MIP *formulation for Anderson's model.* In order to find the optimal estimation of the second stage for solving Anderson's formula in eq. (3.35), after the estimates $\hat{f}'_h s$ and $\hat{\pi}'_h s$ are given, the optimal $\lambda'_{jh} s$ is the final goal. For estimating the $\lambda'_{jh} s$, Gallagher *el al.* [26, 1997] proposed a MIP formulation. Same notation used in Anderson's formula in last section is applied here. The model ensures that the proportion of training data points and total data points n_g of group g in region R_h is less than or equal to a pre-specified percentage, α_{hg} $(0 < \alpha_{hg} < 1)$, for $h, g \in \gamma$ and $h \neq g$.

The original formulation of the approach is a nonlinear mixed integer programming model given by

$$\min_{L_{hgj}, y_{gj}, \lambda_{ih}, u_{gj}} \sum_{g\in\gamma} \sum_{j\in N_g} u_{ggj} \tag{3.36}$$

$$\text{s.t. } L_{hgj} = \pi_h \widehat{p}_h(x^{gj}) - \sum_{i\in\gamma\setminus\{h\}} \lambda_{ih} \widehat{p}_i(x^{gj}), \quad h, g \in \gamma,\ j \in N_g \tag{3.37}$$

$$y_{gj} = \max\{0, L_{hgj} : h = 1, \ldots, G\}, \qquad g \in \gamma,\ j \in N_g \tag{3.38}$$

$$y_{gj} - L_{ggj} \leq M\left(1 - u_{ggj}\right), \qquad g \in \gamma,\ j \in N_g \tag{3.39}$$

$$y_{gj} - L_{hgj} \geq \varepsilon(1 - u_{hgj}), \qquad h, g \in \gamma,\ h \neq g,\ j \in N_g \tag{3.40}$$

$$\sum_{j\in N_g} u_{hgj} \leq \lfloor \alpha_{hg} n_g \rfloor, \qquad h, g \in \gamma,\ h \neq g \tag{3.41}$$

$$-\infty < L_{hgj} < \infty, \qquad h, g \in \gamma,\ j \in N_g$$

$$y_{gj} \geq 0,\ u_{gj} \in \{0, 1\}, \qquad g \in \gamma,\ j \in N_g$$

$$\lambda_{ih} \geq 0, \qquad i \in N_g,\ h \in \gamma.$$

The above nonlinear mixed integer programming model can be transformed to an equivalent linear mixed integer model. The transformation is made by replacing the constraint in eq. (3.38) with the following constraints:

$$y_{gj} \geq L_{hgj}, \qquad h, g \in \gamma,\ j \in N_g,$$

$$\widetilde{y}_{hgj} - L_{hgj} \leq M(1 - v_{ghj}), \quad h, g \in \gamma,\ j \in N_g,$$

$$\widetilde{y}_{hgj} \leq \pi_h \widehat{p}_h(x^{gj}) v_{hgj}, \qquad h, g \in \gamma,\ j \in N_g,$$

$$\sum_{h\in G} v_{hgj} \leq 1, \qquad g \in \gamma,\ j \in N_g,$$

$$\sum_{h\in G} \widetilde{y}_{hgj} = y_{gj}, \qquad g \in \gamma,\ j \in N_g,$$

where $\widetilde{y}_{hgj} \geq 0$ and $v_{ghj} \in \{0, 1\}$, for $h, g \in \gamma$, $j \in N_g$. The constraints in eq. (3.37) define the decision variable L_{hgj} as a function value of L_h at x^{gj} (that is, $L_{hgj} = L_h(X^{gj})$. The variable y_{gj} in eq. (3.38) gives a result that x^{gj} lies in region R_h if and only if, $y_{gj} = L_{hgj}$. The binary variable u_{hgj} is used to indicate whether or not x^{gj} lies in region R_h. The constraints in eq. (3.39) together with the objective function ensure that $u_{ggj} = 1$ if and only if the jth entity from group g is correctly allocated to group g. The constraints in eqs. (3.40)–(3.41) ensure that at most $\lfloor \alpha_{hg} n_g \rfloor$ data points of group g are allocated to group h, $h \neq g$. Note that the condition of indicator variables, $u_{hgj} = 0$, $h \neq g$, implies that $x^{gj} \notin R_h$ by eq. (3.40), but the converse need not hold. As a result, the number of misclassifications may be overcounted. To force the converse hold, (that is $u_{hgj} = 1$ if and only if $x^{gj} \in R_h$,

$\forall h, g \in \gamma$), one can include the following constraints: $y_{gj} - L_{hgj} \leq M(1 - u_{hgj})$ for $h, g \in \gamma$, $j \in N_g$. However, the addition of such constraints substantially increases the solution times and the actual amount of overcounting is minimal. M and ε are large and small positive constants, respectively. Since this MIP formulation is very difficult to solve, especially it involves $2GN$ binary variables. There is a preprocessing strategy suggested in [26] by aggregating variables and constraints. Special branching strategies for solving the MIP model is also suggested in [26]. Those strategies include branching on the smallest indexed fractional-valued binary variable, branching on the most infeasible fractional-valued binary variable, pseudo reduced-cost branching schemes, and strong branching [4, 13].

3.7.2. LP *formulation for Anderson's model.* In order to estimate the $\lambda'_{jh}s$ for solving Anderson's formula in the second stage in eq. (3.35), Lee *el al.* [50, 2003] proposed the Linear Programming (LP) model that minimizes a penalty function in order to allocate each training entity to its correct group or to the reserved-judgement region. Note that same notation used in the MIP approach and Anderson's formula is consistent here. The method is given by

$$\min_{L_{hgj}, \omega_{gj}, y_{gj}, \lambda_{ih}} \sum_{g \in \gamma} \sum_{j \in N_g} (c_1 \omega_{gj} + c_2 y_{gj}) \tag{3.42}$$

$$\text{s.t. } L_{hgj} = \pi_h \widehat{p}_h(x^{gj}) - \sum_{i \in \gamma \setminus \{h\}} \lambda_{ih} \widehat{p}_i(x^{gj}), \quad h, g \in \gamma,\ j \in N_g, \tag{3.43}$$

$$L_{ggj} - L_{hgj} + \omega_{gj} \geq 0, \quad h, g \in \gamma,\ h \neq g,\ j \in N_g, \tag{3.44}$$

$$L_{ggj} + \omega_{gj} \geq 0, \quad g \in \gamma,\ j \in N_g, \tag{3.45}$$

$$-L_{hgj} + y_{gj} \geq 0, \quad h, g \in \gamma,\ j \in N_g, \tag{3.46}$$

$$-\infty < L_{hgj} < \infty, \quad h, g \in \gamma,\ j \in N_g,$$

$$\omega_{gj} \geq 0,\ y_{gj} \geq 0, \quad g \in \gamma,\ j \in N_g,$$

$$\lambda_{ih} \geq 0, \quad i \in N_g,\ h \in \gamma.$$

The constraints in eq. (3.43) define the decision variable L_{hgj} as a function value of L_h for x^{gj} (that is, $L_{hgj} = L_h(X^{gj})$. If the optimal solution yields $\omega_{gj} = 0$, for some (g, j) pair, the constraints in eqs. (3.44)-(3.45) imply that $L_{ggj} = \max\{0, L_{hgj} : h \in \gamma\}$. Thus, when $\omega_{gj} = 0$, it means that the jth entity from group g is correctly classified. If $y_{gj} = 0$ is the case for some (g, j) pair, then the constraints in eq. (3.46) implies that $L_{ggj} = \max\{0, L_{hgj} : h \in \gamma\} = 0$. Hence, the jth entity from group g is placed in the reserved-judgement region. If both ω_{gj} and y_{gj} are positive, the jth entity from group g is misclassified. The optimization solver is attempting either to correctly classify training data points ($\omega_{gj} = 0$), or to place them in the reserved-judgement region ($y_{gj} = 0$). The optimizer's emphasis can be realized by varying the weights c_1 and c_2. It is possible for both ω_{gj} and y_{gj} to be zero. One should decide how to interpret in such situation. Recall the optimal rule in eq. (3.34), which constrains that if x belongs to the reserved judgement region ($h = 0$) then it gives the function value $L_0(x) = 0$.

3.8. Bayesian-based mathematical program. A Bayesian-based mathematical program conformed with the Bayesian decision theoretic approach is proposed by Asparouhov and Danchev [5, 1997]. The model can be formally defined as follows. Denote $c \in \mathbb{R}$ as a cut-off value, $x \in B^m$ as a vector of m binary values, and $\omega \in \mathbb{R}^m$ is a decision variable having m-dimensional vector of real numbers.

A preprocessing needs to be performed so that if $x^T\omega \le c$, the entity x belong to class 1; otherwise it belongs to class 2. Suppose we have a set of n individuals. n_1 of them are from class 1 and n_2 from class 2 ($n = n_1 + n_2$). Let s be a non empty multinomial cell. Denote n_{is} as the number of design set observation from the class i, where $i = 1, 2$, falling in this cell s. There are 2^m number of multinomial cells. Each cell is unique and all observations that belong to it have exactly the same values of the m binary variables. Denote M as a sufficiently large positive real number. In addition to having a geometric interpretation, this formulation is inspired from Bayesian decision theoretic approach and having prior probabilities, $n_i/n\ \forall i$, incorporated. Experimental studies in [5] suggest this Bayesian-based model can give better performance than other contemporary linear discriminant models. The Bayesian-based classification formulation is given by

$$\min_{\omega, z_s, c} \sum_s \left(|n_{1s} - n_{2s}|z_s + \min(n_{1s}, n_{2s})\right) \tag{3.47}$$

$$\text{s.t. } \mathbf{x}_s^T\omega - Mz_s \le c \qquad \text{if} \quad n_{1s} \ge n_{2s}, \tag{3.48}$$

$$\mathbf{x}_s^T\omega + Mz_s \ge c + \varepsilon \qquad \text{if} \quad n_{1s} < n_{2s}, \tag{3.49}$$

$$n_{1s} + n_{2s} \ne 0, \tag{3.50}$$

$$z_s \in \{0, 1\}, \ \omega \in \mathbb{R}^m, \ c \in \mathbb{R}. \tag{3.51}$$

4. Clustering

Clustering is a division of data points into groups of similar objects and the search for clusters is an *unsupervised* learning method which plays an outstanding role in data mining applications such as scientific data exploration, information retrieval, and text mining. In general, clustering is used for the exploration of inter-relationships among a collection of patterns, by organizing them into disjoint and homogeneous clusters. It is called unsupervised learning because unlike classification (known as *supervised learning*), there is no a priori labelling information of some patterns available for categorizing others and inferring the cluster structure of the whole data. Intra-connectivity is a measure of the density of connections between the instances (data points) of a single cluster. A high intra-connectivity indicates a good clustering arrangement because the data points grouped within the same cluster are highly dependent on each other. Inter-connectivity is a measure of the connectivity between distinct clusters. A low degree of interconnectivity is desirable because it indicates that individual clusters are largely independent of each other. Every data point in the data set is represented using the same set of attributes that can be in *continuous*, *categorical* or *binary* form.

A problem with the clustering methods is that the interpretation of the clusters may be difficult. In addition, the algorithms will always assign the data point to clusters even if there were no clusters in the data set. Therefore, if the goal is to make inferences about its cluster structure, it is essential to analyze whether the data set exhibits a clustering tendency. In a real-world application there may be errors (or noises and artifacts) in the collected data set due to inaccurate measurement or due to missing values; therefore, a pre-processing may be needed such as choosing a strategy for handling missing attribute values. On the other hand, the choice of the specific learning algorithm used is another critical step. The issue of relating the learning algorithms to the type of data and to the nature of the problem to be solved still remains an open and fundamental problem [43].

The goal of clustering is to partition the raw data into groups (clusters) that share a common, yet unknown, characteristic property. Similarity is a key property in any clustering method. The difficulty arises from the fact that the process is unsupervised. That is neither the property nor the expected number of groups (clusters) are known ahead of time. The search for the optimal number of clusters is parametric in nature and the optimal point in an "error" vs. "number of clusters" curve is usually identified by a combined objective that weighs appropriately accuracy and number of clusters.

Clustering is a difficult problem due to many factors such as effective similarity measures, criterion functions, algorithms and initial conditions. These factors come into play in devising a well designed clustering technique for a given clustering problem. Moreover, it is well known that no clustering method can sufficiently handle all sorts of cluster structures (*shape, size* and *density*). Various problems such as improving the quality of clusters with pre- or post-processing the data, outlier detection, missing value problem and feature selection are studied in practical applications.

Generally, clustering algorithms can be categorized into *partitioning methods (center-based methods), hierarchical methods, density-based methods, grid-based methods, and model-based methods.* An excellent survey of clustering techniques can be found in [43]. Some more recent works in clustering techniques can also be found in [49]. In each of the clustering methods presented below, there are many papers that describe relevant work. Various applications of the clustering methods and algorithms in many fields can be found in [35]. In the following sections, briefly, it is mentioned about the clustering techniques (as reviewed in [49]) in the literature but the reader can find more details in [43, 49].

4.1. Partitioning methods. Partitioning methods group data points into clusters which are disjoint subsets of similar data points. Since checking all possible subsets is computationally infeasible, certain greedy heuristics are used in the form of *iterative optimization.* Specifically, this means that different relocation schemes iteratively reassign data points between the k clusters. Depending on how the cluster representatives are constructed, iterative optimization partitioning algorithms are subdivided into two major subcategories, *centroid* and *medoid* algorithms. The centroid algorithms represent each cluster by using the gravity center of the data points. The medoid algorithms represent each cluster by means of the data points closest to the gravity center. The most well-known centroid algorithm in the literature is the k-means method [38]. The k-means method partitions the data set into k clusters (subsets) such that all points in a given cluster are closest to the same center. A simple procedure of k-means algorithm can be described as follows. It first randomly selects k of the data points to represent initial clusters, all remaining data points are assigned to their closer center. K-means then computes the new centers by taking the mean of all data points belonging to the same cluster. This operation is iterated until there is no change in the cluster centers. If k is not known a priori, various values of k can be evaluated until the most suitable one is found. The effectiveness of this method as well as of others relies heavily on the objective function used in measuring the distance between data points. The difficulty is in finding a distance measure that works well with all types of data. There are several approaches to define the distance between the data points [43].

The k-modes algorithm [40] is a recent partitioning algorithm and uses the simple matching coefficient measure to deal with categorical attributes. The k-prototypes algorithm [40], through the definition of a combined dissimilarity measure, further integrates the k-means and k-modes algorithms to allow for clustering data points described by mixed attributes. More recently, in [19] another generalization of conventional k-means clustering algorithm has been presented. This new one applicable to ellipse-shaped data clusters as well as ball-shaped ones without dead-unit problem, but also performs correct clustering without pre-determining the number of clusters.

In traditional clustering approaches each data point belongs to one and only one cluster. Hence, the clusters in a *hard clustering* are disjoint. *Soft or Fuzzy clustering* extends this notion to associate each data point with every cluster using a membership function. Larger membership values indicate a higher tendency of the assignment of the data point to the cluster. One widely used algorithm is the Fuzzy C-Means (FCM) algorithm [58], which is based on k-means. In a recent study another soft clustering algorithm has been presented in [8, 41]. The method is based on the joint distance function(JDF), a weighted harmonic mean of the distances in question, that approximates the data in its lowest contours.

Another approach to partitioning methods is to take a conceptual point of view that identifies the cluster with a certain model whose unknown parameters have to be found. Specifically, *probabilistic models* assume that the data comes from a mixture of several populations (*mixture of distributions*) whose distributions and priors are targeted to be found. Most of the probabilistic models are based on the Expectation-Maximization (EM) algorithm [54] where log-likelihood of the data points drawn from a given mixture model. They assume an underlying probability model with parameters that describe the probability that an instance belongs to a certain cluster. The strategy in this algorithm is to start with initial guesses for the mixture model parameters. These values are then used to calculate the cluster probabilities for each data point. These probabilities are in turn used to re-estimate the parameters, and the process is repeated. One clear advantage of probabilistic methods is the interpretability of the constructed clusters. A drawback of such algorithms is that they tend to be computationally expensive. Another problem found in this approach is called the overfitting.

4.2. Hierarchical methods. *Hierarchical* clustering builds a cluster hierarchy or, in other words, a tree of clusters which is also known as a *dendrogram*. Every cluster node contains child clusters (sibling clusters) partition the points covered by their common parent. Hierarchical clustering methods are categorized into two major methods. One is the *agglomerative* (bottom-up) which forms the clusters in a bottom-up fashion until all data points belong to the same cluster. The other is the *divisive* (top-down), which splits up the data set into smaller cluster in a top-down fashion until each cluster contains only one data point [42, 47].

Both divisive and agglomerative algorithms can be represented by dendrograms and they are known for their quick termination. Other advantages of these algorithms include:

(1) they do not require the number of clusters to be known in advance;
(2) they provide a complete hierarchy of clusters;

(3) they provide good result visualizations; and
(4) a "flat" partition can be derived afterwards (using a cut through the dendrogram).

However, both methods suffer from their inability to perform adjustments once the splitting or merging decision is made. For agglomerative hierarchical techniques, the criterion is typically to merge the "closest" pair of clusters, where "close" is defined by a specified measure of cluster proximity. There are three definitions of the closeness between two clusters: *single-link, complete-link* and *average-link*. The single-link similarity between two clusters is the similarity between the two most similar data points, one of which appears in each cluster. Single link is good at handling non-elliptical shapes, but is sensitive to noise and outliers. The complete-link similarity is the similarity between the two most dissimilar data points, one from each cluster. Complete link is less susceptible to noise and outliers, but can break large clusters, and has trouble with convex shapes. The average-link similarity is a compromise between the two. Some of the hierarchical clustering algorithms are: Balanced Iterative Reducing and Clustering using Hierarchies—BIRCH [66], Clustering Using Representatives—CURE [34], and CHAMELEON [46]. More recently, a novel incremental hierarchical clustering algorithm (GRIN) for numerical data sets is presented in [18]. The reader can find the details about these algorithms and a nice comparison in [49, 11].

4.3. Density-based clustering. An open set in the Euclidean space can be divided into a set of its connected components. The implementation of this idea for partitioning of a finite set of points requires concepts of *density, connectivity* and *boundary*. They are closely related to a point's nearest neighbours. A cluster, defined as a connected dense component, grows in any direction that density leads. The *density-based clustering* algorithms basically try to find clusters based on density of data points in a region. The key idea of density-based clustering is that for each data point of a cluster the neighborhood of a given radius (*Eps*) has to contain at least a minimum number of data points (*MinPts*). Since the clusters grow in any direction that density leads the density-based algorithms are capable of discovering clusters of arbitrary shapes which also provides a natural protection against outliers.

One of the most well known density-based clustering algorithms is the DBSCAN [21]. It separates data points into three classes: *core point, border point,* and *noise point.* A core point is a point interior to a cluster. A border point is a point that does not have enough points in its neighborhood, but falls within the neighborhood of a core point. A noise point is a point that is not a core point or a border point. An incremental version of DBSCAN (incremental DBSCAN) is presented in [22]. In addition, another clustering algorithm (GDBSCAN) generalizing the density-based algorithm DBSCAN is presented in [49]. GDBSCAN can cluster data points to both, their numerical and their categorical attributes. Moreover, in [65] the PDBSCAN, a parallel version of DBSCAN is presented. Furthermore, DBCLASD (Distribution-Based Clustering of Large Spatial Data sets) eliminates the need for *MinPts* and *Eps* parameters [64]. DBCLASD incrementally augments an initial cluster by its neigh boring points as long as the nearest neighbour distance set of the resulting cluster still fits the expected distance distribution. While the distance set of the whole cluster might fit the expected distance distribution, this does not necessarily hold for all subsets of this cluster. Another density-based algorithm is the DENCLUE [39] where the basic idea is to model the overall point density analytically as the sum of influence functions of the data points. The influence function can be seen as a function, which describes the impact of a data point within its neighborhood. Then, by determining the maximum of the overall density function can identify clusters. Fast Density-Based Clustering (FDC) algorithm is presented in [67] for density-based clustering defined by the density-linked relationship. The clustering in this algorithm is defined by an equivalence relationship on the objects in the database. The complexity of the algorithm is linear to the size of the database, which is much faster than that of the algorithm DBSCAN.

4.4. Grid-based clustering. Another way of dealing with the concepts of density, connectivity, and boundary used in the previous section is to inherit the topology from the underlying attribute space [11]. To limit the search combinations, multirectangular segments are considered. A segment, that can also be in form of a cube, cell or region, is a direct Cartesian product of individual attribute sub-ranges (contiguous in case of numerical attributes). Since some binning is usually adopted for numerical attributes, methods of partitioning space are frequently called as *grid-based methods*. In general, the attention is shifted from data points to space partitioning. Data partitioning is induced by points' membership in segments resulted from space partitioning, while space partitioning is based on grid-characteristics accumulated from input data. One advantage of this indirect partitioning is that accumulation of grid-data makes grid-based clustering techniques independent of data ordering. Whereas, partitioning methods and incremental algorithms (see Section 4.1) are very sensitive with respect to data ordering. Comparing with the density-based partitioning methods, grid-based clustering methods work with attributes of different types. In contrast density-based partitioning methods work best with numerical attributes.

Grid-based clustering algorithms first quantize the clustering space into a finite number of cells (hyper-rectangles) and then perform the required operations on the quantized space. Cells that contain more than certain number of points are treated as dense and the dense cells are connected to form the clusters. Some of the grid-based clustering algorithms are: STatistical INformation Grid-based method—STING [63], WaveCluster [59], and CLustering In QUEst—CLIQUE [2].

4.5. Model-based clustering. Beyond fitting a particular fixed mixture model, AUTOCLASS proposed in [17] extends the search to different models and different k distributions. To do this AUTOCLASS heavily relies on Bayesian methodology, in which a model complexity is reflected through certain coefficients (priors) in the expression for the likelihood previously dependent only on parameters' values. This approach starts from a random initialization of the parameters, incrementally adjusts them in an attempt to find their maximum likelihood estimates. Moreover, in [56] it is assumed that, there is a hidden variable in addition to the observed or predictive attributes. This unobserved variable reflects the cluster membership for every case in the data set. Therefore, the data-clustering problem is also an example of supervised learning from incomplete data due to the existence of such a hidden variable [44]. Their approach for learning has been called Recursive Bayesian Multinets (RBMNs). Another model based method is Self-Organizing Map (SOM) algorithm [48]. It can be thought as two layers of neural network where each neuron of the network are themselves cluster centers, represented by n-dimensional weight vector; but to accommodate interpretation the map units can be combined to form bigger clusters. The SOM algorithm is trained iteratively. In each training step, one sample vector from the input data set is chosen randomly, and the distance between the input vector and all the weight vectors of SOM is calculated using a distance measure. After finding the best–matching unit (the neuron whose weight vector is closest to the input vector), the weight vectors of SOM are updated so that the best–matching unit is moved closer to the input vector in the input space. This makes the SOM algorithm very robust. Outliers in the data set can be easily detected from the map, since its distance in the input space from other units is large. The SOM algorithm also can deal with missing data values.

5. Mathematical Programming for Clustering

Clustering techniques considered in this section produce a partition of the data points into a specified number of groups by either *minimizing* or *maximizing* some numerical criterion. The analysis here is confined to *distance based cluster analysis* in which a distance measure between the data points is available, and *the volume based cluster analysis* in which the Mahalanobis distance is used as a distance measure instead of Euclidean distance. Differences between the methods in this section occurs because of both the variety of clustering criteria that might be optimized and the various optimization algorithms that might be used. In these methods it is assumed that the number of groups has been fixed before starting the analysis.

The basic idea behind the optimization methods is that associated with each partition of the N data points into the required number of clusters, K, is an index $r(N, K)$, the value of which measures some aspect of the *sufficiency* of this particular partition [23]. For some indices high values are associated with a desirable cluster solution, whereas for others a low value is sought. A variety of clustering criteria have been suggested in the literature. Some operate on the basis of the inter-individual dissimilarities, others employ the original data matrix.

5.1. Clustering criteria from dissimilarity matrix. Most cluster analysis methods rely upon dissimilarities (or similarities) between data points. The concepts of *homogeneity* and *separation* can be employed to develop indices. *Homogeneity* means that data points within the same cluster should resemble one another and *separation* that data points in different clusters should differ one from the other. An informative partition should produce clusters such that the data points within a group have a cohesive structure and clusters are well isolated from each other. This approach is particularly useful for defining cluster criteria on the basis of dissimilarity matrix $\mathcal{D}$, where the elements d_{ij} measuring the dissimilarity between the ith and jth data points. Such similarities satisfy the properties $d_{ij} \geq 0$, $d_{jj} = 0$, and $d_{ij} = d_{ji}$ for $i, j = 1, 2, \ldots, N$. They need not satisfy the triangle inequality. A variety of cluster criteria have been suggested, based on the d_{ij} that minimize the lack of homogeneity or maximize the separation of the clusters [57, 7, 37]. Homogeneity of cluster K_m can be measured by

$$h_1(K_m) = \sum_{i=1}^{N_m} \sum_{j=1, j\neq i}^{N_m} (d_{i,j}),$$

$$h_2(K_m) = \max_{i,j=1,\ldots,N_m; j\neq i} [(d_{i,j})^n],$$

$$h_3(K_m) = \min_{j=1,\dots,N_m} \left[\sum_{i=1}^{N_m} (d_{i,j})^n \right].$$

The first index, $h_1(K_m)$, is the sum of all the (squared) dissimilarities between two data points from cluster m; the second, $h_2(K_m)$, is simply the maximum of the latter. When $n = 1$ and the dissimilarities are metric, $h_2(K_m)$ can be thought of as the diameter of the cluster. Index $h_3(K_m)$ measures the minimum sum of the (squared) dissimilarities between all the data points in cluster K_m and a single cluster member. For $n = 1$ in metric dissimilarities, the index is also referred to as the *star index* because the graph is formed by connecting all the objects of the group with one reference object. The smallest sum of distances is achieved when the reference object is located at the center of the *star*. In this connection the star center can be interpreted as a representative object of the cluster as called the medoid (analogous to calling the cluster mean the centroid). Separation of cluster K_m can be measured by

$$s_1(K_m) = \sum_{i=1}^{N_m} \sum_{k \neq m} \sum_{j=1}^{N_k} (d_{im,jk})^n,$$

$$s_2(K_m) = \min_{\substack{i=1,\dots,N_m \\ j=1,\dots,N_m \\ k \neq m}} [(d_{im,jk})^n].$$

Similarly to the first two homogeneity criteria, $s_1(K_m)$ measures the sum over the (squared) dissimilarities between a data point in the group and a data point outside the group, and $s_2(K_m)$ is simply the minimum of the latter. Having chosen an index measuring a cluster's homogeneity or separation, cluster criteria can be defined by suitable aggregation over clusters, for example:

$$r_1(N, K) = \sum_{m=1}^{K} h_m, \tag{5.1}$$

$$r_2(N, K) = \max_{m=1,\dots,K} [h(m)], \quad \text{or} \tag{5.2}$$

$$r_3(N, K) = \min_{m=1,\dots,K} [h(m)]. \tag{5.3}$$

The criterion in eq. (5.1) reflects the average homogeneity, whereas the criteria in Eqs. (5.2)–(5.3) measure the homogeneity of the worst and best group, respectively. When dealing with the homogeneity criteria a cluster solution is sought that minimizes the cluster criterion $r(N, K)$; when considering separation indices, the aim is to maximize $r(N, K)$. Note that the cluster criteria might also be defined as a combination of homogeneity and separation measures.

5.2. Clustering criteria from continuous data. The most commonly used clustering criteria derived from a $N \times p$ matrix, $\mathbf{X}$, of continuous data make use of a decomposition of the $p \times p$ dispersion matrix, $\mathbf{T}$, given by

$$\mathbf{T} = \sum_{m=1}^{K} \sum_{i=1}^{N_m} (\mathbf{x}_{mi} - \overline{\mathbf{x}})(\mathbf{x}_{mi} - \overline{\mathbf{x}})', \tag{5.4}$$

where $\mathbf{x}_{mi}$ is the p-dimensional vector of observations of the ith data point in cluster K_m and $\overline{\mathbf{x}}$ is the p-dimensional vector of overall sample means for each variable.

This total dispersion matrix can be partitioned into the within-cluster dispersion matrix

$$\mathbf{W} = \sum_{m=1}^{K} \sum_{i=1}^{N_m} (\mathbf{x}_{mi} - \overline{\mathbf{x}}_m)(\mathbf{x}_{mi} - \overline{\mathbf{x}}_m)' \tag{5.5}$$

where $\overline{\mathbf{x}}_m$ is the p-dimensional vector of sample means within cluster K_m, and the between-cluster dispersion matrix

$$\mathbf{B} = \sum_{m=1}^{K} N_m(\overline{\mathbf{x}}_m - \overline{x})(\overline{\mathbf{x}}_m - \overline{x})' \tag{5.6}$$

such that

$$\mathbf{T} = \mathbf{W} + \mathbf{B}. \tag{5.7}$$

For univariate data ($p = 1$), eq. (5.7) represents the division of the total sum of squares of a variable into the within- and between-clusters sum of squares. In the univariate case a natural criterion for grouping would be to choose the partition corresponding to the minimum value of the within-group sum of squares or, equivalently, the maximum value of the between-cluster sum of squares.

In the multivariate case ($p > 1$) the derivation of a clustering criterion from eq. (5.7) is not so clear-cut as the univariate case, and several alternatives have been suggested.

5.2.1. *Minimization of trace* ($\mathbf{W}$). We first consider to minimize the sum of the within-cluster sums of squares, over all the variables, that is

$$\min\{\text{trace}(\mathbf{W})\}, \tag{5.8}$$

which is equivalent to

$$\max\{\text{trace}(\mathbf{B})\}. \tag{5.9}$$

This can be shown to be equivalent to minimizing the sum of the squared Euclidean distances between data points and their cluster mean which is used in K-means algorithms. The criterion can also be derived on the basis of the distance matrix:

$$\mathbf{E} = \sum_{m=1}^{K} \frac{1}{2N_m} \sum_{i=1}^{N_m} \sum_{j=1, j\neq i}^{N_m} d_{i,j}^2, \tag{5.10}$$

where $d_{i,j}$ is the Euclidean distance between the ith and jth data point in cluster K_m. Thus the minimization of trace($\mathbf{W}$) is equivalent to the minimization of the homogeneity criterion $h_1(K_m)/N_m$ for Euclidean distances and $n = 2$ [23].

5.2.2. *Minimization of* det($\mathbf{W}$). The differences in cluster mean vectors are based on the ratio of the determinants of the total and within-cluster dispersion matrices. Large values of $\det(\mathbf{T})/\det(\mathbf{W})$ indicate that the cluster mean vectors differ. Thus, a clustering criterion can be constructed as the maximization of this ratio;

$$\min\left\{\frac{\det(\mathbf{T})}{\det(\mathbf{W})}\right\}. \tag{5.11}$$

Since all partitions of N data points into K clusters, $\mathbf{T}$ remains the same, this problem is equivalent to

$$\min\{\det(\mathbf{W})\}. \tag{5.12}$$

5.2.3. *Maximization of trace* $(\mathbf{BW}^{-1})$. A further criterion considered is a combination of dispersion matrices:

$$\max\left\{\text{trace}\left(\frac{\mathbf{B}}{\mathbf{W}}\right)\right\}. \tag{5.13}$$

This criterion is obtained from the product of the between-clusters dispersion matrix and the inverse of the within-clusters dispersion matrix. This function is also a further test criterion used in the context of multivariate analysis of variance, with large values of trace$(\mathbf{BW}^{-1})$ indicating that the cluster mean vectors differ.

5.2.4. *Comparison of the clustering criteria.* Of the three clustering criteria mentioned above, the criterion in eq. (5.13) is perhaps the one most commonly used. However it suffers from some serious problems [23]. Firstly, the method is scale-dependent. Different solutions may be obtained from the raw data and from the data standardized in some way. Clearly this is of considerable practical importance because of the need for standardization in many applications. Another problem with the use of this criterion is that it may impose a *spherical* structure on the observed clusters even when the natural clusters in the data are of other shapes. The alternative criteria in eqs. (5.8) and (5.13) are not affected by scaling which is the main motivation behind of these criteria. Moreover, the criterion in eq. (5.12) which has been widely used does not restrict clusters to being spherical. It can also identify elliptical clusters. On the other hand, this criteria assumes that all clusters in the data set have the same shape (i.e., same orientation). Finally, both the criteria in Eqs. (5.8) and (5.12) produce clusters that contain roughly equal numbers of data points.

5.2.5. *Optimization algorithms.* After choosing a suitable clustering criterion, the question needs to be answered is how to find a partition into k clusters that optimizes the criterion. Although in theory the value of the criterion can be calculated for each possible partition and choose a partition that gives an optimal value for the criterion, in practice the task is not so straightforward. The complete enumeration of every possible partition is simply not possible [23]. This problem has led to the design of algorithms to search for the optimum value of a clustering criterion. Certain greedy heuristics are used in the form of *iterative optimization.* Specifically, this means different relocation methods that iteratively reassign points between the k clusters. Since in iterative improvements would be too expensive if pair-wise distances or similarities between data points are used, using unique cluster representatives resolves the problem that the computation of objective function becomes linear in N (and in the number of clusters).

Hill-climbing method is a solution-improvement search algorithm. Conceptually, it rearranges existing partitions and keeps the new one only if it provides an improvement. The algorithm works as follows. The first step is to find some initial partition of all n data points into k clusters. Then it calculates the change in clustering criterion produced by moving each data point from its own to another group. Finally, it only makes the change that leads to the greatest improvement in the value of the clustering criterion. These three steps are repeated in an iterative fashion until no move of a data point improves the clustering criterion.

An initial partition can be obtained in a number of different ways. It might be specified on the basis of prior knowledge or it might be the result of a previous application of another cluster method. Alternatively, an initial partition might be chosen at random. Different initial partitions may lead to different local optima of the clustering criterion, although with well-structured data it is reasonable to expect convergence to the same, global optimum from most starting partitions [23]. One of the earliest hill-climbing algorithms consisted of iteratively updating a partition by simultaneously relocating each object to the group to whose mean it was closest and then recalculating the group means [6]. It can also be shown that this is equivalent to the criterion in eq. (5.8). Depending on how cluster representatives are constructed, such iterative optimization partitioning algorithms are subdivided into k-medoids and k-means methods (in Section 4.1). In practice, there are differences in the implementations of the hill-climbing algorithm. Most implementations of the k-means differ in whether the objects are relocated simultaneously or singly.

5.3. Volume based clustering. To overcome the difficulties like clustering with equal size or spherical shapes described in previous section, we can use the Mahalanobis distance instead of the Euclidean distance [61]. For example, if the covariance Σ is known, then the similarity within that cluster, with center $\mathbf{c}$ would be measured by $\|\mathbf{x} - \mathbf{c}\|_{\Sigma^{-1}}$. This measure is scale invariant and can deal with asymmetric, non-spherical clusters. However, the challenge in employing the Mahalanobis distance lies on the calculation of the covariance matrix of each cluster. Even if each cluster had the same variability and spread, covariance matrix is not known a priori and the sample covariance of the points will provide with an erroneous solution. A promising alternative for scale-invariant metric of cluster quality is minimum volume ellipsoids, where data points are allocated into clusters so that the volumes of the covering ellipsoids for each cluster is minimized. The problem of finding the minimum volume ellipsoid can be formulated as a semidefinite programming problem and an efficient algorithm for solving the problem has been proposed by [62].

The problem of minimum volume ellipsoid clustering in [61] is as follows: given n points in $\mathbb{R}^d : \mathbf{x}_1, \mathbf{x}_2, \ldots, \mathbf{x}_n$, allocate these points into K clusters that minimizes the total volume of the minimum volume covering ellipsoid for each cluster. To state this formally, suppose F_j are the set of indices of the data points assigned to cluster $j, j = 1, \ldots, K$, and $E_j, E_j \subset \mathbb{R}^d$, is the ellipsoid that contains all the points $a_i \in F_j$.

The parameters determining an ellipsoid E_j are $\mathbf{c}_j$, its center, and $\mathbf{Q}_j, \mathbf{Q}_j \in \Sigma^d_{++}$ that defines its shape given by

$$E_j = \{\mathbf{x} \mid (\mathbf{x} - \mathbf{c}_j)^T \mathbf{Q}(\mathbf{x} - \mathbf{c}_j) \leq 1\}.$$

However the shape is defined by

$$E_j = \{\mathbf{x} \mid (\mathbf{M}_j\mathbf{x} - \mathbf{z}_j)^T (\mathbf{M}_j\mathbf{x} - \mathbf{z}_j) \leq 1\},$$

where $\mathbf{M}_j = \mathbf{Q}_j^{\frac{1}{2}}$ and $\mathbf{z}_j = \mathbf{Q}_j^{\frac{1}{2}}\mathbf{c}_j$.

Since the volume of E_j is proportional to $\det(\mathbf{M}_j)^{-1}$, we wish to find the optimal allocation $F_j, j = 1, \ldots, K$, which solves the following problem:

$$\begin{aligned}
&\min - \sum_{j=1} K \det(\mathbf{M}_j) \\
&\text{s.t. } (\mathbf{M}_j\mathbf{x}_i - \mathbf{z}_j)^T (\mathbf{M}_j\mathbf{x}_i - \mathbf{z}_j) \leq 1, \quad i \in F_j,\ j = 1, \ldots, K, \\
&\bigcup_j F_j = 1, \ldots, n.
\end{aligned}$$

Two different approaches,

(1) pure branch-and-bound and
(2) convex relaxation branch-and-bound were proposed in [61] to solve the above problem.

References

1. P. L. Abad and W. J. Banks, *New* LP *based heuristics for the classification problem*, European J. Oper. Res. **67** (1993), 88–100.
2. R. Agrawal, J. Gehrke, D. Gunopulos, and P. Raghavan, *Automatic subspace clustering of high dimensional data for data mining applications*, Proc. ACM SIGMOD Internat. Conf. on Management of Data, 1998, pp. 94–105.
3. J. A. Anderson, *Constrained discrimination between k populations*, J. Royal Statist. Soc. Ser. B **31** (1969), 123–139.
4. D. Applegate, R. E. Bixby, V. Chvatal, and W. Cook, *The travelling salesman problem*, Technical report, Rice University, Houston, TX, 1994.
5. O. Asparouhov and S. Danchev, *Discrimination and classification in the presence of binary variables*, Biocybernetics and Biomedical Engrg. **17** (1997), no. 1-2, 25–39.
6. G. H. Ball and D. J. Hall, *A clustering technique for summarizing multivariate data*, Behavioural Science **12** (1967), 153–155.
7. L. Belbin, *The use of non-hierarchical allocation methods for clustering large sets of data*, Austral. Comput. J. **19** (1987), 32–41.
8. A. Ben-Israel and C. Iyigun, *Probabilistic distance clustering*, J. Classification, 2007 (to appear).
9. K. P. Bennett and J. Blue, *A support vector machine approach to decision trees*, Neural Netw. Proc. IEEE World Congress on Computational Intelligence **3** (1998), 2396–2401.
10. K. P. Bennett and O. L. Mangasarian, *Robust linear programming discrimination of two linearly inseparable sets*, Optim. Methods Softw. **1** (1992), 23–34.
11. P. Berkhin, *Survey of clustering data mining techniques*, Accrue Soft., Inc., 2002.
12. M. Better, F. Glover, and M. Samorani, *Multi-hyperplane formulations for classification and discrimination analysis*, Student Paper Award of the Decision Analysis Society, Working paper, University of Colorado, Colorado, 2006.
13. R. E. Bixby, W. Cook, A. Cox, and E. K. Lee, *Computational experience with parallel mixed integer programming in a distributed environment*, Research Monograph CRPC-TR95554, Center for Research on Parallel Computation, Rice University, TX, 1995.
14. P. S. Bradley, O. L. Mangasarian, and W. N. Street, *Feature selection via mathematical programming*, INFORMS J. Comput. **10** (1998), 209–217.
15. C. Burges, *Tutorial on support vector machines for pattern recognition*, Data Min. Knowl. Discov. **2** (1998), 121–167.
16. W. A. Chaovalitwongse, Y. J. Fan, and R. C. Sachdeo, *Novel optimization models for multidimensional time series classification: application to the identification of abnormal brain activity*, Oper. Res., 2006 (submitted).
17. P. Cheeseman and J. Stutz, *Bayesian classification (autoclass): theory and results*, Advances in Knowledge Discovery and Data Mining (U. M. Fayyad, G. Piatetsky-Shapiro, P. Smith, and R. Uthurusamy, eds.), AAAI/MIT Press, Cambridge, MA, 1996, pp. 153–180.
18. C. Y. Chen, S. C. Hwang, and Y. J. Oyang, *An incremental hierarchical data clustering algorithm based on gravity theory*, Proc. 6th Pacific-Asia Conference on Advances in Knowledge Discovery and Data Mining, vol. 2336, 2002, Springer-Verlag, London, UK, pp. 237–250.
19. Y. M. Cheung, *k^*-means: a new generalized k-means clustering algorithm*, Pattern Recognition Lett. **24** (2003), 2883–2893.
20. R. O. Duda, P. E. Hart, and D. G. Stork, *Pattern classification*, 2nd ed., Wiley-Interscience, New York, 2001.
21. M. Ester, H. P. Kriegel, J. Sander, and X. Xu, *A density-based algorithm for discovering clusters in large spatial data sets with noise*, Proc. 2nd Internat. Conf. on Knowledge Discovery and Data Mining, 1996, pp. 226–231.

22. M. Ester, H. P. Kriegel, J. Sander, M. Wimmer, and X. Xu, *Incremental clustering for mining in a data warehousing environment*, Proc. 24th Internat. Conf. on Very Large Data Bases, VLDB, 1998, pp. 323–333, 24–27.
23. B. Everitt, S. Landau, and M. Leese, *Cluster analysis*, 3rd. ed., Edward Arnold, London; copublished in the Americas by Halsted Press [John Wiley & Sons, Inc.], New York, 1993.
24. E. Freed and F. Glover, *Simple but powerful goal programming models for discriminant problems*, European J. Operations Research **7** (1981), no. 1, 44–60.
25. ———, Resolving certain difficulties and improving the classification power of the LP discriminant analysis procedure, Decision Sci. **17** (1986), 589–595.
26. R. J. Gallagher, E. K. Lee, and D. A. Patterson, *Constrained discriminant analysis via* 0/1 *mixed integer programming*, Ann. Oper. Res. **74** (1997), 65–88.
27. J. J. Glen, *Integer programming methods for normalization and variable selection in mathematical programming discriminant analysis models*, J. Oper Res. Soc. **50** (1999), 1043–1053.
28. ———, *An iterative mixed integer programming method for classification accuracy maximizing discriminant analysis*, Comput. Oper. Res. **30** (2003), no. 2, 181–198.
29. ———, *Mathematical programming models for piecewise-linear discriminant analysis*, J. Oper. Res. Soc. **56** (2005), 331–341.
30. F. Glover, *Improved linear programming models for discriminant analysis*, Decis. Sci. **21** (1990), no. 4, 771–785.
31. ———, *Improved linear and integer programming models for discriminant analysis*, Creative and Innovative Approaches to the Science of Management, RGK Foundation Press, 1993, pp. 187–215.
32. F. Glover, S. Keene, and B. Duea, *A new class of models for the discriminant problem*, Decis. Sci. **19** (1988), 269–280.
33. R. L. Grossman, C. Kamath, P. Kegelmeyer, V. Kumar, and E. E. Namburu, *Data mining for scientific and engineering applications*, Kluwer Academic Publ., 2001.
34. S. Guha, R. Rastogi, and K. Shim, CURE: *an efficient clustering algorithm for large databases*, Proc. 1998 ACM SIGMOD Internat. Conf. on Management of Data (Seattle, USA), ACM Press, New York, NY, USA 1998, 73–84.
35. J. Han and M. Kamber, *Data mining*: *concepts and techniques*, 2nd ed., Morgan Kaufmann Publ. Inc. San Francisco, CA, USA, 2006.
36. D. J. Hand, H. Mannila, and P. Smyth, *Principle of data mining*, MIT Press, Cambridge, MA, USA, 2001.
37. P. Hansen and B. Jaumard, *Cluster analysis and math programming*, Math. programming **79** (1997), 191–215.
38. J. Hartigan, *Clustering algorithms*, Wiley Ser. Probab. Math. Statist., John Wiley & Sons, New York-London-Sydney, 1975.
39. A. Hinneburg and D. Keim, *A general approach to clustering in large databases with noise*, J. Knowl. Inform. Syst. **5** (2003), no. 4, 387–415.
40. Z. Huang, *Extensions to the k-means algorithm for clustering large data sets with categorical values*, Data Min. Knowl. Discov. **2** (1998), 283–304.
41. C. Iyigun and A. Ben-Israel, *Probabilistic distance clustering with cluster sizes*, Probab. Engrg. Inform. Syst., 2007 (to appear).
42. A. Jain and R. Dubes, *Algorithms for clustering data*, Prentice-Hall, Englewood Cliffs, NJ, 1988.
43. A. K. Jain, M. N. Murty, and P. J. Flynn, *Data clustering*: *a review*, ACM Computing Surveys **31** (1999), no. 3, 264–323.
44. F. Jensen, *An introduction to Bayesian networks*, UCL Press, London, 1996.
45. E. A. Joachimsthaler and A. Stam, *Mathematical programming approaches for the classification problem in two-group discriminant analysis*, Multivariate Behavioral Research **25** (1990), no. 4, 427–454.
46. G. Karypis, E.-H. (Sam) Han, and V. Kumar, *Chameleon*: *hierarchical clustering using dynamic modeling*, Computer **32** (1999), no. 8, 68–75.
47. L. Kaufman and P. Rousseeuw, *Finding groups in data*: *an introduction to cluster analysis*, John Wiley & Sons, New York, NY, 1990.
48. T. Kohonen, *Self-organizing maps*, 2nd extended ed., Springer Ser. Inform. Sci., vol. 30, Springer, Berlin, Heidelberg, New York, 1997.

49. S. B. Kotsiantis and P. E. Pintelas, *Recent advances in clustering: a brief survey*, WSEAS Trans. Inform. Sci. Appl. **1**, no. 1, 73–81.
50. E. K. Lee, R. J. Gallagher, and D. A. Patterson, *A linear programming approach to discriminant analysis with a reserved-judgement region*, INFORMS J. Comput. **15** (2003), no. 1, 23–41.
51. O. L. Mangasarian, *Mathematical programming in data mining*, Data Min. Knowl. Discov. **42** (1997), no. 1, 183–201.
52. ———, *Linear and nonlinear separation of pattern by linear programming*, Oper. Res. **13** (1965), 444–452S.
53. ———, *Multisurface method for pattern separation*, IEEE Trans. Inform. Theory **14** (1968), 801–807.
54. G. J. McLachlan and T. Krishnan, *The* EM *algorithm and extensions*, Wiley Ser. Prob. Stat., John Wiley & Sons, New York, NY, 1997.
55. P. M. Pardalos, V. Boginski, and A. Vazacopoulos, *Data mining in biomedicine*, Springer, New York, 2007.
56. J. Pena, J. Lozano, and P. Larranaga, *Learning recursive Bayesian multinets for data clustering by means of constructive induction*, Mach. Learning **47** (2002), 63–89.
57. J. Rubin, *Optimal classification into groups: an approach for solving the taxonomy problem*, J. Theoret. Biology **15** (1967), 103–144.
58. M. Sato, Y. Sato, and L. Jain, *Fuzzy clustering models and applications* (J. Kacprzyk, ed.), Physica-Verlag, 1997.
59. C. Sheikholeslami, S. Chatterjee, and A. Zhang, *Wavecluster: a-multiresolution clustering approach for very large spatial data set*, Proc. 24th Internat. Conf. on Very Large Data Bases (August, 1998), Morgan Kaufmann Publishers Inc., San Francisco, CA, USA, 1998, pp. 428–439.
60. H. Shimodaira, K. I. Noma, M. Nakai, and S. Sagayama, *Dynamic time-alignment Kernel in support vector machine*, Adv. Neural Inform. Processing Syst. **14** (2001), no. 2, 921–928.
61. R. Shioda and L. Tuncel, *Clustering via minimum volume ellipsoids*, Comput. Optim. Appl., 2006 (to appear).
62. P. Sun and R. M. Freund, *Computation of minimum volume covering ellipsoids*, Oper. Res. **52** (2004), no. 5, 690–706.
63. W. Wang, J. Yang, and R. R. Muntz, STING: *a statistical information grid approach to spatial data mining*, Proc. 23rd Internat. Conf. on Very Large Data Bases (Greece, 1997) (M. Jarke, M.-J. Carey, K.-R. Dittrich, F.-H. Lochovsky, P. Loucopoulos, and M.-A. Jeusfeld, eds.), Morgan Kaufmann Publ. Inc., San Francisco, CA, USA, pp. 186–195.
64. X. Xu, M. Ester, H. P. Kriegel, and J. Sander, *A nonparametric clustering algorithm for knowledge discovery in large spatial data sets*, Proc. IEEE Internat. Conf. on Data Engineering, IEEE Computer Soc. Press, 1998.
65. X. Xu and J. Jager and H. P. Kriegel, *A fast parallel clustering algorithm for large spatial data sets*, Data Min. Knowl. Discov. **3** (1999), 263–290.
66. T. Zhang, R. Ramakrishnan, and M. Linvy, BIRCH: *an efficient data clustering method for very large data sets*, Proc. 1996 ACM SIGMOD Internat. Conf. on Management of Data (Montréal, 1996), ACM Press, New York, NY, USA, 1996, pp. 103–114.
67. B. Zhou, D. W. Cheung, and B. Kao, *A fast algorithm for density-based clustering in large database*, Proc. 3rd Pacific-Asia Conf. on Methodologies for Knowledge Discovery and Data Mining, London, UK, 1999, Springer-Verlag, pp. 338–349.

Department of Industrial and Systems Engineering, Rutgers University, 96 Frelinghuysen Road, CoRE Building, Room 201, Piscataway, NJ 08854, USA.
E-mail address: `yjfan@eden.rutgers.edu`

Rutgers Center for Operations Research, Rutgers University, 640 Bartholomew Road, Piscataway, NJ 08854-8003, USA
E-mail address: `iyigun@rutcor.rutgers.edu`

Department of Industrial and Systems Engineering, Room 201 CoRE Building, Rutgers University, 96 Frelinghuysen Rd., Piscataway, NJ 08854, USA.
E-mail address: `wchaoval@rci.rutgers.edu`

Centre de Recherches Mathématiques
CRM Proceedings and Lecture Notes
Volume **45**, 2008

Nonlinear Skeletons of Data Sets and Applications—Methods Based on Subspace Clustering

Pando G. Georgiev

ABSTRACT. In many practical problems the data (given by the columns of a data matrix) lie, up to some precision, on a union of m affine subspaces, called *affine skeleton* of the data. Finding the best possible skeleton when $m = 1$ can be performed easily by Principal Component Analysis (PCA). When $m > 1$, however, this is a global optimization problem, which appears to be hard for bigger m. Its solution reveals an internal structure of the data set, namely, under mild assumptions, the data matrix can be decomposed uniquely (up to some unimportant ambiguities) as multiplication of a *mixing* matrix and a *source* matrix, which, more importantly, is *sparse*. This a kind of Blind Signal Separation problem based on identifiability conditions involving sparseness, which we describe precisely. We develop a subspace clustering algorithm, which is a generalization of the k-plane clustering algorithm, and is suitable for separation of sparse mixtures with bigger sparsity. We present "kernelization" of this idea, leading to the notion of *nonlinear skeletons*, when we work in Reproducing Kernel Hilbert Spaces (RKHS). Nonlinear skeleton identification in RKHS is a fusion of two ideas: clustering and Kernel PCA. We propose the idea that binary classification tasks can be performed by finding nonlinear skeletons of the training points in the both classes. We demonstrate our algorithms by examples.

1. Introduction

The notion of skeleton of data arises in a natural way when we want to approximate a data set by a simple one which could be described easily. The simplest example of skeleton is presented by Principle Component Analysis, which finds a subspace with appropriate dimension, which fits the data in the best way. In this paper we extend this idea to the case when the data vectors are approximated by elements in a union of finitely many affine subspaces, called *skeleton* of the given data set. One motivation for developing algorithms for finding skeletons of data sets is presented by the so called Sparse Blind Signal Separation problem [7], which exploits the fact that under mild sparseness conditions of the source matrix, the data vectors lie on a finite union of subspaces. So, finding these subspaces (skeleton of the data set) is crucial for separation of the mixture. We develop a subspace clustering algorithm, which is a generalization of the k-plane clustering algorithm [2],

This is the final form of the paper.

and is suitable for separation of sparse mixtures with bigger sparsity. We present "kernelization" of this idea, leading to the notion of *nonlinear skeletons*, when we work in Reproducing Kernel Hilbert Spaces.

Another motivation for skeleton identification algorithms is presented by binary classification tasks. We develop the idea that such tasks can be performed by finding nonlinear skeletons of the training points in the both classes.

2. Affine Hyperplane Skeletons of a Finite Set of Points

The solution $\{(\mathbf{n}_i^0, b_i^0)\}_{i=1}^k$ of the minimization problem:

$$\text{minimize } f(\{(\mathbf{n}_i^T, b_i)\}_{i=1}^k) = \sum_{j=1}^N \min_{1\le i\le k} |\mathbf{n}_i^T \mathbf{x}_j - b_i|^l \tag{2.1}$$

$$\text{subject to } \|\mathbf{n}_i\| = 1, \quad b_i \in \mathbb{R}, \ i = 1, \dots, k, \tag{2.2}$$

defines *affine hyperplane* $k^{(l)}$*-skeleton* of $\mathbf{X} \in \mathbb{R}^{m\times N}$, introduced for $l = 1$ in [11] and for $l = 2$, in [2]. It consists of a union of k *affine* hyper-planes

$$H_i = \{\mathbf{x} \in \mathbb{R}^m : \mathbf{n}_i^{0^T} \mathbf{x} = b_i^0\}, \quad i = 1, \dots, k,$$

such that the sum of minimum distances raised to power l, from every point $\mathbf{x}_j$ to them is minimal. Consider the following minimization problem:

$$\text{minimize } \sum_{j=1}^N \min_{1\le i\le k} |\widetilde{\mathbf{n}}_i^T \widetilde{\mathbf{x}}_j|^l \tag{2.3}$$

$$\text{subject to } |\widetilde{\mathbf{n}}_i\| = 1, \quad i = 1, \dots, k, \tag{2.4}$$

where $\widetilde{\mathbf{n}}_i = (\mathbf{n}_i, b_i)$, $\widetilde{\mathbf{x}}_j = (\mathbf{x}_j, -1)$. Its solution $\widetilde{\mathbf{n}}_i$ defines *hyperpane skeleton* in $\mathbb{R}^{m+1}$, consisting of a union of k hyperplanes

$$\widetilde{H}_i = \{\widetilde{\mathbf{x}} \in \mathbb{R}^{m+1} : \widetilde{\mathbf{n}}_i^T \widetilde{\mathbf{x}} = 0\}, \quad i = 1, \dots, k,$$

such that the sum of minimum distances raised to power l, from every point $\widetilde{\mathbf{x}}_j$ to them is minimal. The restriction of $\widetilde{H}_i$ to $\mathbb{R}^m$ gives H_i. The usefulness of working in $\mathbb{R}^{m+1}$ can be seen for $l = 2$, if we denote by $\widetilde{\mathbf{X}}_i$ the ith cluster

$$\widetilde{\mathbf{X}}_i = \{\widetilde{\mathbf{x}}_j : \min_{1\le p\le k} |\widetilde{\mathbf{n}}_p^T \widetilde{\mathbf{x}}_j|^2 = |\widetilde{\mathbf{n}}_i^T \widetilde{\mathbf{x}}_j|^2\} \tag{2.5}$$

for some given $\{\widetilde{\mathbf{n}}_i^T\}_{i=1}^k$. Then the following minimization problem

$$\text{minimize } \mathbf{v}_i^T \widetilde{\mathbf{X}}_i \widetilde{\mathbf{X}}_i^T \mathbf{v}_i \tag{2.6}$$

$$\text{subject to } \mathbf{v} \in \mathbb{R}^{m+1}, \quad \|\mathbf{v}_i\| = 1, \tag{2.7}$$

will produce a solution $\{\mathbf{v}_i^0\}_{i=1}^k$ (consisting of eigenvectors corresponding to the minimal eigenvalues) for which f in (2.1) will not increase, i.e. $f(\{\mathbf{v}_i^0\}_{i=1}^k) \le f(\{\widetilde{\mathbf{n}}_i\}_{i=1}^k)$.

So we arrived to the following analogue of the classical k-means clustering algorithm.

Hyperplane clustering algorithm. [See the pseudo code of the subspace clustering algorithm below in the particular case when $r_i = 1$.]

Apply iteratively the following two steps until convergence:

(1) Cluster assignment: forming $\widetilde{\mathbf{X}}_i$, $i = 1, \ldots, k$ by (2.5),
(2) Cluster update: solving the eigenvalue problem (2.6), (2.7).

Such kind of analogue of the classical k-means clustering algorithm, (working in $\mathbb{R}^m$ instead in $\mathbb{R}^{m+1}$) was introduced by Bradley and Mangasarian in [2] and called k-plane clustering algorithm.

Since our hyperplane clustering algorithm is equivalent to the k-plane clustering algorithm, it has the same properties, i.e. termination after finite number of steps at a cluster assignment which is locally optimal, i.e. f in (2.1) cannot be decreased by either reassigning of a point to a different cluster plane, or by defining a new cluster plane for any of the clusters (see [2, Theorem 3.7]).

Affine subspace skeleton. The solution of the following minimization problem:

$$\text{minimize } F(\{\mathbf{U}_i\}_{i=1}^k) = \sum_{j=1}^N \min_{1\le i\le k} \sum_{s=1}^{r_i} |\mathbf{u}_{i,s}^T \mathbf{x}_j - b_{i,s}|^l \tag{2.8}$$

$$\text{subject to } \|\mathbf{u}_{i,s}\| = 1, \quad b_{i,s} \in \mathbb{R},\ i = 1, \ldots, k, \tag{2.9}$$

$$s = 1, \ldots, r_i,\ \mathbf{u}_{i,p}^T \mathbf{u}_{i,q} = 0,\ p \neq q, \tag{2.10}$$

where $\mathbf{U}_i = \{\mathbf{u}_{i,s}\}_{s=1}^{r_i}$, will be called *affine subspace skeleton.* It consists of a union of k affine subspaces $S_i = \{x \in \mathbb{R}^m : \mathbf{u}_{i,s}^T \mathbf{x} = b_{i,s}, s = 1, \ldots, r_i\}$ each with codimension r_i, $i = 1, \ldots, k$, such that the sum of minimum distances raised to power l, from every point $\mathbf{x}_j$ to them is minimal. Consider the following minimization problem:

$$\text{minimize } \widetilde{F}(\{\widetilde{\mathbf{U}}_i\}_{i=1}^k) = \sum_{j=1}^N \min_{1\le i\le k} \sum_{s=1}^{r_i} |\widetilde{\mathbf{u}}_{i,s}^T \widetilde{\mathbf{x}}_j|^l \tag{2.11}$$

$$\text{subject to } \|\widetilde{\mathbf{u}}_{i,s}\| = 1, \quad b_{i,s} \in \mathbb{R},\ i = 1, \ldots, k, \tag{2.12}$$

$$s = 1, \ldots, r_i,\ \widetilde{\mathbf{u}}_{i,p}^T \widetilde{\mathbf{u}}_{i,q} = 0,\ p \neq q, \tag{2.13}$$

where $\widetilde{\mathbf{U}}_i = \{\widetilde{\mathbf{u}}_{i,s}\}_{s=1}^{r_i}$, $\widetilde{\mathbf{u}}_{i,s} = (\mathbf{u}_{i,s}, b_{i,s})$, $\widetilde{\mathbf{x}}_j = (\mathbf{x}_j, -1)$. Its solution $\widetilde{\mathbf{n}}_{i,s}$ defines a *subspace skeleton* in $\mathbb{R}^{m+1}$, consisting of a union of k subspaces

$$\widetilde{S}_i = \{\widetilde{\mathbf{x}} \in \mathbb{R}^{m+1} : \widetilde{\mathbf{n}}_{i,s}^T \widetilde{\mathbf{x}} = 0, s = 1, \ldots, r_i\}, \quad i = 1, \ldots, k,$$

such that the sum of minimum distances raised to power l, from every point $\widetilde{\mathbf{x}}_j$ to them is minimal. The restriction of $\widetilde{S}_i$ to $\mathbb{R}^m$ gives S_i (a non-trivial fact).

Similarly to the hyperspace clustering algorithm, we can write the following analogue of the classical k-means clustering algorithm for finding the subspace skeleton, when $l = 2$.

Subspace clustering algorithm. [See the pseudocode below]

It consists of two steps applied alternatively until convergence:

Algorithm 1: Subspace clustering algorithm.

Data: samples $\mathbf{x}(1), \dots, \mathbf{x}(T)$

Result: estimated k subspaces $\widetilde{S}_i \subset \mathbb{R}^{n+1}, i = 1, \dots, k$ given by the orthonormal bases $\{\widetilde{\mathbf{u}}_{i,s}\}_{s=1}^{r_i}$ of their orthogonal complements $\widetilde{S}_i^{\perp}$.

Initialize $\varepsilon > 0$ and randomly $\widetilde{\mathbf{u}}_{i,s} = (\mathbf{u}_{i,s}, b_{i,s})$ with $|\widetilde{\mathbf{u}}_{i,s}| = 1$, $i = 1, \dots, k$, $s = 1, \dots, r_i$.

do (*Until the difference of* $\widetilde{F}$ *calculated in two subsequent estimates of* $\{\widetilde{\mathbf{U}}_i\}_{i=1}^k$ *is less than* ε)

    Cluster assignment.

    for $t \leftarrow 1, \dots, T$ **do**

        Add $\widetilde{\mathbf{x}}(t) = (\mathbf{x}(t), -1)$ to cluster $\widetilde{\mathbf{X}}_i$ (a matrix), where i is chosen to minimize the distance of $\widetilde{\mathbf{x}}(t)$ to the subspace S_i: $\sum_{s=1}^{r_i} |\mathbf{u}_{i,s}^T \mathbf{x}_j - b_{i,s}|^2$.

    end

    Cluster update.

    for $i \leftarrow 1, \dots, k$ **do**

        Calculate the i-th cluster correlation $\mathbf{C} := \widetilde{\mathbf{X}}_i \widetilde{\mathbf{X}}_i^T$.

        Choose eigenvectors $\mathbf{v}_s, s = 1, \dots, r_i$ of $\mathbf{C}$ corresponding to the r_i minimal eigenvalues.

        Set $\widetilde{\mathbf{u}}_{i,s} \leftarrow \mathbf{v}_s / |\mathbf{v}_s|$, $s = 1, \dots, r_i$, $\widetilde{\mathbf{U}}_i = \bigcup_{s=1}^{r_i} \widetilde{\mathbf{u}}_{i,s}$.

    end

end

(1) Cluster assignment: forming $\widetilde{\mathbf{X}}_i, i = 1, \dots, k$ by: for given k orthonormal families of vectors $\{\widetilde{\mathbf{n}}_{i,s}\}_{s=1}^{r_i}$, $i = 1, \dots, k$, put

$$\widetilde{\mathbf{X}}_i = \left\{ \widetilde{\mathbf{x}}_j : \sum_{s=1}^{r_i} |\widetilde{\mathbf{u}}_{i,s}^T \widetilde{\mathbf{x}}_j|^2 = \min_{1 \le i \le k} \sum_{s=1}^{r_i} |\widetilde{\mathbf{u}}_{i,s}^T \widetilde{\mathbf{x}}_j|^2 \right\}. \tag{2.14}$$

(2) Cluster update: for every $i = 1, \dots, k$, solving the minimization problem

$$\text{minimize} \sum_{s=1}^{r_i} \operatorname{trace}(\mathbf{V}_i^T \widetilde{\mathbf{X}}_i \widetilde{\mathbf{X}}_i^T \mathbf{V}_i) \tag{2.15}$$

$$\text{under orthogonality constraints} \mathbf{V}_i^T \mathbf{V}_i = \mathbf{I}_{r_i}, \tag{2.16}$$

where $\mathbf{v}_i$ has dimensionality $(n + 1 \times r_i)$.

For any i, it will produce a solution $\mathbf{V}_i^0$—a matrix with columns equal to the eigenvectors corresponding to the r_i smallest eigenvalues of the matrix $\widetilde{\mathbf{X}}_i \widetilde{\mathbf{X}}_i^T$ (a non-trivial fact) and $\widetilde{F}$ in (2.14) will not increase, i.e. $\widetilde{F}(\{\mathbf{V}_i^0\}_{i=1}^k) \le \widetilde{F}(\{\widetilde{\mathbf{U}}_i\}_{i=1}^k)$. This step is realized simply by eigenvalue decomposition of $\widetilde{\mathbf{X}}_i \widetilde{\mathbf{X}}_i^T$.

This algorithm terminates in a finite number of steps (similarly to the Bradley–Mangasarian algorithm) to a solution which is locally optimal, i.e. $\widetilde{F}$ in (2.14) cannot be decreased by either reassigning of a point to a different cluster subspace, or by defining a new cluster subspace for any of the clusters.

3. Applications to Blind Signal Separation

A goal of the Blind Signal Separation (BSS) is the recovering of underlying source signals of some given set of observations obtained by an unknown linear mixture of the sources. In order to decompose the data set, different assumptions on the sources have to be made. The most common assumption nowadays is statistical independence of the sources, which leads to the field of *Independent Component Analysis* (ICA), see for instance [3, 8] and references therein. ICA is very successful in the linear *complete* case, when as many signals as underlying sources are observed, and the mixing matrix is non-singular. In [4] it is shown that the mixing matrix and the sources are identifiable except for permutation and scaling. In the *overcomplete* or *underdetermined* case, less observations than sources are given. It can be seen that still the mixing matrix can be recovered [6], but source identifiability does not hold. In order to approximatively detect the sources, additional requirements have to be made, usually sparsity of the sources. We refer to [9, 14, 16, 17] and reference therein for some recent papers on sparsity and underdetermined ICA ($m < n$).

Definition 1. A vector $\mathbf{v} \in \mathbf{R}^m$ is said to be *k-sparse* if $\mathbf{v}$ has at least k zero entries. A matrix $\mathbf{S} \in \mathbf{R}^{m \times n}$ is said to be *k-sparse* if each column of it is k-sparse.

The goal of *Blind Signal Separation* of level k (k-BSS) is to decompose a given m-dimensional random vector $\mathbf{X}$ into

$$\mathbf{X} = \mathbf{AS} \tag{3.1}$$

with a real $m \times n$-matrix $\mathbf{A}$ and an $n \times N$-dimensional k-sparse matrix $\mathbf{S}$. $\mathbf{S}$ is called the *source matrix*, $\mathbf{X}$ the *mixtures* and $\mathbf{A}$ the *mixing matrix*. We speak of *complete*, *overcomplete* or *undercomplete* k-BSS if $m = n$, $m < n$ or $m > n$ respectively.

More generally, when a solution to the above problem doesn't exist, we formulate *least square BSS* problem as follows:

- find a best approximation of $\mathbf{X}$ by $\mathbf{AS}$, such that $\mathbf{S}$ is k-sparse, i.e.

$$\text{minimize } \|\mathbf{X} - \mathbf{AS}\| \tag{3.2}$$

$$\text{under constraints } \mathbf{A} \in \mathbb{R}^{m \times n}, \mathbf{S} \in \mathbb{R}^{n \times N} \quad \text{and} \quad \mathbf{S} \text{ is } k\text{-sparse.} \tag{3.3}$$

In the following without loss of generality we will assume $m \leq n$: the undercomplete case can be reduced to the complete case by projection of $\mathbf{X}$.

3.1. Identifiability conditions. The following identifiability conditions are extension and refinement of those in [7]. The extension is for the case of bigger sparsity i.e. when the source matrix is at least $(n - m + 2)$-sparse. The refinement can be seen, for instance, when $n \geq m+2$ and the source matrix is $(n-m+1)$-sparse. The provided examples illustrate both cases. Assume that:

(H1) the indeces $\{1, \ldots, N\}$ are divided in two groups $\mathcal{N}_1$ and $\mathcal{N}_2$ such that
  (a) for any index $i \in \{1, \ldots, n\}$ there exist $k_i \geq m$ different sets of indeces $I_{i,j} \subset \{1, \ldots, n\}$, $j = 1, \ldots, k_i$ each of which contains i, such that $|I_{i,j}| \leq m - 1$,
  (b) for any set $I_{i,j}$ from (a) there exist $m_{i,j} > |I_{i,j}|$ vectors $\mathbf{s}_1, \ldots, \mathbf{s}_{m_{i,j}}$ from the set $\mathbf{S}_1 := \{\mathbf{S}(:, j), j \in \mathcal{N}_1\}$, such that each of them has zero elements in places with indeces not in $I_{i,j}$, the rank of the set $\{\mathbf{s}_1, \ldots, \mathbf{s}_{m_{i,j}}\}$ is full (i.e. equal to $|I_{i,j}|$), and
  (c) $\bigcap_{j=1}^{k_i} I_{i,j} = \{i\}$.

(H2) Every m vectors from the group $\{\mathbf{S}(:,j), j \in \mathcal{N}_2\}$ are linearly independent.

Theorem 3.1 (Matrix identifiability). *Let the assumptions* (H1) *and* (H2) *be satisfied. Then, in the set of all matrices with dimension* $m \times n$ *there is a subset of a full measure* $\mathcal{A}$ *such that, if* $\mathbf{X} = \mathbf{AS}$ *and* $\mathbf{A} \in \mathcal{A}$, *then* $\mathbf{A}$ *is uniquely determined by* $\mathbf{X}$ *up to permutation and scaling of the columns.*

PROOF. Sketch of the proof The condition (H1) (c) implies that for any column $\mathbf{A}_i$ of the mixing matrix there exists less that $|I_{i,j}|$ detectable subspaces in the data which intersection is this column. One more subspace is reserved for "confirmation" that their intersection is indeed a column of the mixing matrix. Their intersection is stable under small perturbation of the columns, so "most" of the matrices will exclude false intersection (not corresponding to a column). "Detectable subspace" here means that it contains at least $|I_{i,j}| + 1$ data vectors (columns of $\mathbf{X}$) different from zero and any $|I_{i,j}|$ from them are linearly independent (follows from (H1) (b)), so such a subspace can be detected by an algorithm. Condition (H2) ensures that there is no false intersection of subspaces (not corresponding to a column of $\mathbf{A}$). □

Algorithm for identification of the mixing matrix.

(1) Cluster the columns $\{\mathbf{X}(:,j) : j \in \mathcal{N}_1\}$ in k groups $\mathcal{H}_i, i = 1, \ldots, k$ such that the span of the elements of each group $\mathcal{H}_i$ produces r_i-codimensional subspace and these r_i-codimensional subspaces are different.
(2) Calculate any basis of the orthogonal complement of each of these r_i-codimensional subspaces.
(3) Find all possible groups such that each of them is composed of the elements of at least m bases in 2), and the vectors in each group lie on a hyperplane. The number of these hyperplanes gives the number of sources n. The normal vectors to these hyperplanes are estimations of the columns of the mixing matrix $\mathbf{A}$ (up to permutation and scaling).

In practical realization all operations in the above algorithm are performed up to some precision $\varepsilon > 0$.

Source recovery algorithm.

(1) Repeat for $j = 1$ to N:
(2) (a) Identify the subspace $\mathcal{H}_i$ containing $\mathbf{x}_j := \mathbf{X}(:,j)$, or, in practical situation with presence of noise, identify $\mathcal{H}_i$ to which the distance from $\mathbf{x}_j$ is minimal and project $\mathbf{x}_j$ onto $\mathcal{H}_i$ to $\widetilde{\mathbf{x}}_j$;
 (b) if $\mathcal{H}_i$ is produced by the linear hull of column vectors $\mathbf{a}_{p_1}, \ldots, \mathbf{a}_{p_{m-r_i}}$, then find coefficients $\mathbf{L}_{j,l}$ such that $\widetilde{\mathbf{x}}_j = \sum_{l=1}^{m-r_i} \mathbf{L}_{j,l} \mathbf{a}_{p_l}$. These coefficients are uniquely determined generically (i.e. up to measure zero) for $\widetilde{\mathbf{x}}_j$ (see [7], Theorem 3).
 (c) Construct the solution $\mathbf{s}_j = \mathbf{S}(:,j)$: it contains $\mathbf{L}_{j,l}$ in the place p_l for $l = 1, \ldots, m - r_i$, the other its components are zero.

It is clear that such found couple $(\mathbf{A}, \mathbf{S})$ is a solution to the least square BSS problem defined by (3.2).

3.2. Computer simulation examples.

First example. We created artificially four source signals, sparse of level 2, i.e. each column of the source matrix contains at least 2 zeros (shown in Fig. 1). They are mixed with a square nonsingular randomly generated matrix **A**:

$$\mathbf{A} = \begin{pmatrix} 0.4621 & 0.6285 & 0.5606 & 0.4399 \\ 0.2002 & 0.4921 & 0.5829 & 0.3058 \\ 0.8138 & 0.4558 & 0.2195 & 0.2762 \\ 0.2899 & 0.3938 & 0.5457 & 0.7979 \end{pmatrix}.$$

The mixed sources are shown in Fig. 2, the recovered sources by our algorithm are shown in Fig. 3. The created signals are superposition of sinusoidal signals, so it is clear that they are dependent and ICA methods could not separate them (lack of space prevent us to show the results of applying ICA algorithms). The estimated mixing matrix with our subspace clustering algorithm is

$$\mathbf{M} = \begin{pmatrix} 0.6285 & 0.4399 & 0.4621 & 0.5606 \\ 0.4921 & 0.3059 & 0.2002 & 0.5829 \\ 0.4558 & 0.2762 & 0.8138 & 0.2195 \\ 0.3938 & 0.7979 & 0.2899 & 0.5457 \end{pmatrix}.$$

Second example. Now we create six signals with level of sparsity 4, i.e. each column of the source matrix has 4 zeros. We mixed them with a randomly generated matrix with dimension (3×6):

$$\mathbf{H} = \begin{pmatrix} 0.8256 & 0.3828 & 0.4857 & 0.4053 & 0.7720 & 0.2959 \\ 0.2008 & 0.7021 & 0.0197 & 0.5610 & 0.6182 & 0.6822 \\ 0.5273 & 0.6004 & 0.8739 & 0.7218 & 0.1476 & 0.6686 \end{pmatrix}.$$

The estimated mixing matrix with our subspace hyperplane clustering algorithm is

$$\mathbf{M} = \begin{pmatrix} 0.2959 & 0.4857 & 0.8256 & 0.4053 & 0.3828 & 0.7720 \\ 0.6822 & 0.0197 & 0.2008 & 0.5610 & 0.7021 & 0.6182 \\ 0.6686 & 0.8739 & 0.5273 & 0.7218 & 0.6004 & 0.1476 \end{pmatrix}.$$

Here the number of hyperplanes presented in the data are 9, while the number of all possible hyperplanes (according to the identifiability conditions in [7]) when all combination of 4 zeros in the columns of the source matrix are present, are 15.

4. Reproducing Kernel Hilbert Spaces

Recall one of the possible constructions of Reproducing Kernel Hilbert Spaces (RKHS) (see [1, 5] for more details). Let $\mathbf{X} \subset \mathbb{R}^m$ and $K : \mathbf{X} \times \mathbf{X} \to \mathbb{R}$ be a continuous and symmetric function such that $\{K(\mathbf{x}_i, \mathbf{x}_j)\}_{i,j=1}^n$ is positive definite for every $\{\mathbf{x}_i\}_{i=1}^n \subset \mathbf{X}$ and every n, i.e. $\sum_{i,j=1}^n c_i c_j K(x_i, x_j) \geq 0$ for any $n \in N$ and any choice of $x_i \in X$ and $c_i \in \mathbb{R}$ $(i = 1, \dots, n)$. Note $K(x, x) \geq 0$ for all x. The mapping **K** is called *positive definite kernel.*

Let H_0 be the linear span of the functions $\{K(\mathbf{x}, .), \mathbf{x} \in \mathbf{X}\}$:

$$H_0 = \left\{ f : \mathbf{X} \to \mathbb{R} : f(.) = \sum_{i=1}^n \alpha_i K(\mathbf{x}_i, .), \mathbf{x}_i \in \mathbf{X}, \alpha_i \in \mathbb{R}, n \in \mathbb{N} \right\}.$$

Define an inner product in H_0 by

$$\langle f, g \rangle = \sum_{i=1}^{n} \sum_{j=1}^{n'} \alpha_i \beta_j K(\mathbf{x}_i, \mathbf{x}'_j) \tag{4.1}$$

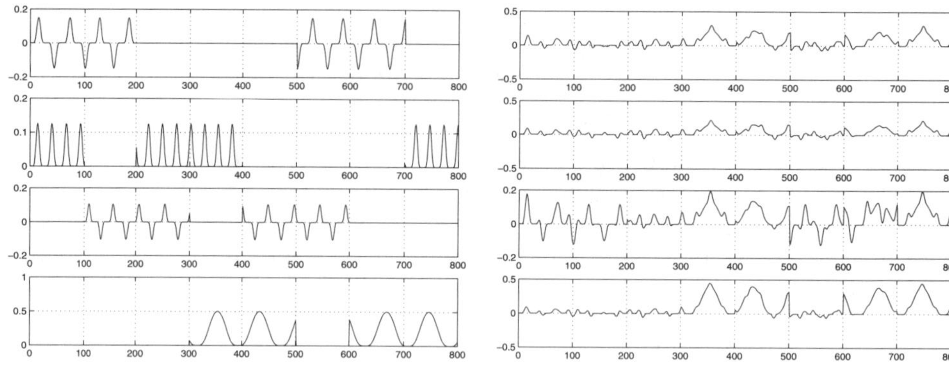

FIGURE 1. Example 1: original source signals (left) and mixed ones (right).

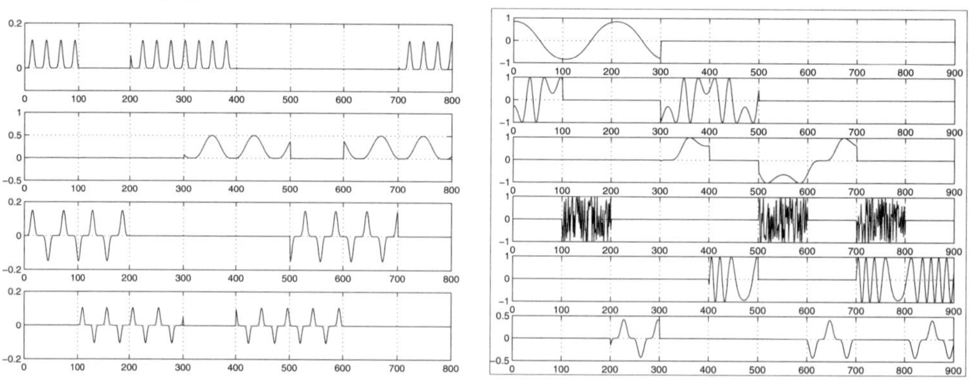

FIGURE 2. Left: Example 1: separated signals. Right: Example 2—original source signals.

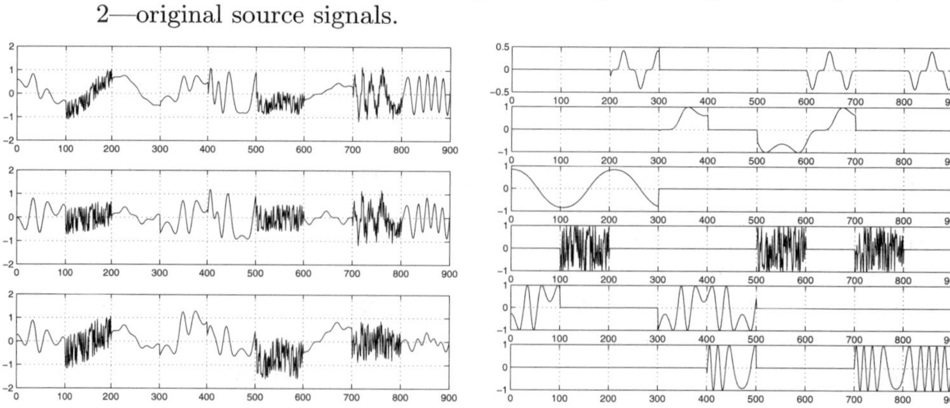

FIGURE 3. Example 2: mixed signals (left) and separated signals (right).

where $f, g \in H_0$, $f(.) = \sum_{i=1}^{n} \alpha_i K(\mathbf{x}_i, .), g(.) = \sum_{i=1}^{n'} \beta_i K(\mathbf{x}'_i, .)$.

The *reproducing property*

$$f(\mathbf{x}) = \langle f, K(\mathbf{x}, .)\rangle \quad \forall \mathbf{x} \in \mathbf{X}$$

follows from (4.1).

The induced norm $\|.\|_K$ in H_0 is $\|f\|_K = \sqrt{\langle f, f\rangle}$. It is indeed a norm, since if $\|f\| = 0$, then $f(\mathbf{x}) = \langle f, K(\mathbf{x}, .)\rangle \leq \|f\| \|K(\mathbf{x}, .)\| = 0\ \forall \mathbf{x}$.

The completion of $(H_0, \|.\|_K)$ is called *Reproducing Kernel Hilbert Space.*

The so called *kernel trick*

$$K(\mathbf{x}, \mathbf{y}) = \langle \Phi(\mathbf{x}), \Phi(\mathbf{y})\rangle \quad \text{for every} \quad \mathbf{x},\ \mathbf{y} \in \mathbf{X}$$

where $\Phi(\mathbf{x}) = K(\mathbf{x}, .)$ is called *feature map*, follows again from (4.1).

We give two typical examples of positive definite kernels. The first is the Gaussian kernel $K : \mathbb{R}^n \times \mathbb{R}^n \to \mathbb{R}$ defined by $K(\mathbf{x}, \mathbf{x}') = \exp(-\|\mathbf{x} - \mathbf{x}'\|^2/c^2)$ $(c > 0)$.

The second is the polynomial kernel of degree p, $K : \mathbb{R}^n \times \mathbb{R}^n \to \mathbb{R}$ defined by $K_p(\mathbf{x}, \mathbf{x}') = (\langle \mathbf{x}, \mathbf{x}'\rangle + 1)^p$. In the case when for $m = 2$, $p = 3$, we have (for $\mathbf{x} = (x, y), \mathbf{x}' = (x', y')$),

$$\Phi(\mathbf{x}) = \big(x^3, y^3, \sqrt{3}x^2y, \sqrt{3}xy^2, \sqrt{3}x^2, \sqrt{3}y^2, \sqrt{3}xy, \sqrt{3}x, \sqrt{3}y, 1\big),$$
$$\langle \Phi(\mathbf{x}), \Phi(\mathbf{x}')\rangle = K\big((x, y), (x', y')\big).$$

5. Nonlinear Skeletons

The notion of subspace skeleton developed in previous sections can be extended easily to the notion "nonlinear skeleton" as follows: in previous definitions we change $\mathbf{x}$ with $\Phi(\mathbf{x})$ and work in a Reproducing Kernel Hilbert space H defined by a kernel K such that $\Phi(\mathbf{x}) = K(\mathbf{x}, .)$. For instance, the notion of hyperplane skeleton of a data set $\{\mathbf{x}_j\}_{j=1}^N$ is extended to nonlinear hyperplane skeleton by the following definition:

$$\mathrm{NS}(\mathbf{X}) = \big\{\mathbf{x} \in \mathbb{R}^m : \langle \Phi(\mathbf{x}), \mathbf{h}_i\rangle = 0 \text{ for some } i \in \{1, \dots, k\}\big\}$$

where $\{\mathbf{h}_i\}_{i=1}^k \subset H$ is solution of the following minimization problem:

$$\text{minimize} \quad \sum_{j=1}^{N} \min_{1\leq i\leq k} \langle \Phi(\mathbf{x}_j), \mathbf{h}_i\rangle^2 \tag{5.1}$$

$$\text{subject to } \|\mathbf{h}_i\| = 1, \quad i = 1, \dots, k. \tag{5.2}$$

Clustering on subspaces in RKHS. Consider two cases:

(a) *Primal Kernel Subspace Clustering Algorithm*: the feature map Φ is known and the feature space (RKHS) is finite dimensional.

Then the subspace clustering algorithm (Algorithm 1) is performed with the only change $\mathbf{x}$ with $\Phi(\mathbf{x})$.

(b) *Dual Kernel Subspace Clustering Algorithm*: the kernel K is known only. Then the following two steps are performed until convergence:
   (1) Cluster assignment.
   (2) Cluster update: this step resemble a kernel PCA step.

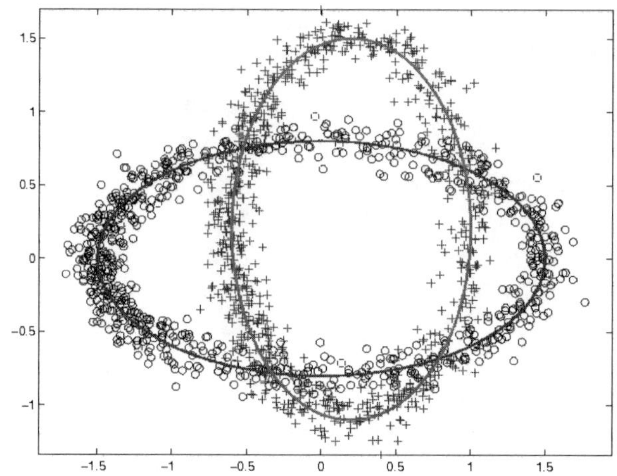

FIGURE 4. Nonlinear skeleton of degree 2.

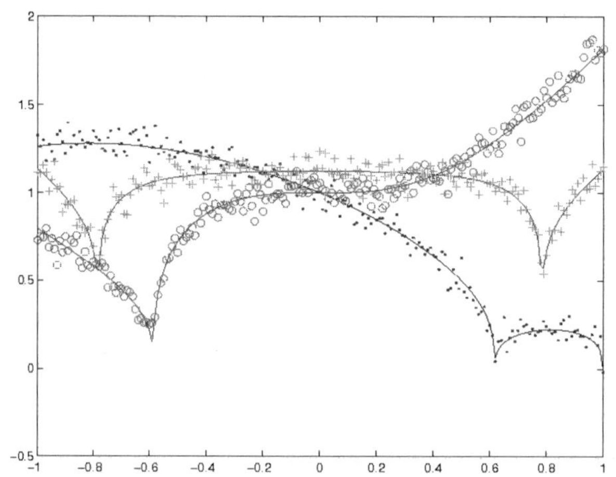

FIGURE 5. Nonlinear skeleton of degree 3.

The subspace clustering algorithm (Algorithm 1) is changed as follows:

The justification of the formula in the Cluster Assignment step is the fact that the eigenvector $\mathbf{u}_{i,s}$ of the correlation matrix $\mathbf{Y}_i\mathbf{Y}_i^T$ corresponding to the eigenvalue $\lambda_{i,s}$ is expressed by the the eigenvector $\mathbf{v}_{i,s}$ of the kernel matrix $\mathbf{Y}_i^T\mathbf{Y}_i$ as

$$\mathbf{u}_{i,s} = \lambda_{i,s}^{-1/2} \sum_{p=1}^{l_i} (\mathbf{v}_{i,s})_p \Phi(\mathbf{x}_{i,p}),$$

see for instance, [13, p. 142].

This algorithm again terminates in finitely many steps to a solution which is locally optimal, i.e. it finds k clusters $\mathbf{X}_i, i = 1, \dots, k$ such that $\min_{1\le l\le k}\langle \Phi(\mathbf{x}_j), \mathbf{h}_l\rangle = \langle \Phi(\mathbf{x}_j), \mathbf{h}_i\rangle$ if $\mathbf{x}_j \in \mathbf{X}_i$, where $\{\mathbf{h}_i\}_{i=1}^k$ is a local solution of (5.1), (5.2). For small values of k (less than 7), this algorithm terminates usually in clusters which are

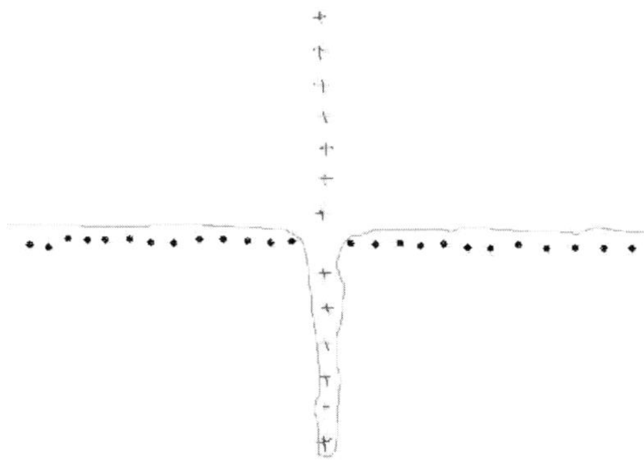

FIGURE 6. Binary classification by classical SVM classifier. New data points from both classes, converging to the intersection of their skeletons (which in this case are lines) cannot be classified correctly, while the proposed skeleton classifier classifies them perfectly.

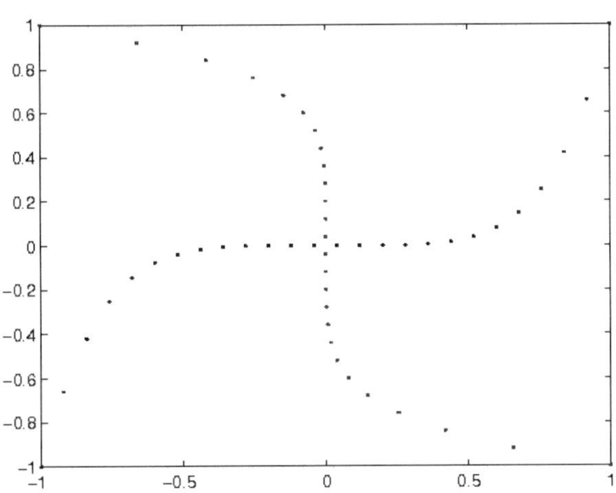

FIGURE 7. Nonlinear cross. The skeletons of the two classes are level sets of polynomials of two variables of degree 2. They are hyperplanes in $\mathbb{R}^6$ after applying the feature map $\Phi : (x, y) \mapsto (x^2, y^2, \sqrt{2}xy, \sqrt{2}x, \sqrt{2}y, 1)$. The same reasoning like in previous example shows the advantage of the proposed skeleton classifier.

globally optimal. A big challenge is to design a global optimization algorithm for subspace (and kernel subspace) clustering.

Algorithm 2: Dual kernel subspace clustering algorithm.

Data: samples $\mathbf{x}_1, \ldots, \mathbf{x}_T, \widetilde{\mathbf{x}}$, natural numbers $k, r_i, i = 1, \ldots, k$
Result: estimated k clusters $\mathbf{X}_i, i = 1, \ldots, k$ such that
$\min_{1 \le l \le k} \langle \Phi(\mathbf{x}_j), \mathbf{h}_l \rangle = \langle \Phi(\mathbf{x}_j), \mathbf{h}_i \rangle$ if $\mathbf{x}_j \in \mathbf{X}_i$, where $\{\mathbf{h}_i\}_{i=1}^k$ is a solution of (5.1), (5.2) and estimate a cluster with index $\tilde{i}$ in which $\widetilde{\mathbf{x}}$ is classified.

Initialize $\varepsilon > 0$ and randomly generate clusters $\mathbf{X}_i, i = 1, \ldots, k$.
do (*Until there is no change of the current clusters*)
  Cluster update.
  for $i \leftarrow 1, \ldots, k$ **do**
    Calculate the i-th kernel matrix $\mathbf{K} := \mathbf{Y}_i^T \mathbf{Y}_i$, where $\mathbf{Y}_i$ is a matrix with columns $\Phi(\mathbf{z}) : \mathbf{z} \in \mathbf{X}_i$
    Choose unit eigenvectors $\mathbf{v}_{i,s}, s = 1, \ldots, r_i$ of $\mathbf{K}$ corresponding to the r_i minimal eigenvalues $\lambda_{i,s}, s = 1, \ldots, r_i$.
  end
  Cluster assignment.
  for $j \leftarrow 1, \ldots, T$ **do**
    Add $\mathbf{x}_j$ to cluster $\mathbf{X}_i$, where i is chosen to minimize the distance from $\Phi(\mathbf{x}_j)$ to the subspace with codimension r_i which fits in the best way the columns of $\mathbf{Y}_i$; i.e. i is chosen such that

$$\text{(5.3)}\quad \min_{1 \le l \le k} \sum_{s=1}^{r_l} \lambda_{l,s}^{-1/2} \sum_{p=1}^{l_i} (\mathbf{v}_{l,s})_p K(\mathbf{x}_{i,p}, \mathbf{x}_j) = \sum_{s=1}^{r_i} \lambda_{i,s}^{-1/2} \sum_{p=1}^{l_i} (\mathbf{v}_{i,s})_p K(\mathbf{x}_{i,p}, \mathbf{x}_j),$$

    where l_i is the number of elements in the cluster $\mathbf{X}_i$ and $\Phi(\mathbf{x}_{i,p}) = \mathbf{Y}_i(:, p)$ (the columns of the matrix $\mathbf{Y}_i$).
  end
end
Estimate a cluster with index $\tilde{i}$ in which $\widetilde{\mathbf{x}}$ is classified: choose index $\tilde{i}$ for which a minimum in (5.3) is attained when $\mathbf{x}_j$ is replaced with $\widetilde{\mathbf{x}}$.

Examples. Figure 4 shows artificially created data which are fitted by two ellipses. We used the primal kernel subspace clustering algorithm with polynomial kernel of degree 2. Figure 5 shows another set of artificially created data which are fitted by three level sets of polynomials of degree three. We used again the primal kernel subspace clustering algorithm with polynomial kernel of degree 3.

6. Skeleton classifiers

Here we present the idea that binary classification tasks can be performed by finding nonlinear skeletons of the training points in the both classes. Figure 6 shows a possible separation of the two classes belonging to two lines (forming a cross) by standard support vector machine (SVM) classifier [15, 12]. It is clear that this classifier cannot classify correctly new points belonging to these lines and converging to their intersecting point (not shown).

Now we propose that the skeletons of the two classes can serve as a classifier (called skeleton classifier) in sense that a new point belongs to a class A if it is at a nearest distance to the skeleton of the class A. Then it is clear that the new points from the above example that cannot be classified correctly by SVM (converging to the intersection of the two skeletons, which are two lines) can be classified perfectly by the proposed skeleton classifier. In this example the skeleton is affine hyperplane skeleton (union of hyperplanes). We have to note that a similar (but more complicated) idea was proposed in [10].

Figure 7 shows a nonlinear perturbation of the previous data set (nonlinear cross, two level sets of polynomials of degree 2). After mapping of the points in both classes by the feature map $\Phi : (x, y) \mapsto (x^2, y^2, \sqrt{2}xy, \sqrt{2}x, \sqrt{2}y, 1)$, the data set belong to two hyperplanes in $\mathbb{R}^6$ (which are also the image of the skeletons of the two classes after applying the feature map Φ). The same reasoning like in previous example shoes that next points from both classes (belonging near to the corresponding skeletons and converging to the intersection of nonlinear skeletons) cannot be classified by SVM classifier, while can be classified perfectly by the proposed skeleton classifier.

Several issues concerning skeleton classifiers remain to be investigated in future research: statistical properties, stability, applications to real data, etc.

7. Conclusion

We develop the idea of nonlinear skeleton and present subspace clustering algorithms for finding such a skeleton, based on eigenvalue decomposition. Finding the affine skeletons of data sets has a direct application to Blind Signal Separation problems for de-mixing of unknown mixture of source signals under mild sparsity assumptions. We present identifiability conditions for sparse BSS problems, allowing less hyperplanes in the data points for full recovery of the original sources and the mixing matrix. We propose a new classifier—a skeleton classifier, which classifies the points in accordance with their nearness to the skeleton of a corresponding class. Presented examples suggest that the proposed skeleton classifier could be better compared with the classical SVM classifier.

References

1. N. Aronszajn, *Theory of reproducing kernels*, Trans. Amer. Math. Soc. **68** (1950), 337–404.
2. P. S. Bradley and O. L. Mangasarian, *k-plane clustering*, J. Global Optim. **16** (2000), no. 1, 23–32.
3. A. Cichocki and S. Amari, *Adaptive blind signal and image processing*, John Wiley, Chichester, 2002.
4. P. Comon, *Independent component analysis—a new concept?*, Signal Process. **36** (1994), 287–314.
5. F. Cucker and S. Smale, *On the mathematical foundation of learning*, Bull. Amer. Math. Soc. (N.S.) **39** (19??), no. 1, 1–49.
6. J. Eriksson and V. Koivunen, *Identifiability and separability of linear* ICA *models revisited*, 4th Internat. Conf. on ICA and BSS, ICA-2003 (Japan, 2003), PUBL???, PUBLADDR ???, 2003, pp. 23–27.
7. P. Georgiev, F. Theis and A. Cichocki, *Sparse component analysis and blind source separation of underdetermined mixtures*, IEEE Trans. Neural Networks **16** (2005), no. 4, 992–996.
8. A. Hyvärinen, J. Karhunen, and E. Oja, *Independent component analysis*, John Wiley & Sons, NY, 2001.
9. T.-W. Lee, M. S. Lewicki, M. Girolami, and T. J. Sejnowski, *Blind sourse separation of more sources than mixtures using overcomplete representations*, IEEE Signal Process. Lett. **6** (1999), no. 4, 87–90.
10. O. L. Mangasarian and E. W. Wild, *Multisurface proximal support vector machine classification via generalized eigenvalues*, IEEE Trans. Pattern Analysis and Machine Intelligence **28** (2006), no. 1, 69–74.

11. A. M. Rubinov, N. Soukhoroukova, and J. Ugon, *Minimization of the sum of minima of convex functions and its application to clustering*, Continuous Optimization, Current trends and modern applications (V. Jeyakumar and A. Rubinov, eds.), Springer, Berlin, 2005, pp. 409–434.
12. B. Scholkopf and A. Smola, *Learning with kernels*, MIT Press, Cambridge, MA, 2002.
13. J. Shawe-Taylor and N. Cristianini, *Kernel methods for pattern analysis*, Cambridge University Press, Cambridge, UK, 2004.
14. F. J. Theis, E. W. Lang, and C. G. Puntonet, *A geometric algorithm for overcomplete linear* ICA, Neurocomputing **56** (2004), 381–398.
15. V. N. Vapnik, *The nature of statistical learning theory*, 2nd ed., Springer, NY, 2000.
16. K. Waheed and F. Salem, *Algebraic overcomplete independent component analysis*, 4th Internat. Conf. on ICA and BSS, ICA-2003 (Japan, 2003), PUBL???, PUBLADDR ???, 2003, pp. 1077–1082.
17. M. Zibulevsky and B. A. Pearlmutter, *Blind source separation by sparse decomposition in a signal dictionary*, Neural Comput. **13** (2001), no. 4, 863–882.

Computer Science Department, University of Cincinnati, Cincinnati, OH 45221, USA.

E-mail address: pgeorgie@ececs.uc.edu

Centre de Recherches Mathématiques
CRM Proceedings and Lecture Notes
Volume **45**, 2008

Current Classification Algorithms for Biomedical Applications

Mario R. Guarracino, Salvatore Cuciniello, Davide Feminiano, Gerardo Toraldo, and Panos M. Pardalos

ABSTRACT. In this study we report the advances in supervised learning methods that have been devised to analyze medical data sets. As mining of data sets produced by medical equipments is becoming an increasingly challenging task, due to the size of the databases and the gradient of their update, new methods need to provide classification models that can handle the complexity of the problems. We start describing standard methods and we show how kernel methods, incremental learning algorithms and feature reduction techniques, applied to standard classification techniques, can be successfully used to discriminate biological and medical data sets. Among existing methods, we describe those that have their foundations in the statistical learning theory and have been successfully applied to the field. We provide numerical experiments based on publicly available data sets, and discuss results in terms of classification accuracy. Finally, we draw conclusions and outline future research directions.

1. Introduction

Supervised learning refers to the capability of a system to learn from a set of input/output couples, which is called the *training set*. The trained system is able to provide an answer (output) for a new question (input). The term *supervised* originates from the fact that the desired output for the training set of points is given by an external teacher. Supervised learning systems can find applications in the prognosis and diagnosis of different illnesses: differentiation of breast cancers induced by the BRCA1 mutations versus the BRCA2 and sporadic mutations [17], prediction of patient outcome after prostatectomy prostate cancer [32], classification of malignant gliomas survival based on gene expression and histological analysis [27], clinical outcome of breast cancer [34], recurrence of hepatocellular carcinoma after curative resection [18], discrimination of tumor versus normal colon tissues [2],

---

2000 *Mathematics Subject Classification.* Primary: 68Q32; Secondary: 68T10.
The authors are grateful to Dr. Hervé Philippe for his help in the analysis of the `rpl2e` data.
This work has been partially funded by Italian Research Council and Centro Ricerche *Enrico Fermi*.
This is the final form of the paper.

classification of Acute Myeloid versus Lymphoblastic Leukemia [13]. Among supervised methods there are K-Nearest Neighbours (KNN) and Fisher Discriminant Analysis (FDA) [9], Support Vector Machines (SVM) [12], Regularized Generalized Eigenvalue Classifier (ReGEC) [14], and Neural Networks [29].

Unsupervised learning starts with a collection of multivariate data sets and produces groupings of samples, or combination of variables, based upon information inherent in the data set without additional outside information. Among the unsupervised methods, Eisen*el al.* [10] successfully applied hierarchical clustering to genomic problems, Bradley *et al.* [4] proposed local maximum clustering for gene expression data set analysis. In [20] self-organizing maps (SOM) are proposed and Tavazoie*el al.* [33] applied K-means clustering and its variations to genetic network architectures.

In almost every application area, data sets are ever growing and continuously updated. Moreover, many applications on massive data sets analysis are emerging [1], which require efficient computational procedures to face the dynamics of large databases. Keeping an effective continuous updating of the learning system by including new knowledge that becomes available is a very challenging task. A key point with respect to this would be to understand how a new training point may influence the classification function, in order to design effective solutions to be used when the data set updating is required, in place of a complete retraining, whose computational burden could result prohibitive. For this purpose several methods, especially in the kernel-based nonlinear classification case, have been proposed to reduce the size of the training set, and thus, the related kernel: Reduced SVM (RSVM) [21, 23], local incremental learning of SVM (LISVM) [31], Incremental and decremented SVM [5, 8], incremental ReGEC (I-ReGEC) [6] and LibSVM [22].

In the last years, mining information from huge data sets produced by the Human Genome project [11], and other related projects using microarray technology, represent a very challenging task to work with. The classification of gene expression data sets is interesting for many reasons. Indeed, data sets have the unusual feature of comprising a very large number of variables and publicly available data sets contain gene expression data sets for tens of thousands genes, which are updated regularly. This tendency is going to bring in the next years the need for classification algorithms that can handle such complexity.

To solve the problem, feature selection or transformation methods have been introduced, for instance Principal Component Analysis (PCA) [28], Singular Value Decomposition (SVD) [36] and feature selection derived by Golub [13] to reduce the *curse of dimensionality.*

In this work we start describing SVM, ReGEC, FDA and KNN. These are useful for the classification of data sets with a small number of samples and features. Then, we show the limited capabilities of those algorithms through numerical examples. This is mainly due to the fact that a classification problem can involve a large number of training points. Then, to describe the advantages of Chunking and incremental methods, we show performances of LibSVM and I-ReGEC. Finally, we detail and provide accuracy results for feature transformation and reduction techniques applied to those algorithms.

The notation used in the paper is as follows. A scalar will be denoted by y, while a vector by $\mathbf{x}$ and its components by $\mathbf{x}^i$. Different vectors are identified by $\mathbf{x}_i$, where i is the index of a vector set.

Scalar product of two vectors $\mathbf{x}_i$ and $\mathbf{x}_j$ in $\mathbb{R}^n$ will be denoted by $\mathbf{x}_i'\mathbf{x}_j$, 2-norm of $\mathbf{x}$ will be denoted by $\|\mathbf{x}\|$ and the unit vector will be denoted by $\mathbf{e}$. The transpose of a matrix C is C^T.

The present work is organized as follows: in Section 2, we describe and test standard classification methods. In Section 3, we discuss the advantages of Chunking and incremental methods, giving some more results. In Section 4, we detail and provide accuracy results for feature transformation and reduction techniques applied to those algorithms. In Section 5, current techniques are reported. Finally, in Section 6, conclusions are drawn and future research directions are highlighted.

2. Standard Classification Methods

The problem of classifying objects into classes is as old as the interest of human beings in trying to discern difference and similarities. Many efforts have been made for centuries to define good systems of classification. Ancient classification examples can be found in ancient repositories such as the Library of Alexandria in Egypt. An ancient field of application of classification methods is the solution of mathematical problems. Indeed, some Babylonian tables of the Old Babylonian Empire (1900 b.C.—1600 b.C.) report one of the earliest known classification theorem relative to some geometric measurements related to the Pitagora's theorem. A significant system of classification, applied to non-mathematical information, was attempted into IV century a.C. by Aristotle, subdividing animals in groups by such criteria as reproduction method or lack of red blood. Classification systems have been used to subdivide animals and plants into groups, and, more generally, to handle diversity and similarity in the natural world.

In the next section we analyze methods that have been devised during the last century and have been successfully implemented on computing machineries.

2.1. Support vector machine.

SVMs are the state-of-the-art supervised classification methods and they have been widely accepted in many application areas. A SVM finds a hyperplane that separates the elements belonging to two different classes S^+ and S^- [35, 7]. The separating hyperplane $\mathbf{w}'\mathbf{x} + b = 0$ is usually chosen to maximize the *margin* between the two classes. The margin can be defined as the minimum distance of all of the elements of either class to the hyperplane and it is equal to $2/\|\mathbf{w}\|$. These elements with the minimum distance from both classes are called *support vectors* and are the only elements needed to train the classifier.

Figure 1 shows the hyperplane that separates the points of the two classes, the margin μ, and the support vectors (ringed points).

Let consider a data set composed of n pairs $(\mathbf{x_i}, y_i)$ where $\mathbf{x_i} \in \mathbb{R}^m$ is the feature vector that characterizes the point $\mathbf{x_i}$, and $y_i \in \{-1, 1\}$, is the class label. Then, the solution to the following quadratic linearly constrained problem is the optimal hyperplane with the maximum margin:

$$\min f(\mathbf{w}) = \frac{\mathbf{w}'\mathbf{w}}{2} \qquad (2.1)$$
$$\text{s.t. } y_i(\mathbf{w}'\mathbf{x}_i + b) \geq 1.$$

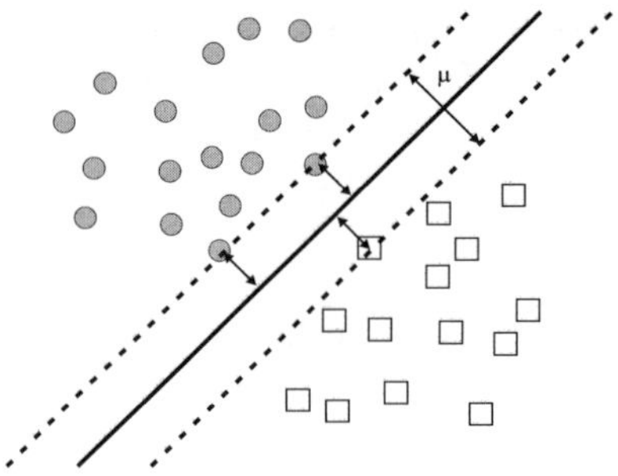

FIGURE 1. SVM classification.

A dual representation of this problem is as follows:

$$\max f(\boldsymbol{\alpha}) = \tfrac{1}{2}\boldsymbol{\alpha}'Q\boldsymbol{\alpha} - \mathbf{e}'\boldsymbol{\alpha} \quad \text{s.t. } \alpha_i \geq 0,\ i = 1, \dots, n, \quad \mathbf{y}'\boldsymbol{\alpha} = 0, \tag{2.2}$$

and the classification function is:

$$f(\mathbf{x}) = \sum_{i=1}^{n} \alpha_i y_i x_i \mathbf{x}, \tag{2.3}$$

where $\boldsymbol{\alpha}$ is a vector of training examples $(\mathbf{x}_i, y_i)$ and $Q_{i,j} = y_i y_j (\mathbf{x}_i \mathbf{x}_j)$.

In the classification of a large number of points, usually a very small number of support vectors are sufficient to define the optimal separating hyperplane. This is shown in Fig. 1. Therefore a good hyperplane can easily be found using a subset of points from both classes whose memberships are known and the test points can be efficiently classified using the function $f(\mathbf{x})$ in (2.3).

Formulation (2.2) has a feasible solution only when the data are linearly separable. To obtain a feasible solution in the more common cases of real data sets, slack variables are introduced and a penalty term in the objective function is added to allow some misclassification. Furthermore, the relationship between features and class labels can be nonlinear. In this case, a hyperplane could give an inaccurate classification. For this purpose, an implicit nonlinear embedding of the data sets to a higher dimensional space is required. This nonlinear mapping can be done implicitly by kernel functions, which represent the inner product of the elements in a nonlinear space. Kernel functions are represented as:

$$K(\mathbf{x}_i, \mathbf{x}_j) = \langle \phi(\mathbf{x}_i) - \phi(\mathbf{x}_j), \phi(\mathbf{x}_i) - \phi(\mathbf{x}_j) \rangle, \tag{2.4}$$

where $\phi(\mathbf{x})$ is the embedding function.

Some commonly used kernel functions are as follows.

$$\begin{array}{ll} \text{linear} & K(\mathbf{x}_i, \mathbf{x}_j) = \mathbf{x}_i'\mathbf{x}_j \\ \text{polynomial} & K(\mathbf{x}_i, \mathbf{x}_j) = (\mathbf{x}_i'\mathbf{x}_j + 1)^d \\ \text{Gaussian} & K(\mathbf{x}_i, \mathbf{x}_j) = \exp\left(-\dfrac{\|\mathbf{x}_i - \mathbf{x}_j\|^2}{\sigma}\right). \end{array}$$

In Fig. 2 we show points of two classes which are not linearly separable and a function ϕ projecting them in a space where they are separable.

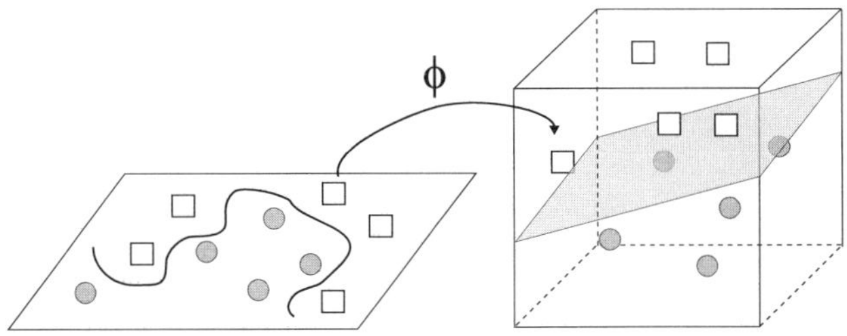

FIGURE 2. Nonlinearly separable points projected in a higher dimensional space.

Replacing the inner product $\mathbf{x}_i'\mathbf{x}_j$ in $Q_{i,j} = y_i y_j(\mathbf{x}_i\mathbf{x}_j)$, with the kernel function $K(\mathbf{x}_i, \mathbf{x}_j)$ we obtain $Q_{i,j} = y_i y_j K(\mathbf{x}_i\mathbf{x}_j)$.

2.2. Fisher discriminant analysis. Another classification technique, based on linear separation between classes, is the Fisher discriminant analysis. The idea is to find a direction that maximizes the scattering of data sets among classes and minimizes the scattering within each class. That direction permits to maximize class separation and to regroup points of each class, as depicted in Fig. 3.

This is the equation describing the scatter matrices, which are proportional to the covariance matrices:

$$S_B = \sum_c N_c(\boldsymbol{\mu_c} - \bar{\mathbf{x}})(\boldsymbol{\mu_c} - \bar{\mathbf{x}})' \tag{2.5}$$

$$S_W = \sum_c \sum_{i \in c} (\boldsymbol{\mu_c} - \mathbf{x}^i)(\boldsymbol{\mu_c} - \mathbf{x}^i)', \tag{2.6}$$

where

$$\boldsymbol{\mu_c} = \frac{1}{N_c} \sum_{i \in c} \mathbf{x}^i; \quad \bar{\mathbf{x}} = \frac{1}{N} \sum_i \mathbf{x}^i = \frac{1}{N} \sum_c N_c \boldsymbol{\mu_c}, \tag{2.7}$$

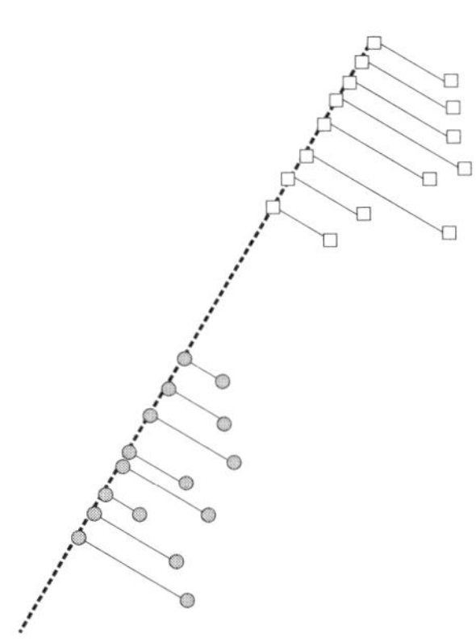

FIGURE 3. Scattering of data sets among classes and distances between points of the same class.

and N_c is the number of elements in class c. In order to obtain the solution, the Fisher discriminant analysis maximizes the class separability criterion:

$$J(\mathbf{w}) = \frac{\mathbf{w}' S_B \mathbf{w}}{\mathbf{w}' S_W \mathbf{w}}. \tag{2.8}$$

This is the Raleigh quotient of the generalized eigenvalue problem $S_B\mathbf{w} = \lambda S_W$. Since S_B is a symmetric positive definite matrix, the stationary points are reached in correspondence of the eigenvectors, and the objective function values in those points equal the corresponding eigenvalues. Thus, to maximize (2.8), it is sufficient to evaluate the eigenvector corresponding to the largest eigenvalue.

The solution of this problem is usually obtained by transforming it in a quadratic programming task. Since the total scatter matrix is such that $S_T = S_B + S_W$, the objective function (2.8) may be rewritten as:

$$J(\mathbf{w}) = \frac{\mathbf{w}' S_T \mathbf{w}}{\mathbf{w}' S_W \mathbf{w}} - 1. \tag{2.9}$$

The objective function is invariant with respect to re-scalings of the vectors w, (2.9) is equivalent to the following constrained optimization problem:

$$\begin{aligned} &\min_{\mathbf{w}\neq 0} -\tfrac{1}{2}\mathbf{w}' S_B \mathbf{w} \\ &\text{s.t. } \mathbf{w}' S_W \mathbf{w} = 1. \end{aligned} \tag{2.10}$$

Although the nonlinear formulation of the method does not naturally arise as in SVM, it is possible to obtain a kernel Fisher LDA.

2.3. Nearest neighbour. The idea behind Nearest Neighbour classification is very simple: a point is assigned to the class of its nearest neighbour. The method can be generalized to take into account k neighbours of a given point (k *Nearest Neighbour*—kNN), as depicted in Fig. 4. In this case, the class label is chosen equal to the label belonging to the greater part of neighbours. Furthermore, this notion can be extended to take into account different discriminating functions D_i for each class i. For the Parzen version of the algorithm, a point $\mathbf{y}$ belongs to the class i

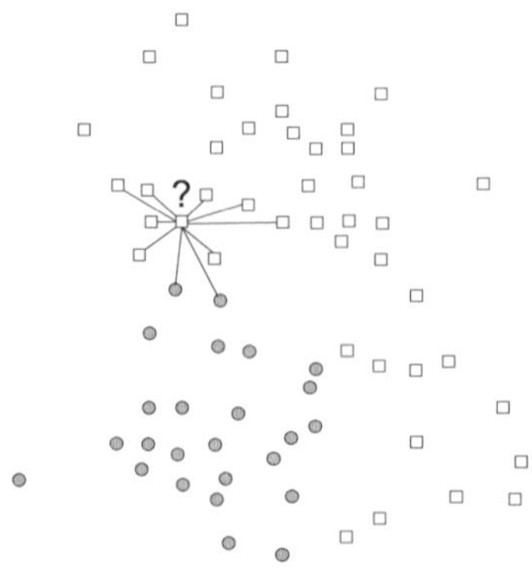

FIGURE 4. The class of white point is derived from its distances from k nearest points.

which maximizes the discriminating function:

$$D_i(\mathbf{y}) = \frac{1}{N^{(i)}} \sum_{j=1}^{N^{(i)}} \phi_i(\mathbf{y} - \mathbf{x}_j^{(i)}), \tag{2.11}$$

where $N^{(i)}$ is the number of elements of class i and the non-parametric kernel ϕ_i can be different for each class.

2.4. Regularized generalized eigenvalue classifier. There are also efficient algorithms that exploit the special structure of a slightly different optimization problem, such as the Generalized Proximal SVMs (GEPSVM) [24], in which the binary classification problem can be formulated as a generalized eigenvalue problem. This formulation differs from SVMs because, rather than finding one hyperplane that separates the two classes, it finds two hyperplanes that approximate the two classes. In its original formulation, such approach requires the solution of two different eigenvalue problems; more recently a classifier that uses a new regularization technique, known as Regularized General Eigenvalue Classifier (ReGEC), has been proposed. It requires the solution of a single eigenvalue problem to find both hyperplanes [14].

Let consider two matrices $A \in \mathbb{R}^{n \times m}$ and $B \in \mathbb{R}^{k \times m}$, that represent the two classes, each row being a point in the feature space. Mangasarian *et al.* [24] propose to classify these two sets of points A and B using two hyperplanes in the feature space, each closest to one set of points, and furthest from the other.

In Fig. 5, the hyperplanes generated by GEPSVM for two simple data sets are shown. The first hyperplane is the closest to the set of points in A and the furthest from those in B and the second hyperplane is closest set of points in B and the furthest from those in A. Furthermore, Fig. 5 (left) shows that GEPSVM can linearly discriminate sets that cannot be separated by a single hyperplane.

In order to satisfy the previous condition for the points in A, the two hyperplanes

$$K(\mathbf{x}, C)\mathbf{u}_1 - \gamma_1 = 0, \quad K(\mathbf{x}, C)\mathbf{u}_2 - \gamma_2 = 0 \tag{2.12}$$

FIGURE 5. Separation obtained with generalized eigenvalues.

can be obtained by solving the following two regularized optimization problems:

$$\min_{\mathbf{u},\gamma\neq 0} \frac{\|K(A,C)\mathbf{u}-\mathbf{e}\gamma\|^2+\delta\|\left[\begin{smallmatrix}\mathbf{u}\\ \gamma\end{smallmatrix}\right]\|^2}{\|K(B,C)\mathbf{u}-\mathbf{e}\gamma\|^2}, \tag{2.13}$$

and

$$\min_{\mathbf{u},\gamma\neq 0} \frac{\|K(B,C)\mathbf{u}-\mathbf{e}\gamma\|^2+\delta\|\left[\begin{smallmatrix}\mathbf{u}\\ \gamma\end{smallmatrix}\right]\|^2}{\|K(A,C)\mathbf{u}-\mathbf{e}\gamma\|^2}, \tag{2.14}$$

where $C^T = \begin{bmatrix} A^T & B^T \end{bmatrix}$ and δ is the regularization parameter.

The number of eigenvalue problems can be reduced from two to one, with the (*ReGEC*), proposed by Guarracino *el al.* [14], which requires the solution of the following generalized eigenvalue problem:

$$\min_{\mathbf{u},\gamma\neq 0} \frac{\|K(A,C)\mathbf{u}-\mathbf{e}\gamma\|^2+\delta\|\widetilde{K}_B\mathbf{u}-\mathbf{e}\gamma\|^2}{\|K(B,C)\mathbf{u}-\mathbf{e}\gamma\|^2+\delta\|\widetilde{K}_A\mathbf{u}-\mathbf{e}\gamma\|^2}. \tag{2.15}$$

Here $\widetilde{K_A}$ and $\widetilde{K_B}$ are diagonal matrices with the diagonal entries from the matrices $K(A,C)$ and $K(B,C)$. The new regularization leads to a regularized problem which provides accuracy results comparable to the ones obtained by solving equations (2.13) and (2.14).

The eigenvectors related to the minimum and the maximum eigenvalues obtained from the solution of (2.15) provide the proximal planes P_i, $i = 1,2$ to classify the new points. The distance of a point $\mathbf{x}$ from hyperplane P_i is:

$$\mathrm{dist}(\mathbf{x},P_i) = \frac{|K(\mathbf{x},C)\mathbf{u}-\gamma|}{\|\mathbf{u}\|}, \tag{2.16}$$

and the class of a point $\mathbf{x}$ is determined as

$$\mathrm{class}(\mathbf{x}) = \mathrm{argmin}_{i=-1,1}\{\mathrm{dist}(\mathbf{x},P_i)\}. \tag{2.17}$$

2.5. Standard classification methods results. The different methods have been implemented with Matlab 7.1. Results are calculated using an Intel Pentium 4 CPU 2.8GHz, 512MB RAM running Windows XP. Matlab function *eig* for the solution of the generalized eigenvalue problem is used as computational kernel of ReGEC. Matlab functions *svmtrain* and *svmclassify* for the solution of support vector machine are used. Classification algorithms KNN and KLDA are included in the open-source software package *MatlabArsenal* [37]. Results regarding their performance in terms of classification accuracy are presented.

Tests have been performed training the algorithms on a random sample of 90% of the data set and validating the accuracy on the remaining 10%. Tests have been repeated 100 times.

The data sets are: breast cancer (*BRCA*2) [17], high-grade glioma (*Nutt*) [27], Thyroid [3], Wisconsin Prognostic Breast Cancer (*WPBC*) [3] and Banana [26]. In Table 1, for each data sets are reported, name, dimension of the training and test sets, distribution of points between classes and percentage of null classification.

In Table 2, classification accuracy percentage is evaluated for SVM, KNN, KLDA and ReGEC. We note that results are better for data sets with less features, such as Banana and Thyroid, whereas, for those with thousands of features, results are close to null classification. The large number of features not only increases the space and time complexity, but also leads to overfitting and underfitting, as it will be explained in more detail in the following. Classification accuracy and performance

TABLE 1. Data sets characteristics.

Data set	train	test	m	1	-1	null (%)
BRCA2	20	2	3226	8	11	34
Nutt	45	5	12625	28	22	44
WPBC	99	11	32	41	69	31
Thyroid	140	75	5	65	150	30
Banana	400	4900	2	2376	2924	45

TABLE 2. Classification accuracy of standard methods (%).

Data set	SVM	KNN	KLDA	ReGEC
BRCA2	83.00	75.50	62.50	73.00
Nutt	54.00	71.80	54.00	46.00
WPBC	60.27	50.82	63.64	60.18
Thyroid	94.72	95.64	70.25	92.76
Banana	89.15	86.36	73.34	84.44

can be improved either by reducing the number of considered features, or defining a smaller subset of points for the training set.

3. Incremental Methods

The last few decades have been characterized by an increasing amount of data sets produced by medical trials. Furthermore, medical trials have been conducted in collaboration among different centers, which have produced many large data sets that are continuously updated during time. This leads to the necessity to device methods that can update classification models as soon as new data sets are made available, without reconsidering the entire data set. The basic idea of those methods is to take into account a small subset of the training set that retains as much information as possible. As soon as new data sets are available, they are tested against the previously selected points.

3.1. Chunking techniques. LibSVM is a simple and fast implementation of SVM. It uses a heuristic to find the hyperplane of optimal separation with the smallest number of nonzero coefficients. This produces a decrease in CPU time, although the solution provided might be a suboptimal one. LibSVM does not supply to SVM the entire data set: data sets are subdivided in chunks and SVM is executed for each chunk. This process is repeated until the size of the input data sets is reduced to two points. The problem solved by SVM is:

$$\min f(\boldsymbol{\alpha}) = \tfrac{1}{2}\boldsymbol{\alpha}'Q\boldsymbol{\alpha} - e'\boldsymbol{\alpha} \tag{3.1}$$

$$\text{s.t. } 0 \leq \alpha_i \leq C, \quad i = 1, \ldots, n \tag{3.2}$$

$$\mathbf{y}'\boldsymbol{\alpha} = 0,$$

where C is the upper bound of the all variables and $Q_{i,j} = y_i y_j K(\mathbf{x}_i, \mathbf{x}_j)$. Other methods update α at each step of an iterative procedure. In here, we consider the Sequential Minimal Optimization (SMO) [30], that has only two active elements at each step. A description of the SMO-type decomposition method can be found in [22].

3.2. Incremental subset selection. Incremental subset selection consists in constructing a small set of points that retains the information of the entire training set, providing comparable accuracy results. A kernel built from a smaller subset is computationally more efficient in predicting new elements, compared to the one that uses the entire training set. Furthermore, a smaller set of points reduces the probability of over-fitting the data sets. Finally, as new points are available, the cost to retrain the algorithm decreases if the influence of those new points on classification is evaluated only with respect to that subset, rather than the entire training set.

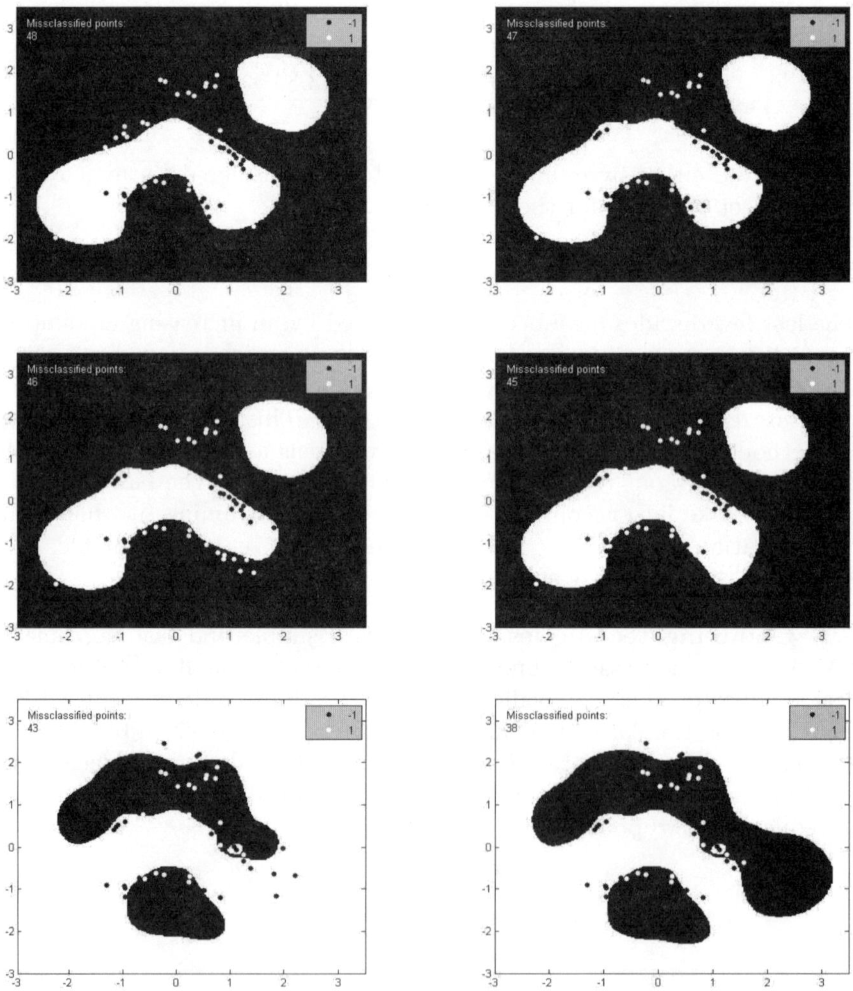

FIGURE 6. Classification surface during the I-ReGEC algorithm.

The algorithm takes an initial set of points $C \supset C_0 = A_0 \cup B_0$ and the entire training set C as input, where A_0 and B_0 are sets of points in C_0 which belong to the two classes A and B respectively. We refer to C_0 as the *incremental subset*. Let $\Gamma_0 = C \setminus C_0$ be the initial set of points that can be included in the incremental subset. ReGEC classifies all the points in the training set C using the kernel from C_0. Let P_{A_0} and P_{B_0} be the hyperplanes found by ReGEC, R_0 be the classification accuracy and M_0 be the points which are misclassified. Then, among the points in $\Gamma_0 \cap M_0$, the one that is farthest from its respective hyperplane is selected, i.e.

$$\mathbf{x}_i = \mathbf{x} : \max_{\mathbf{x} \in \{\Gamma_0 \cap M_0\}} \{\text{dist}(\mathbf{x}, P_{\text{class}(\mathbf{x})})\}, \tag{3.3}$$

where class($\mathbf{x}$) returns A or B depending on the class of $\mathbf{x}$. This point is the candidate point to be included in the incremental subset. This choice is based on the idea that a point very far from its plane either is needed in the incremental subset to improve accuracy, or it is an outlier. We update the incremental set as $C_1 = C_0 \cup \{\mathbf{x}_1\}$. Then, we classify the entire training set C using the points in C_1 to build the kernel. Let the classification accuracy be R_1. If $R_1 > R_0$, then we keep the new subset; otherwise we reject the new point, that is $C_1 = C_0$. In both cases $\Gamma_1 = \Gamma_0 \setminus \{\mathbf{x}_1\}$. The algorithm repeats until $|\Gamma_k| = 0$, at some kth iteration. The initial points are the training points closest to the centroids determined by a simple *k-means* algorithm applied to each class.

In [6] it has been shown that the k-means based selection criteria give the best performance in terms of stability and accuracy, with respect to random selection of initial points. In Fig. 6, the classification surfaces for Banana data set build during I-ReGEC algorithm with Banana data set, for a choice of $k = 5$, $\sigma = 0.2$ and $\delta = 10^{-5}$ are shown. The algorithm starts with a $2k$ initial points and it defines a classification surface with a small subset of points. When the algorithm adds new points to the incremental subset, the classification surface is modified and the number the points of the train, which are missclassified, decreases. At the beginning there are 48 missclassified points, whereas at the end of the algorithm, only 38 are left.

3.3. Incremental methods results. In Table 3, classification accuracy percentage is evaluated for LibSVM and I-ReGEC. We show that classification accuracy well compares with previous methods, although we are using only a subset as training set. LibSVM always gives better results in comparison with SVM. The same holds for I-ReGEC: for all cases classification accuracy has increased when compared with ReGEC. In particular, I-ReGEC performs better then ReGEC for Banana and Thyroid data sets, which have larger training sets. It was not possible to compute the libSVM classification for the Banana data set due to memory limit.

TABLE 3. Classification accuracy of incremental methods(%).

Data set	Lib-Svm	I-ReGEC
BRCA2	83.00	77.50
Nutt	75.40	65.20
WPBC	63.64	60.45
Thyroid	88.65	94.41
Banana	NaN	89.15

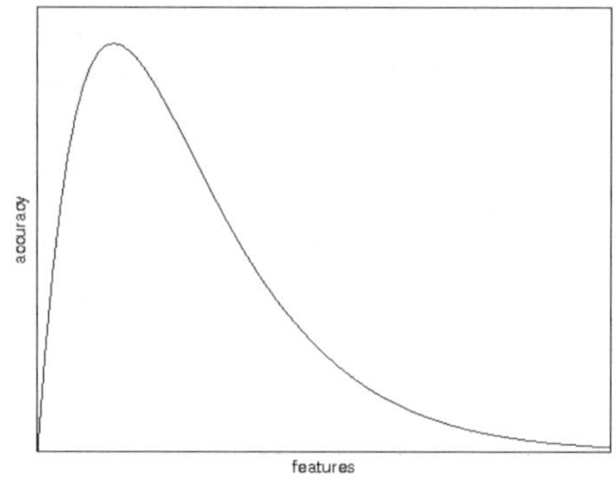

FIGURE 7. Number of features versus classifier accuracy.

4. Feature Reduction Algorithms

Features reduction methods find a *features subset* F highly correlated with a class [16]. Feature reduction is useful for different reasons: it improves the performances of the training procedure decreasing its execution time and avoiding problems related to the curse of dimensionality. In [15] it is shown that the growth of the number of features for a fixed number of samples degrades the classifier performances as described in Fig. 7. Furthermore, data sets may be affected by noise and redundancy due to instruments and experimental conditions. The features subset F should contain only features needed to have a good accuracy of the training procedure. Features fall into one of three categories: *strongly relevant*, *weakly relevant* and *irrelevant*. *Strongly relevant* features are those necessary to obtain a good classification model, while *weakly relevant* are not always useful and irrelevant features are not important [19]. Feature reduction techniques can be divided into two classes: *feature transformation* and *feature selection*. Feature selection addresses the problem of searching a minimal set of features that maximizes the discrimination among classes. If there are n features, there are 2^n possible features subsets. When n in very large, it is impossible to find the optimal features subset, therefore a suboptimal solution needs to be found. Feature transformation methods obtain a new set of features as linear combination of the original ones, that is, they linearly project points in a space of lower dimension.

4.1. Principal component analysis. PCA is an orthogonal linear transformation method that transforms data sets into a new coordinate system. It can be used as a statistical method to reduce multidimensional data sets to lower dimensions for analysis. PCA involves the computation of the eigenvalue decomposition of the covariance matrix $C \in \mathbb{R}^{n \times n}$, whose elements are:

$$C_{ij} = \mathrm{cov}(\mathbf{x}_i, \mathbf{x}_j), \tag{4.1}$$

where cov indicates the covariance of features $\mathbf{x}_i$ and $\mathbf{x}_j$.

The term *principal component* refers to the main direction along which data sets have maximum variance. Each principal component is described by a vector, namely by an eigenvector of the covariance matrix. Assuming the most important characteristics are those which contribute most to data set variance, the dimensionality reduction can be done by projecting the information in the space spanned by the eigenvectors which correspond to the largest eigenvalues. This means that it is possible to preprocess a data set with PCA and use its compact representation, in the space spanned by few principal components, to train a classifier. Obviously, the less components we consider, the more we can loose in terms of information about the data sets. However, if the discarded eigenvectors have relatively negligible eigenvalues, it is possible to summarize data sets without a particular loss of relevant information.

4.2. Singular value decomposition. The singular value decomposition is a linear algebraic method to factorize a matrix as product of three matrices. It reveals the rank of the matrix, thus allowing the detection of relevant features and discarding noise and redundancy.

The Singular Value Decomposition (SVD) of a matrix $X \in \mathbb{R}^{m \times n}$ with rank $r \leq \min\{m, n\}$, and $k = \min\{m, n\}$ is the factorization

$$X = U\Sigma V^T, \tag{4.2}$$

with

$$\Sigma = \mathrm{diag}(\sigma_1, \dots, \sigma_k) \in \mathbb{R}^{m \times n}, \tag{4.3}$$

where $\sigma_1 \geq \cdots \geq \sigma_r > 0 = \sigma_{r+1} = \cdots \sigma_k$ are the *singular values* of X, and U and V unitary matrices respectively of dimension $m \times m$ and $n \times n$. The columns of U and V are respectively the *left* and the *right singular vectors* for the corresponding singular values. Given (4.2), the following relation holds

$$X^T X = V\Sigma^2 V^T, \quad XX^T = U\Sigma\Sigma^T U^T, \tag{4.4}$$

which implies:

$$\sigma_i = \sqrt{\lambda_i(X^T X)} = \sqrt{\lambda_i(XX^T)}, \quad i = 1, \dots, n \tag{4.5}$$

i.e. the non-zero singular values of X are the square root of the non-zero eigenvalues of $X^T X$ (XX^T) and the left (right) singular values are the eigenvectors of XX^T ($X^T X$). This shows the equivalence of PCA and SVD. It can be shown that the matrix

$$X^{(l)} = \sum_{k=1}^{l} \mathbf{u}_k \mathbf{s}_k \mathbf{v}_k^T \tag{4.6}$$

is the closest rank-l matrix approximation of X. The term 'closest' means that $X^{(l)}$ minimizes the sum of the squares of the difference between the elements of X and $X^{(l)}$, $\min_{\mathbf{x}_{ij}^{(l)}} \sum_{i,j} |\mathbf{x}_{ij} - \mathbf{x}_{ij}^{(l)}|^2$. Equation (4.6) allows to use SVD for dimensional reduction. Consider the matrix X with features on columns, the right singular matrix contains an eigenvectors basis for the space. Taking the matrix V_l made with the first l columns of V, with $l \leq m$, the approximate representation of X in the subspace spanned by V_p^T is $V_l^T X$. The choice of the number l of eigenvectors to be considered, can be based on the percentage of information related to the first l eigenvectors defined as

$$p_l = \sum_{k=1}^{l} \sigma_k^2 \bigg/ \sum_{k=1}^{r} \sigma_k^2, \tag{4.7}$$

TABLE 4. Classification accuracy with feature reduction %.

Data set	SVD SVM	SVD *K*NN	SVD *K*LDA	SVD ReGEC
BRCA2	62.50	80.50	62.50	86.00
Nutt	54.00	71.80	54.00	72.00
WPBC	60.27	47.82	60.27	60.27
Thyroid	94.72	95.64	70.25	92.76

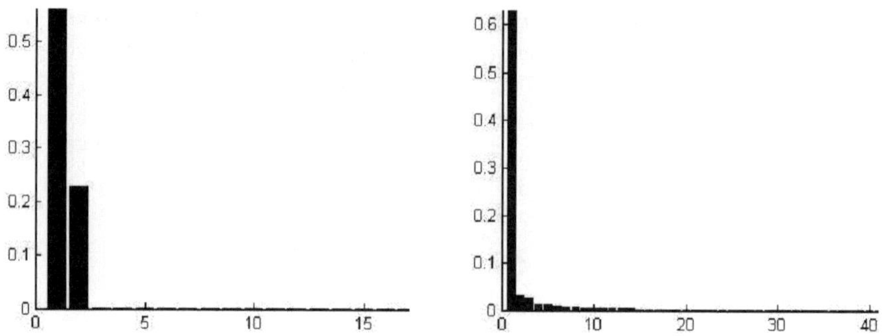

FIGURE 8. Singular value histogram on BRCA2 and Nutt data set.

In order to use SVD as a pre-processing stage to reduce the number of features of the train matrix, we consider only the first l columns of V, where p_l is computed according to (4.7), to give a satisfactory value.

4.3. Feature reduction algorithms results. Table 4 shows accuracy results obtained by various methods when the data set is preprocessed with SVD. Since the data set Banana consists of only two features, it is not considered in the following.

We note an improvement of the execution time and a slight improvement in classification accuracy. In particular, we note an improvement with BRCA2 and Nutt, which are data sets with many features, using ReGEC. In all cases we obtain at least the same accuracy because the information are contained in just a few features: in Fig. 8 we show singular value histogram on BRCA2 and Nutt data set from which it is possible to observe that only a few features gather up an almost complete information. The number of features is evaluated accordingly to (4.7), setting $p_l = .90$. For BRCA2 data set the percentage of used features is 0.68% and for Nutt data set only the 0.40%.

5. Current Methods

In many biomedical applications, the number of points and features in the training set grows continuously. In the previous pages, we have shown some methods to reduce the number of features and training points. For genomic data sets, it is often necessary only to apply the reduction of the features, because data sets provided by microarray technology refers to the comparison of thousands of genes on a small number of samples. In the next future it will be possible to compare a greater number of samples and therefore it will be necessary to device methods that at the same time reduce the number of points and features.

5.1. ILDC-ReGEC algorithm. We recall that computational complexity of the eigenvalue problem increases with the number of points and features in the training set.

The feature selection algorithm presented in the following is a hybrid method: it applies Golub's procedure for feature selection and then it controls whether to eliminate features accordingly to the built classifier. The original Golub's method selects an equal number of genes with positive and negative correlation coefficients, while in this version a new rule is defined in order to determine the number of the features. The function *FeatureSelection*, shown in the Algorithm 1, calculates the $F(\mathbf{x}_j)$ score for each gene and it sorts the features for decreasing values of $F(\mathbf{x}_j)$.

The algorithm sorts the values of $F(\mathbf{x}_j)$ in descending order and it returns the set of samples C^0 with only those genes whose sum of the $F(\mathbf{x}_j)$ is equal to a fixed percentage α of the total sum. In other words, the algorithm returns the genes $\mathbf{x}_{j_k}$

Algorithm 1 ILDC-ReGEC (C_0, C, α)

$C^0 =$ FeatureSelection (C, α)
$\Gamma_0 = C^0 \setminus C_0^0$
$\{R_0, M_0\} = 4$ Classify (C^0, C_0^0)
$k = 1$
$j = 0$
while $|\Gamma_k| > 0$ **do**
  $\mathbf{x}_k = \mathbf{x} : \max_{\mathbf{x} \in \{M_k \cap \Gamma_{k-1}\}} \{\text{dist}(\mathbf{x}, P_{\text{class}(\mathbf{x})})\}$
  $\{R_k^j, M_k^j\} =$ Classify $(C^j, \{C_{k-1}^j \cup \{\mathbf{x}_k\}\})$
  if $R_k^j >= R_{k-1}^j$ **then**
    $C_k^j = C_{k-1}^j \cup \{\mathbf{x}_k\}$
    FLAG = False
    repeat
      $C_k^{j+1} =$ LastFeature (C_k^j);
      $\{\overline{R}_k^j, M_k^j\} =$ Classify (C_k^j, C_k^{j+1});
      if $\overline{R}_k^j > R_k^j$ **then**
        $R_k^j = \overline{R}_k^j$
        $j = j + 1$
      else
        FLAG=True
      end if
    until FLAG
  end if
  $\Gamma_k = \Gamma_{k-1} \setminus \{\mathbf{x}_k\}$
  $k = k + 1$
end while

TABLE 5. Classification accuracy of current methods (%).

Data set	SVD-LibSVM	SVD-I-ReGEC	ILDC-ReGEC
BRCA2	80.00	84.50	86.00
Nutt	75.54	72.20	77.80

with higher values of $F(\mathbf{x}_j)$ such that:

$$\sum_{k=1}^{m_\alpha} F(\mathbf{x}_{j_k}) = \alpha \sum_{j=1}^{m} F(\mathbf{x}_j), \tag{5.1}$$

with $0 \leq \alpha \leq 1$ and $m_\alpha \leq m$.

After selecting the subset of features, ReGEC is applied in order to define the incremental subset. The execution of ReGEC becomes faster, and therefore overall execution time decreases.

As stated in Section 2.4, at each step, ILDC-ReGEC tries to add a new point to the incremental subset. When a new point is added, ILDC-ReGEC checks whether it is possible to delete some features.

Let R_k be the classification accuracy with all considered features and with $\overline{R}_K$ the one obtained omitting one feature. If $\overline{R}_K >= R_k$ then it deletes last feature and the process is repeated. The procedure is restarted until further accuracy improvement is obtained.

The algorithm is summarized in Algorithm 1.

5.2. Current methods results. In Table 5 classification accuracy results of IDLC-ReGEC are compared with the algorithms LibSVM and I-ReGEC, which use SVD as a preprocessing step to reduce the number of features.

We test only BRCA2 and Nutt, which have thousands of features. We note that LibSVM is still accurate when SVD is applied. Classification accuracy of I-ReGEC improves on both data sets and IDLC-ReGEC is better than the other two methods. ILDC-ReGEC selects only the 4.19% of the features and the 21.77% of the points of BRCA2; on Nutt, it selects only the 1.37% of the features and the 17.3% of the points.

6. Conclusions and Future Work

The use of standard classification methods with large training sets is not always straightforward. To improve results, we describe some techniques that reduce the number of samples of the training set. When these techniques are applied to data sets with many samples (Thyroid and Banana), they provide acceptable results. This idea permits to train the algorithm with just a few points. In case of high dimensional data sets, classification accuracy does not improve much, and therefore we introduce methods that reduce both points and features. We firstly use SVD as filter method to reduce the number of features and then we perform the selection of training points. This strategy is not always useful. In some cases, classification accuracy is not even better. We obtain an improvement only with ILDC-ReGEC, which uses a different strategy to select features. In this case we obtain improvements for classification accuracy using a very small number of points and features.

In future, it will be interesting to investigate how prior knowledge acquired by experts can be used in the classification models. Some attempts have already been done by [25]. In that case, prior knowledge is taken into account by the model in terms of constrains on the underlying optimization problem, but no clue is given on how to identify regions where such knowledge provides more insight. We will investigate how to determine regions in which further knowledge is required to improve accuracy of classification models.

References

1. J. Abello, P. M. Pardalos, and M. G. C. Resende (eds.), *Handbook of massive data sets*, Kluwer Academic Publishers, Norwell, MA, USA, 2002.
2. A. Alon, N. Barkai, D. A. Notterman, K. Gish, S. Ybarra, D. Mack, and A. J. Levine, *Broad patterns of gene expression revealed by clustering analysis of tumor and normal colon tissues probed by oligonucleotide arrays*, Proc. Natl. Acad. Sci. USA **96** (1999), 6745–6750.
3. C. L. Blake and C. J. Merz, Uci *repository of machine learning databases*, `www.ics.uci.edu/~mlearn/MLRepository.html`, 1998.
4. P. Bradley, O. Mangasarian, and W. N. Street, *The local maximun clustering method and its application in microarray gene expression data analysis*, EURASIP J. Appl. Signal Proc. **1** (2004), 51–61.
5. G. Cauwenberghs and T. Poggio, *Incremental and decremental support vector machine learning*, Advances in Neural Information Processing Systems (NIPS'2000), MIT Press, Cambridge, MA, vol 13, pp. 409–415.
6. C. Cifarelli, M. R. Guarracino, O. Seref, S. Cuciniello, and P. M. Pardalos, *Incremental classification with generalized eigenvalues*, Technical report, University of Florida, 2006.
7. N. Cristianini and J. Shawe-Taylor, *An introduction to support vector machines*, Cambridge University Press, Cambridge, UK, 2000.
8. D. Domeniconi and C. Gunopulos, *Incremental support vector machine construction*, 1st IEEE International Conference on Data Mining (ICDM'01), IEEE Computer Soc., Washington, DC, 2001, pp. 589–593.
9. R. O. Duda, P. E. Hart, and D. G. Stork, *Pattern classification*, Wiley-Interscience Publication, 2000.
10. M. Eisen *et al.*, *Cluster analysis and display of genome-wide expression pattern*, Proc. Natl. Acad. Sci. USA **95**, 1998, 14683–14688.
11. K. H. Fasman, A. J. Cuticchia, and D. T. Kingsbury, *The* GDB T *human genoma database*, Nucleic Acid. Res. **22** (1994), no. 17, 3462–3469.
12. T. Fureyet *al.*, *Support vector machine classification and validation of cancer tissue samples using microarray expression data*, Bull. Amer. Math. Soc. **16** (2003), 906–914.
13. T. R. Golub, D. K. Slonim, P. Tamayo, C. Huard, M. Gaasenbeek, J. P. Mesirov, H. Coller, M. Loh, J. R. Downing, M. A. Caligiuri, C. D. Bloomfield, and E. S. Lander, *Molecular classification of cancer: class discovery and class prediction by gene expression monitoring*, Science **286** (1999), 531–537.
14. M. R. Guarracino, C. Cifarelli, O. Seref, and P. M. Pardalos, *A classification algorithm based on generalized eigenvalue problems*, Optim. Methods Softw., 2007 (to appear).
15. I. Guyon and A. Elisseff, *An introduction to variable and feature selection*, J. Machine Learning Res. **3** (2003), 1157–1182.
16. M. Hall, *Correlation-based feature selection for machine learning*, M.Sc., Waikato University, 1999.
17. I. Hedenfalk, D. Duggan, Y. Chen, M. Radmacher, R. Simon, P. Meltzer, B. Gusterson, M. Esteller, M. Raffeld, Z. Yakhini, A. Ben-Dor, E. Dougherty, J. Kononen, L. Bubendorf, W. Fehrle, S. Pttalunga, S. Gruvberger, N. Loman, O. Johannsson, H. Olsson, B. Wilfond, G. Sauter, O. P. Kallioniemi, A. Borg, and J. Trent, *Gene-expression profiles in hereditary breast cancer*, N. Engl. J. Med. **344** (2001), 539–548.
18. N. Iizuka, M. Oka, H. Yamada Okabe, M. Nishida, Y. Maeda, N. Mori, T. Takao, T. Tamesa, A. Tangoku, H. Tabuchi, K. Hamada, H. Nakayama, H. Ishitsuka, T. Miyamoto, A. Hirabayashi, S. Uchimura, and Hamamoto, *Oligonucleotide microarray for prediction of*

early intrahepatic recurrence of hepatocelllaur carcinoma after curative resection, Lancet **361** (2003), 923–929.

19. R. Kohavi and G. H. John, *Wrappers for feature subset selection*, Artificial Intelligence **97** (1997), no. 1-2, 273–324.
20. T. Kohonen, *Self-organizing maps*, Proc. IEEE **78** (1990), no. 9, 1464–1480.
21. O. L. Lee, Y. J. Mangasarian, *Rsvm: reduced support vector machines*, Technical report 00-07, Data Mining Inst., University of Wisconsin, 2000; 1st SIAM Internat. Conf. on Data Mining (April 2001), CD-ROM; `ftp://ftp.cs.wisc.edu/pub/dmi/tech-reports/00-07.ps`.
22. C. J. Lin, C. W. Hsu, and C. C. Chang, *A practical guide to support vector classification*, `http://www.csie.ntu.edu.tw/ cjlin/papers/guide/guide.pdf`, 2004.
23. K. Lin and C. Lin, *A study on reduced support vector machines*, IEEE Trans. Neural Networks **14** (2003), no. 14, 1449–1459.
24. O. L. Mangasarian and E. W. Wild, *Multisurface proximal support vector classification via generalized eigenvalues*, Technical report 04-03, Data Mining Inst., University of Wisconsin, 2004.
25. ———, *Nonlinear knowledge-based classification*, Technical report 06-04, Data Mining Inst., University of Wisconsin, 2006.
26. S. Mika, G. Ratsch, J. Weston, B. Scholkopf, and K. Muller, *Fisher discriminant analysis with kernels*, Neural Networks for Signal Processing IX (Y.-H. Hu, J. Larsen, E. Wilson, and S. Douglas, eds.), IEEE, 1999, pp. 41–48.
27. C. L. Nutt, D. R. Mani, R. A. Betensky, P. Tamayo, J. G. Cairncross, C. Ladd, U. Phl, C. Hartmann, M. E. McLaughlin, T. T. Batchelor, P. M. Black, A. von Deimling, S. L. Pomeroy, T. R. Golub, and D. N. Louis, *Oligonucleotide microarray for prediction of early intrahepatic recurrence of hepatocelllaur carcinoma after curative resection*, Lancet **63** (2003), no. 7, 1602–1607.
28. E. Oja, *A simplified neuron model as a principal component analyzer*, J. Math. Biol. **15** (1982), 267–273.
29. M. O'Neill and L. Song, *Neural network analysis of lymphoma microarray data: prognosis and diagnosis near-perfect*, Optim. Methods Softw. **4** (2003), 28–41.
30. J. Platt, *Sequential minimal optimization: a fast algorithm for training support vector machines*, Technical report MSR-TR-98-14, 1998.
31. L. Ralaivola, *Incremental support vector machine learning: a local approach*, Lecture Notes in Comput. Sci. **2130** (2001), 322–330.
32. D. Singh, P. G. Febbo, K. Ross, D. G. Jackson, J. Manola, C. Ladd, P. Tamayo, A. A. Renshaw, A. V. D'Amico, J. P. Richie, E. S. Lander, M. Loda, P. W. Kantoff, T. R. Golub, and W. R. Sellers, *Gene expression correlates of clinical prostate cancer behavior*, Cancer Cell **1** (2002), no. 1, 203–209.
33. S. Tavazoie, J. D. Hughes, M. J. Campbell, R. J. Cho, and G. M. Church, *Systematic determination of genetic network architecture*, Nature Genetics **22** (1999), no. 3, 281–285.
34. L. J. van't Veer, H. Dai, M. J. Van De Vijver, T. D. He, A. A. M. Hart, M. Mao, H. L. Peterse, K. Van Der Kooy, M. J. Marton, A. T. Witteveen, G. J. Schreiber, R. M. Kerkhoven, C. Roberts, P. S. Linsley, R. Bernards, and S. H. Friend, *Gene expression profiling predicts clinical outcome of breast cancer*, Nature **415** (2002), 536–536.
35. V. Vapnik, *The nature of statistical learning theory*, Springer-Verlag, New York, 1995.
36. M. E. Wall, P. A. Dyck, and T. S. Brettin, Svdman—*Singular value decomposition analysis of microarray data*, Bioinformatics **17** (2001), no. 6, 566–568.
37. R. Yan, *A matlab package for classification algorithms*, `http://finalfantasyxi.inf.cs.cmu.edu/tmp/MATLABArsenal.zip`, 2006.

Area della Ricerca CNR, ICAR-CNR, Via Pietro Castellino, 111, 80131 Naples, Italy.
E-mail address: mario.guarracino@icar.cnr.it

Area della Ricerca CNR, ICAR-CNR, Via Pietro Castellino, 111, 80131 Naples, Italy.
E-mail address: salvatore.cuciniello@icar.cnr.it

Area della Ricerca CNR, ICAR-CNR, Via Pietro Castellino, 111, 80131 Naples, Italy.
E-mail address: malingut@hotmail.it

Area della Ricerca CNR, ICAR-CNR, Via Pietro Castellino, 111, 80131 Naples, Italy; Department of Agricultural Engineering and Agronomy, The University of Naples Federico II. Agronomy School, via Università 100, 80055 Portici, Italy
E-mail address: toraldo@unina.it

Center for Applied Optimization, University of Florida, Gainesville, FL, 32611-6595, USA
E-mail address: pardalos@ufl.edu

Centre de Recherches Mathématiques
CRM Proceedings and Lecture Notes
Volume **45**, 2008

Bilevel Model Selection for Support Vector Machines

Gautam Kunapuli, Kristin P. Bennett, Jing Hu, and Jong-Shi Pang

Abstract. The successful application of Support Vector Machines (SVMs), kernel methods and other statistical machine learning methods requires selection of model parameters based on estimates of the generalization error. This paper presents a novel approach to systematic model selection through bilevel optimization. We show how modelling tasks for widely used machine learning methods can be formulated as bilevel optimization problems and describe how the approach can address a broad range of tasks—among which are parameter, feature and kernel selection In addition, we also discuss the challenges in implementing these approaches and enumerate opportunities for future work in this emerging research area.

1. Introduction

Currently, Support Vector Machines (SVM) [12] and kernel methods enjoy enormous popularity in the machine learning community. There are several reasons: theoretically, their foundations are rooted in Structural Risk Minimization (SRM) [54], which leads to excellent generalization capabilities; practically, they are very flexible owing to the fact that they are modular and can be "kernelized" to capture highly nonlinear relationships [51]; and computationally, they are tractable and can be applied to large high-dimensional data sets which contain several thousand points. The kernel framework can be readily adapted to many learning tasks, e.g. regression, classification, ranking, and novelty detection. One other reason why SVMs may be considered one of the pre-eminent methods is because, with regard to applicability, they have had wide ranging success on a variety of real-world applications. However, the many papers reporting the success of such methods frequently gloss over an important issue: model selection.

In kernel methods, the root learning task is to construct a linear function that minimizes a regularized convex loss function. Nonlinear functions can then be constructed using the so-called "kernel trick". The resulting optimization problem is convex, but it typically contains hyper-parameters that must be selected by the

---

2000 *Mathematics Subject Classification.* Primary: 62F40; Secondary: 90C26.

This work was supported in part by the Office of Naval Research under grant no. N00014-06-1-0014.

This is the final form of the paper.

user. For example, in SVMs, the appropriate kernel function and trade-off parameter between error and regularization must both be selected. Certain loss functions, such as the ε-insensitive loss, require the selection of additional hyper-parameters. There have been many interesting attempts to pick these hyper-parameters; notable among these are approaches that use bounds [10] or attempt to trace the complete regularization path of the SVM [31]. However, the most systematic, commonly used and widely accepted method for selecting these hyper-parameters is still T-fold cross validation (CV).

The focus of this chapter is to demonstrate how T-fold cross-validation model selection for many different learning tasks can be formulated as bilevel optimization problems. We begin with a review of the cross-validation problem. We illustrate the bilevel programming model using support vector regression as an example. Then, we introduce the generic T-fold CV formulation and discuss many possible variations. We briefly review possible methods for solving the models and direct the reader to references that present computational results for one approach applied to support vector regression and classification. Solution of the other resulting problems is left as a significant and compelling challenge to mathematical programming researchers.

1.1. Model selection. The general predictive learning task is to construct a function using present data that performs well on future data. A loss function specific to the learning tasks is used to measure how well the function is performing. Cross validation (CV) is a method of estimating the out-of-sample generalization error of the model for given hyper-parameters. Cross validation leaves out subsets of the training data, trains models on the reduced sets of data, and then tests the resulting models on the left-out data. Cross validation can be applied to arbitrary machine learning problems, gives a very good estimate of generalization error (even for small data sets) which shows a strong correlation with the test error [20]. The CV step is typically followed by a post processing step in which the final model is trained on all the available data, using the "optimal" hyper-parameters given by CV, to build the final model. The efficacy is this model may further be examined by observing its performance on a hold-out test set.

To perform model selection, CV must be embedded within an optimization algorithm. In the most common approach, Grid Search, CV is performed over a grid that discretizes the hyper-parameter space of interest and involves, for T folds, training T models at each grid point. As the number of hyper-parameters grows, so does the number of problems to be solved and cross validation becomes prohibitively expensive. Efficiency can only be achieved at the expense of grid refinement and coarser grids inevitably yield poor models. In fact, even for a small number of parameters, cross validation can still be expensive for high-dimensional data sets. For example, feature selection for high-dimensional data sets leads to a combinatorial explosion of grid points. Such problems are ubiquitous in machine learning e.g., in feature selection [5, 30], kernel construction [37, 48], and multi-task learning [9, 21]. For such high-dimensional problems, greedy strategies such as stepwise regression, backward elimination, filter methods, or genetic algorithms are used. Yet, these heuristic methods, including grid search, have a fundamental deficiency in addition to their practical inefficiency; namely, they are incapable of assuring the overall quality of the produced "solution".

Another drawback in grid search is that the discretization is restricted to examining only a finite set of points. Recent work on determining the full regularization path of support vector machines underscores the fact that regularization parameter is continuous. In particular, the paper [31] argues that the choice of the single regularization parameter, C, is critical and shows that it is quite tractable to compute the SVM solution for all possible values of the regularization parameter C. But as it is well known in optimization, this parametric programming approach for a single parameter is not extendable to models with multiple parameters and certainly is not possible for models with a large number of parameters. Bayesian methods can treat model parameters as random variables but then the challenge becomes the choice of appropriate priors. In the end, out-of-sample testing is still the gold standard for selecting parameters values. From the standpoint of "optimizing model selection" using out-of-sample estimates, there is an urgent need for improved methodologies that combine sound theoretical foundation and robust computational efficiency. This paper proposes one such methodology that is based on the methods of *bilevel optimization.*

In addition to model selection for support vector machines through continuous cross validation, the bilevel approach can also be applied to a wide variety of problems like semi-supervised learning, predicting missing values in the data, kernel selection, multi-task learning and complexity minimization. This is because each of these problems can be formulated as a bilevel program where the overall testing/generalization objective is minimized in the "outer" (or upper) level subject to the learning functions which are optimized in the "inner" (or lower) level. For example, in cross validation, training is performed in the inner level and validation in the outer level. Prior bilevel approaches have been developed and successfully used for inner-level problems with closed form solutions and a single parameter, e.g. the generalized cross-validation method for selecting the ridge parameter in ridge regression [28]. Still, these approaches are limited to a single hyper-parameter and inner-level function with a closed-form solution.

Bilevel model selection offers several advantages over prior approaches. The most obvious advantage is the ability to deal with multi-parametric model selection and deal with them in continuous rather than discrete space. This is possible because of recent advances in bilevel programming in the optimization community, which permit the systematic treatment of models based on different loss and regularization functions and kernels. In addition to being able to incorporate existing methods, the bilevel approach offers a broad framework in which novel regularization methods and generalization measures can be developed. Most significantly, these advantages allow for improved model selection.

1.2. Bilevel programming. In this subsection, we introduce the bilevel methodology by means of a brief historical perspective. Succinctly, bilevel programs are a class of hierarchical optimization problems in variables x and y, with the optimal x being chosen by solving a constrained optimization problem whose constraints themselves are optimization problems in y, or possibly both x and y. In operations research literature, the class of bilevel optimization problems was introduced by Bracken and McGill [6], and applied to defence problems like minimum-cost weapon mix and economic problems like optimal production and marketing decision making models. Their work is closely related to the extensively studied economic problem of the Stackelberg game [53], whose origin predates the work of Bracken and McGill.

Stackelberg used a hierarchical model to describe the market situation where different decision makers try to optimize their decisions based on individually different objectives according to some hierarchy. The Stackelberg game can be considered an extension of the well-known Nash game. In the Nash game, there are T players, each of whom has a strategy set, Y_t, and the objective of player t is chose a strategy, $y_t \in Y_t$, given that the other players have already chosen theirs, to minimize some utility function. Thus, each player chooses a strategy based on the choices of the other players and there is no hierarchy.

In contrast, in the Stackelberg game, there is a hierarchy where a distinctive player, called the *leader* is aware of the choices of the other players, called the *followers*. Thus, the leader, being in a superior position with regard to everyone else can achieve the best objective while forcing the followers to respond to this choice of strategy by solving the Stackelberg game. Consider the case of a single leader and follower. Let X and Y denote the strategy sets for the leader and follower; let $F(x, y)$ and $f(x, y)$ be their utility functions respectively. Based on the selection, x, of the leader, the follower can select the best strategy $y(x) \in Y$ such that $f(x, y)$ is maximized i.e.,

$$y(x) \in \Psi(x) = \arg\max_{y \in Y} f(x, y). \tag{1.1}$$

The leader then computes the best strategy $x \in X$ as (see Fig. 1),

$$x \equiv \max_{x \in X} \{F(x, y) \mid y \in \Psi(x)\}. \tag{1.2}$$

Equations (1.1) and (1.2) can be combined to express the Stackelberg game compactly as

$$\begin{aligned} &\max_{x \in X, y} \; F(x, y) \\ &\text{s. t. } y \in \arg\max_{\eta \in Y} f(x, \eta). \end{aligned} \tag{1.3}$$

Bilevel programs are more general than Stackelberg games in the sense that the strategy sets, also known as admissible sets, can depend on both x and y. This leads us to the general bilevel program formulated by Bracken and McGill:

$$\begin{aligned} &\max_{x \in X, y} \; F(x, y) && \text{outer level} \\ &\text{s. t. } G(x, y) \leq 0, \\ &\quad y \in \left\{ \begin{array}{l} \arg\max_{y \in Y} f(x, y) \\ \text{s. t. } g(x, y) \leq 0 \end{array} \right\}, && \text{inner level} \end{aligned} \tag{1.4}$$

The bilevel program, (1.4), is a generalization of several well-known optimization problems as noted in [18]. If $F(x, y) = -f(x, y)$, we have classical minimax problem; if $F(x, y) = f(x, y)$, we have a realization of the decomposition approach to optimization problems; if the dependence of both problems on y is dropped, we have bicriteria optimization.

Now, a given model selection problem can be recast in the bilevel framework by optimizing the generalization criteria in the outer level while performing the required machine learning task in the inner level (see Figure 1). This reformulation

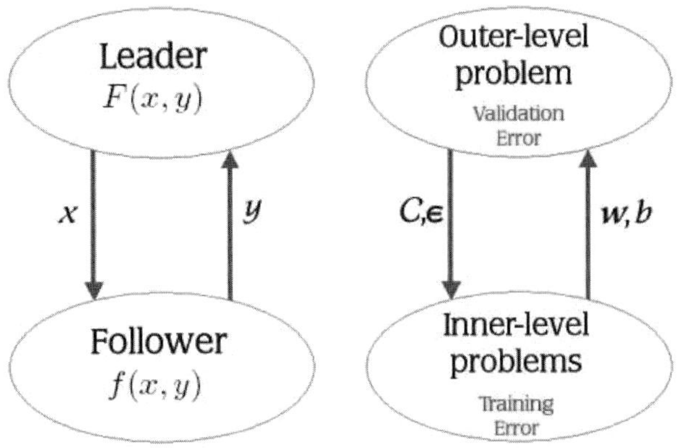

FIGURE 1. The Stackelberg game (left), showing the hierarchy between the leader and the follower; Cross validation modelled as a bilevel program (right), showing the interaction between the parameters, which are optimized in the outer level and the models which are trained in the inner level.

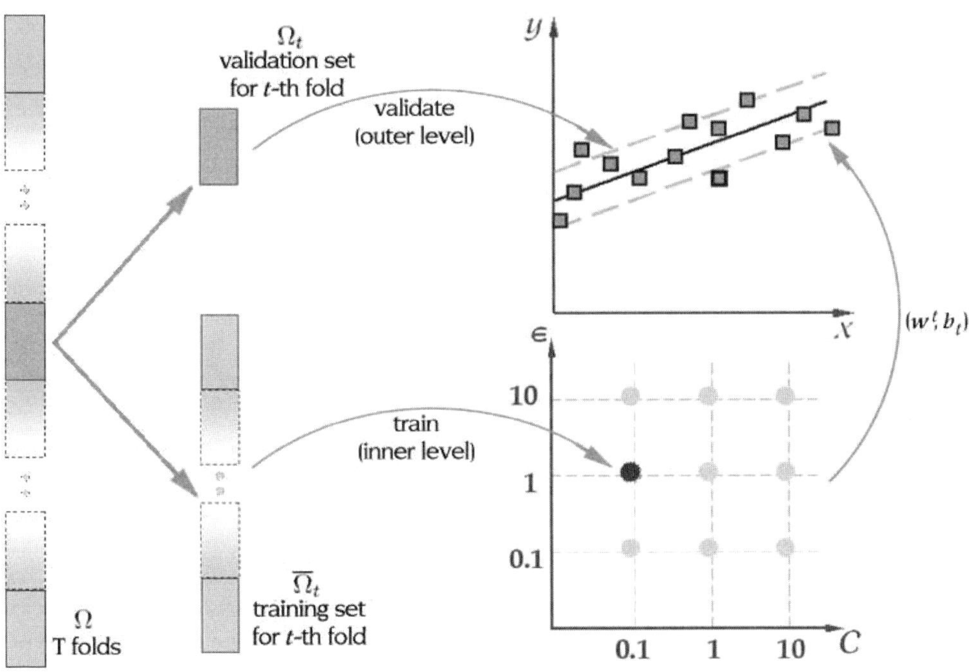

FIGURE 2. T-fold cross validation at a particular grid point, (C_k, ε_k), on a base-10 logarithmic grid.

gives rise to a significant difficulty: many machine learning problems are convex and differentiable, while the bilevel program is non-convex and functions like $\Psi(x)$, which are not even Fréchet differentiable in general, occur in the constraints—for example, in (1.1). There exist techniques to convert bilevel programs into solvable optimization problems such as using implicit function theorems. However, we will restrict our attention to the approach where the inner-level problems are replaced by their Karush–Kuhn–Tucker conditions, a semi-infinite system of inequalities or finite-dimensional variational inequalities. Such an optimization problem is called a *Mathematical Program with Equilibrium Constraints* (MPEC).

The systematic study of the bilevel optimization problem and its MPEC extension attracted the intensive attention of mathematical programmers about a decade ago with the publication of a focused monograph [40], which is followed by two related monographs, [49, 18]. During the past decade, there has been an explosion of research on these optimization problems. See the annotated bibliography [19] which contains many references. In general, bilevel programs/MPECs provide a powerful computational framework for dealing with hyper-parameter identification problems in an optimization setting. As such, they offer a novel paradigm for dealing with the model selection problems described in this chapter. We illustrate the bilevel paradigm on a specific learning task—linear SV regression—and then proceed with the more general formulation.

2. Parameter Selection for Linear SV Regression

Cross validation for support vector regression was the first problem to reformulated as a bilevel program that can perform parameter selection [3]. The learning task is to construct a linear regression function that minimizes the ε-insensitive error regularized by the 2-norm of the weights of the linear function. Thus, SV regression has two hyper-parameters to be picked by T-fold cross validation. More specifically the goal is to find a function, $f(x) = x'w$, such that the the function generalizes well on future data e.g. $f(x) \approx y$ and, typically the generalization error is estimated by cross validation.

2.1. Cross validation. We start with a few words about our notation. The training set consists of ℓ pairs of data and labels, $\{(\mathbf{x}_i, y_i)\}_{i=1}^{\ell} \subset \mathbb{R}^{n+1}$, denoted by Ω; thus, $|\Omega| = \ell$. Let the set of indices for the points in Ω be $\mathcal{N} = \{1, \ldots, \ell\}$. Since we are interested in performing T-fold cross validation, Ω is partitioned into T pairwise disjoint subsets, Ω_t, the *validation sets* within each fold. The sets $\overline{\Omega}_t = \Omega \backslash \Omega_t$ are the *training sets* within each fold. The corresponding index sets for the validation and training sets are $\mathcal{N}_t$ and $\overline{\mathcal{N}}_t$ respectively. The tth training set, $\overline{\Omega}_t$ is used to train the linear function $\mathbf{w}^t \in \mathbb{R}^n$. For simplicity of presentation, we ignore the bias term for now. For compactness of notation, the vectors $\mathbf{w}^t$ are collected, column-wise, into the matrix $\mathbf{W} \in \mathbb{R}^{n \times T}$. Also, given two vectors $\mathbf{r}, \mathbf{s} \in \mathbb{R}^n$, the complementarity condition $\mathbf{r} \perp \mathbf{s}$ is equivalent to $\mathbf{r}'\mathbf{s} = 0$. A vector of ones of arbitrary dimension is denoted $\mathbb{1}$.

The support vector regression (SVR) problem contains two hyper-parameters: the regularization constant, C, and the tube width, ε, for the ε-insensitive loss function, which are to be selected by cross validation. This is typically accomplished using grid search. Grid search involves discretizing the C-ε space, typically on a logarithmic scale of base 2 or 10 (see Fig. 2). At each grid point, the parameters of the regression functions, $\{\mathbf{w}^t\}_{t=1}^T$, are trained on the corresponding training sets, $\overline{\Omega}_t$, using the quadratic program

$$\mathbf{w}^t \in \arg\min_{\mathbf{w}\in\mathbb{R}^n} \left\{ \frac{1}{2}\|\mathbf{w}\|_2^2 + \frac{C}{|\overline{\mathcal{N}}_t|} \sum_{j\in\overline{\mathcal{N}}_t} \max(|\mathbf{x}_j'\mathbf{w} - y_j| - \varepsilon, 0) \right\}. \tag{2.1}$$

Equation (2.1) represents the regression problem of finding a function, $f^\star : \mathbb{R}^n \to \mathbb{R}$, among a given class that minimizes the regularized risk functional,

$$R[f] \equiv \mathcal{P}[f] + \frac{C}{\ell} \sum_{j=1}^{\ell} \mathcal{L}(y_j, f(\mathbf{x}_j)). \tag{2.2}$$

where $\mathcal{L}$ is the ε-insensitive loss function and $\mathcal{P}$ is the classic SVR ℓ_2-norm regularization operator. The regression functions are validated at each grid point by computing the generalization error, such as the mean average deviation (MAD), on the validation sets, Ω_t:

$$\Theta(\mathbf{W}) = \frac{1}{T} \sum_{t=1}^{T} \frac{1}{|\mathcal{N}_t|} \sum_{i\in\mathcal{N}_t} |\mathbf{x}_i'\mathbf{w} - y_i|, \tag{2.3}$$

or alternatively the mean squared error (MSE),

$$\Theta(\mathbf{W}) = \frac{1}{T} \sum_{t=1}^{T} \frac{1}{|\mathcal{N}_t|} \sum_{i\in\mathcal{N}_t} (\mathbf{x}_i'\mathbf{w} - y_i)^2. \tag{2.4}$$

The grid point with the smallest estimate of the generalization error, $\Theta(\mathbf{W})$, yields the choice of "optimal" hyper-parameters; this choice may be further refined by a local search. Assuming that each parameter is discretized to take on d distinct values, T-fold cross validation involves solving $O(Td^2)$ problems, a number that grows quickly as more accuracy is desired or if the number of parameters increases. This, combined with other issues described in Section 1.1, leads us to formulate the bilevel cross-validation model.

2.2. Bilevel cross validation. In a fairly general formulation in which we list only the essential constraints, the model selection bilevel program is to find the hyper-parameters ε and C and the hyperplanes, $\mathbf{w}^t$ in order to

$$\begin{aligned} &\underset{C,\varepsilon,\mathbf{w}^t}{\text{minimize}}\ \Theta(\mathbf{W}) \\ &\text{subject to}\varepsilon, C \geq 0, \\ &\qquad \text{and for } t = 1,\ldots,T, \\ &\qquad \mathbf{w}^t \in \arg\min_{\mathbf{w}\in\mathbb{R}^n} \left\{ \frac{1}{2}\|\mathbf{w}\|_2^2 + \frac{C}{|\overline{\mathcal{N}}_t|} \sum_{j\in\overline{\mathcal{N}}_t} \max(|\mathbf{x}_j'\mathbf{w} - y_j| - \varepsilon, 0) \right\}, \end{aligned} \tag{2.5}$$

where the outer-level objective function, $\Theta(\mathbf{W})$, is an estimate of the generalization error and a performance measure as described in (2.3) and (2.4). Thus, the hyper-parameters are optimized in the outer level while the learning is performed in the inner level. It should be noted that the inner- and outer-level loss functions do not need to match. The inner-level optimization problem is the classic SV regression machine which uses ε-insensitive loss because it produces robust solutions that are sparse in the dual space. The ε-insensitive loss is not typically used for the generalization estimate in cross validation since it contains an unknown parameter ε; instead, we will use the MAD, (2.3), in the analysis below. The bilevel program that uses MSE can easily be formulated with only minor changes.

To solve (2.5), we rewrite each of the T inner-level problems, assuming that the hyper-parameters, C and ε, are fixed. In particular, for the tth problem, we introduce slack variables, $\boldsymbol{\xi}^t \geq 0$, to reformulate the max function in the inner level using the standard trick from linear programming, to give the following convex, quadratic program:

$$\begin{aligned} &\min_{\mathbf{w}^t, \boldsymbol{\xi}^t} \frac{1}{2}\|\mathbf{w}\|_2^2 + \frac{C}{|\overline{\mathcal{N}}_t|} \sum_{j \in \overline{\mathcal{N}}_t} \xi_j^t \\ &\text{s. t.} \\ &\qquad \left.\begin{aligned} &\xi_j^t \geq y_j - \mathbf{x}_j'\mathbf{w} - \varepsilon, \\ &\xi_j^t \geq \mathbf{x}_j'\mathbf{w} - y_j - \varepsilon, \\ &\xi_j^t \geq 0, \end{aligned}\right\}, \quad \forall j \in \overline{\mathcal{N}}_t. \end{aligned} \tag{2.6}$$

This is the SV regression problem that trains $\mathbf{w}^t$ using the training set, $\overline{\Omega}_t$, within the tth fold. Let $\alpha_j^{t,+}$, $\alpha_j^{t,-} \geq 0$ be the Lagrange multipliers for the upper and lower hyperplane constraints respectively. Using these multipliers, we can write down the primal and dual feasibility and complementarity slackness conditions of (2.6) as follows.

$$\left.\begin{aligned} 0 \leq \alpha_j^{t,+} \perp y_j - \mathbf{x}_j'\mathbf{w}^t + \varepsilon + \xi_j^t \geq 0, \\ 0 \leq \alpha_j^{t,-} \perp \mathbf{x}_j'\mathbf{w}^t - y_j + \varepsilon + \xi_j^t \geq 0, \\ 0 \leq \xi_j^t \perp \frac{C}{|\overline{\mathcal{N}}_t|} - \alpha_j^{t,+} - \alpha_j^{t,-} \geq 0, \end{aligned}\right\} \quad \forall j \in \overline{\mathcal{N}}_t. \tag{2.7}$$

There also exists the following first-order condition:

$$\mathbf{w}^t + \sum_{j \in \overline{\mathcal{N}}_t} (\alpha_j^{t,+} - \alpha_j^{t,-})\mathbf{x}_j = 0. \tag{2.8}$$

Equations (2.7) and (2.8) together constitute the well-known Karush–Kuhn–Tucker first-order optimality conditions for (2.6). The outer-level objective also needs to be rewritten as it contains the absolute value function, which is not everywhere differentiable. This is easily achieved by introducing additional variables $\mathbf{z}^t$ and the constraints,

$$-z_i^t \leq \mathbf{x}_i'\mathbf{w}^t - y_i \leq z_i^t, \quad \forall i \in \mathcal{N}_t, \tag{2.9}$$

into the outer level. Here, $\mathbf{z}^t$ measures the mean absolute deviation of each point in the validation set, Ω_t, from the hyperplane, $\mathbf{w}^t$, trained on $\overline{\Omega}_t$. Thus, the overall two-level regression problem can be converted to a one-level problem as shown below

$$
\begin{array}{ll}
\min\limits_{\substack{C,\varepsilon,\mathbf{w}^t,\mathbf{z}^t,\\ \boldsymbol{\alpha}^{t,\pm},\boldsymbol{\xi}^t}} & \dfrac{1}{T}\sum_{t=1}^{T}\dfrac{1}{|\mathcal{N}_t|}\sum_{i\in\mathcal{N}_t} z_i^t \\
\text{s. t.} & \varepsilon, C \geq 0, \\
& \text{and for } t = 1,\ldots,T, \\
& -z_i^t \leq \mathbf{x}_i'\mathbf{w}^t - y_i \leq z_i^t, \qquad \forall i \in \mathcal{N}_t, \\
& \left.\begin{array}{r} 0 \leq \alpha_j^{t,+} \perp y_j - \mathbf{x}_j'\mathbf{w}^t + \varepsilon + \xi_j^t \geq 0, \\ 0 \leq \alpha_j^{t,-} \perp \mathbf{x}_j'\mathbf{w}^t - y_j + \varepsilon + \xi_j^t \geq 0, \\ 0 \leq \xi_j^t \perp \dfrac{C}{|\overline{\mathcal{N}}_t|} - \alpha_j^{t,+} - \alpha_j^{t,-} \geq 0, \end{array}\right\}, \quad \forall j \in \overline{\mathcal{N}}_t, \\
& \mathbf{w}^t + \sum_{j\in\overline{\mathcal{N}}_t}(\alpha_j^{t,+} - \alpha_j^{t,-})\mathbf{x}_j = 0.
\end{array} \tag{2.10}
$$

The optimization problem (2.10) is an instance of a *Linear Program with Equilibrium Constraints* (LPEC); it is a nonlinearly constrained problem and is non-convex in general because of the presence of the complementarity constraints. Several approaches can be used to solve (2.10) including exact penalty methods, integer programming approaches and nonlinear programming techniques, to name a few. We defer a discussion of these techniques until Section 8.

3. Generalized Bilevel Cross validation

The number of machine learning tasks that can be cast into the bilevel cross-validation framework is virtually limitless. These learning tasks determine the objective and constraints used in the inner-level problems and the outer-level objective. The resulting cross-validation problem can be re-formed as a bilevel optimization problem as long as the inner-level problems can be replaced by their corresponding KKT conditions, and the outer-level objective and constraints can be replaced by differentiable counterparts. Keeping this in mind, T-fold cross validation for some general machine learning problem can be written as

$$
\begin{array}{ll}
\underset{f^t,\boldsymbol{\lambda}}{\text{minimize}} & \Theta(f^1|_{\Omega_1},\ldots,f^T|_{\Omega_T};\boldsymbol{\lambda}) \\
\text{subject to} & \boldsymbol{\lambda} \in \Lambda, \\
& \text{and for } t = 1,\ldots,T, \\
& f^t \in \underset{f\in\mathcal{F}}{\arg\min}\left\{\mathcal{P}(f,\boldsymbol{\lambda}) + \sum_{j\in\overline{\mathcal{N}}_t}\mathcal{L}(y_j, f(\mathbf{x}_j),\boldsymbol{\lambda})\right\}.
\end{array} \tag{3.1}
$$

Here, $f^t : (\mathbb{R}^n \times \mathbb{R}) \cap \mathcal{F} \to \mathbb{R}$ is the learning function trained within the tth fold, $\boldsymbol{\lambda} \in \Lambda$ is the set of model selection parameters for the machine learning problem, $\mathcal{P}$ is the regularization operator and $\mathcal{L}$ is the inner-level loss function. This definition admits many variations and well-known machine learning problems. We provide some more examples below.

4. Parameter Selection for Linear SV Classification

In linear SV classification (SVC), the learning task is to construct a decision function that distinguishes one class from another. In this section, we extend the parameter selection idea introduced in the previous section to support vector classification and show how the bilevel formulation can handle a large number of hyper-parameters; this is a review of work that was first introduced in [36]. The inner-level problem is the standard SVC model [12] augmented with additional feature selection constraints. The outer-level objective minimizes the number of points misclassified in each of the T folds.

We again consider a labelled data set, $\Omega = \{(\mathbf{x}_i, y_i)\}_{i=1}^{\ell} \subset \mathbb{R}^{n+1}$, containing ℓ training examples. However, the labels y_i are restricted to ± 1 as we are interested in binary classification. We also introduce the bias into the decision rule: the hyperplane, $\mathbf{x}'\mathbf{w}^t - b_t = 0$, is to be represented by the pair $(\mathbf{w}^t, b_t)$. If $\mathbf{x}'\mathbf{w}^t > 0$, then $\mathbf{x}$ is predicted to be in class 1 by decision rule t, otherwise $\mathbf{x}$ is predicted to be in class -1. As before, the vectors, $\mathbf{w}^t$, are collected column-wise into the matrix $\mathbf{W} \in \mathbb{R}^{n \times T}$. The scalars, b_t, are collected into the vector $\mathbf{b} \in \mathbb{R}^T$.

Using this additional notation, the formulation of cross validation for SV classification as a bilevel program [36] is shown below.

$$
\begin{aligned}
&\underset{C,\overline{\mathbf{w}},\mathbf{w}^t,b_t}{\text{minimize}} \ \Theta(\mathbf{W}, \mathbf{b}) \\
&\text{subject to } C_{\mathbf{lb}} \le C \le C_{\mathbf{ub}}, \\
&\qquad\qquad \overline{\mathbf{w}}_{\mathbf{lb}} \le \overline{\mathbf{w}} \le \overline{\mathbf{w}}_{\mathbf{ub}}, \\
&\qquad\qquad \text{and for } t = 1, \dots, T, \\
&(\mathbf{w}^t, b_t) \in \underset{\substack{-\overline{\mathbf{w}} \le \mathbf{w} \le \overline{\mathbf{w}} \\ b \in \mathbb{R}}}{\arg\min} \left\{ \frac{1}{2}\|\mathbf{w}\|_2^2 + \frac{C}{|\overline{\mathcal{N}}_t|} \sum_{j \in \overline{\mathcal{N}}_t} \max(1 - y_j(\mathbf{x}_j'\mathbf{w} - b), 0) \right\}.
\end{aligned}
\tag{4.1}
$$

It should be noted that the T inner-level problems are nearly identical to the classical support vector classification problem except for the introduction of the additional constraints $-\overline{\mathbf{w}} \le \mathbf{w} \le \overline{\mathbf{w}}$. These constraints can be thought of as a weighted ℓ_∞-norm on $\mathbf{w}$ and provide additional regularization while also performing feature selection. To intuitively see how the box-constrained SV classifier (BoxSVC) performs feature selection, consider the following. An irrelevant feature would induce the corresponding weight in $\overline{\mathbf{w}}$ to be small, which, in turn, forces the respective weights in each fold, $\mathbf{w}^t$, to be small. Therefore, in addition to ensuring consistency across the folds, the box constraint performs capacity control leading to potentially improved overall generalization.

As a consequence, $\overline{\mathbf{w}} \in \mathbb{R}^n$ enters the outer level as a n-dimensional feature selecting hyper-parameter, with the box constraint entering each inner-level problem. Here, n is the dimension of the input space. The bilevel program (4.1) contains $n+1$ hyper-parameters that need to be optimized and model selection entails simultaneous parameter and input-space feature selection. In addition to the box constraints, bounds, $0 < C_{\mathbf{lb}} \le C_{\mathbf{ub}}$ and $0 < \overline{\mathbf{w}}_{\mathbf{lb}} \le \overline{\mathbf{w}}_{\mathbf{ub}}$, have also been introduced on the hyper-parameters in the outer level. This is to cut-off degenerate stationary points that are meaningless in the machine learning context.

4.1. The inner-level problems. Just as in the regression case, we can replace the inner-level problems with their corresponding constraints in order to convert the bilevel program to an MPEC. First, slack variables, $\boldsymbol{\xi}^t \geq 0$, are introduced to reformulate the non-differentiable max function in the inner-level objective giving the following quadratic program to train a decision rule within the tth fold:

$$
\begin{aligned}
&\min_{\mathbf{w}^t, b_t, \boldsymbol{\xi}^t} \frac{1}{2}\|\mathbf{w}^t\|_2^2 + \frac{C}{|\overline{\mathcal{N}}_t|} \sum_{j \in \overline{\mathcal{N}}_t} \xi_j^t \\
&\text{s. t.} \ -\overline{\mathbf{w}} \leq \mathbf{w}^t \leq \overline{\mathbf{w}}, \\
&\left.\begin{aligned} & y_j(\mathbf{x}_j'\mathbf{w}^t - b_t) \geq 1 - \xi_j^t, \\ & \xi_j^t \geq 0, \end{aligned}\right\}, \quad \forall j \in \overline{\mathcal{N}}_t.
\end{aligned} \tag{4.2}
$$

In addition to the Lagrange multipliers, α^t, for the hyperplane constraints and η^t for the non-negativity of ξ^t constraints, we introduce $\boldsymbol{\gamma}^{t,+}$ and $\boldsymbol{\gamma}^{t,-}$ for the lower and upper box constraints. Thus, the KKT complementarity conditions for (4.2) are

$$
\begin{aligned}
&\left.\begin{aligned} & 0 \leq \alpha_j^t \perp y_j(\mathbf{x}_j'\mathbf{w}^t - b_t) - 1 + \xi_j^t \geq 0, \\ & 0 \leq \xi_j^t \perp \frac{C}{|\overline{\mathcal{N}}_t|} - \alpha_j^t \geq 0, \end{aligned}\right\}, \quad \forall j \in \overline{\mathcal{N}}_t, \\
& 0 \leq \boldsymbol{\gamma}^{t,+} \perp \overline{\mathbf{w}} - \mathbf{w}^t \geq 0, \\
& 0 \leq \boldsymbol{\gamma}^{t,-} \perp \overline{\mathbf{w}} + \mathbf{w}^t \geq 0,
\end{aligned}
$$

and the KKT first order conditions are

$$
\begin{aligned}
\mathbf{w}^t - \sum_{j \in \overline{\mathcal{N}}_t} y_j \alpha_j^t \mathbf{x}_j + \boldsymbol{\gamma}^{t,+} - \boldsymbol{\gamma}^{t,-} &= 0, \\
\sum_{j \in \overline{\mathcal{N}}_t} y_j \alpha_j^t &= 0.
\end{aligned} \tag{4.3}
$$

We now look at various outer-level objectives.

4.2. The outer-level objective. The standard performance measure used to validate classification models is the average misclassification error on the validation sets, i.e.,

$$
\Theta(\mathbf{W}, \mathbf{b}) = \frac{1}{T} \sum_{t=1}^{T} \frac{1}{|\mathcal{N}_t|} \sum_{i \in \mathcal{N}_t} [-y_i(\mathbf{x}_i'\mathbf{w}^t - b_t)]_\star, \tag{4.4}
$$

where the inner summand counts the number of misclassifications of points in the validation set within each fold and the outer summand sums the averaged misclassification error over each fold. The objective, (4.4), is formulated in terms of the step function, which, for a vector, $\mathbf{r}_\star$, is defined as

$$
(\mathbf{r}_\star)_i = \begin{cases} 1, & r_i > 0, \\ 0, & r_i \leq 0. \end{cases} \tag{4.5}
$$

The step function, unlike the absolute value or the max functions, is not even continuous. This poses a problem, since now, the outer-level objective is not differentiable. However, the step function can be characterized as the solution to a linear program [41]:

$$\mathbf{r}_\star = \arg\min_{\boldsymbol{\zeta}}\{-\boldsymbol{\zeta}'\mathbf{r} \mid 0 \le \boldsymbol{\zeta} \le \mathbb{1}\}. \tag{4.6}$$

Thus, in order to count the number of misclassifications in the T validation sets, it is necessary to introduce T additional inner-level problems of the type (4.6),

$$\boldsymbol{\zeta}^t \in \arg\min_{0 \le \boldsymbol{\zeta} \le \mathbb{1}} \left\{ \sum_{i \in \mathcal{N}_t} \zeta_i y_i (\mathbf{x}_i'\mathbf{w}^t - b_t) \right\}. \tag{4.7}$$

Then, as in the case of the inner-level problems (4.2), we can replace (4.6) with linear complementarity conditions. This necessitates the introduction of additional Lagrange multipliers, (4.6) also satisfies

$$\begin{aligned} &0 \le \boldsymbol{\zeta} \perp -\mathbf{r} + \mathbf{z} \ge 0, \\ &0 \le \mathbf{z} \perp \mathbb{1} - \boldsymbol{\zeta} \ge 0. \end{aligned} \tag{4.8}$$

Thus, combining (4.3)–(4.3), and (4.7)–(4.8), we can replace the bilevel program with a single-level program, which is also an instance of an MPEC. Note that the variables, $\mathbf{z}^t$, measure the distance of the misclassified validation points from the trained hyperplane within each fold

$$\begin{aligned} \min_{\substack{C,\overline{\mathbf{w}},\mathbf{w}^t,b_t,\mathbf{z}^t,\\ \boldsymbol{\xi}^t,\boldsymbol{\zeta}^t,\boldsymbol{\alpha}^t,b_t,\boldsymbol{\gamma}^{t,\pm}}} \quad & \frac{1}{T}\sum_{t=1}^{T} \frac{1}{|\mathcal{N}_t|} \sum_{i \in \mathcal{N}_t} \zeta_i^t \\ \text{s. t. } & C_{\mathbf{lb}} \le C \le C_{\mathbf{ub}},\ \overline{\mathbf{w}}_{\mathbf{lb}} \le \overline{\mathbf{w}} \le \overline{\mathbf{w}}_{\mathbf{ub}}, \\ & \text{and for } t = 1, \ldots T, \\ & \left.\begin{aligned} &0 \le \zeta_i^t \perp y_i(\mathbf{x}_i'\mathbf{w}^t - b_t) + z_i^t \ge 0, \\ &0 \le z_i^t \perp 1 - \zeta_i^t \ge 0, \end{aligned}\right\}, \qquad \forall i \in \mathcal{N}_t, \\ & \left.\begin{aligned} &0 \le \alpha_j^t \perp y_j(\mathbf{x}_j'\mathbf{w}^t - b_t) - 1 + \xi_j^t \ge 0, \\ &0 \le \xi_j^t \perp \frac{C}{|\overline{\mathcal{N}}_t|} - \alpha_j^t \ge 0, \end{aligned}\right\}, \qquad \forall j \in \overline{\mathcal{N}}_t, \\ & 0 \le \boldsymbol{\gamma}^{t,+} \perp \overline{\mathbf{w}} - \mathbf{w}^t \ge 0, \\ & 0 \le \boldsymbol{\gamma}^{t,-} \perp \overline{\mathbf{w}} + \mathbf{w}^t \ge 0, \\ & \mathbf{w}^t - \sum_{j \in \overline{\mathcal{N}}_t} y_j \alpha_j^t \mathbf{x}_j + \boldsymbol{\gamma}^{t,+} - \boldsymbol{\gamma}^{t,-} = 0, \\ & \sum_{j \in \overline{\mathcal{N}}_t} y_j \alpha_j^t = 0, \end{aligned} \tag{4.9}$$

The outer-level objective, $\Theta(\mathbf{W}, \mathbf{b})$, can be treated not just as performance measure that can estimate the generalization error but also as a loss function. This observation combined with the flexibility of the bilevel approach allows us to formulate several different approaches to model selection.

The most straightforward variation is to use the average distance of the misclassified validation point from the trained decision rule within each fold i.e.,

$$\Theta(\mathbf{W}, \mathbf{b}) = \frac{1}{T}\sum_{t=1}^{T} \frac{1}{|\mathcal{N}_t|} \sum_{i \in \mathcal{N}_t} z_i^t. \tag{4.10}$$

Alternately, the *hinge loss* could be used in the outer-level objective. The hinge loss is a relaxed overestimate of the misclassification loss. Using this loss in the outer level effectively matches the inner- and outer-level loss functions.

$$\Theta(\mathbf{W}, \mathbf{b}) = \frac{1}{T}\sum_{t=1}^{T} \frac{1}{|\mathcal{N}_t|} \sum_{i \in \mathcal{N}_t} \max(1 - y_i(\mathbf{x}_i'\mathbf{w}^t - b_t), 0). \tag{4.11}$$

4.3. ℓ_1-norm regularization. Different norms can be used to regularize the inner-level problem. It is well-known that ℓ_1-norm regularization produces sparser solutions than its ℓ_2 counterpart, hence performing better feature selection if used for regularization. This can easily be incorporated into, say, the classification bilevel formulation (4.1) by replacing $\frac{1}{2}\|\mathbf{w}\|_2^2$ with $\|\mathbf{w}\|_1$, yielding the following inner-level problems:

$$\underset{\substack{-\overline{\mathbf{w}} \le \mathbf{w}^t \le \overline{\mathbf{w}} \\ b_t \in \mathbb{R}}}{\arg\min} \left\{ \|\mathbf{w}\|_1 + \frac{C}{|\mathcal{N}_t|} \sum_{j \in \mathcal{N}_t} \max(1 - y_j(\mathbf{x}_j'\mathbf{w}^t - b_t), 0) \right\}. \tag{4.12}$$

Since the $\|\mathbf{w}^t\|_1$ is not differentiable at 0, we introduce two new variables, $\mathbf{w}^{t,\pm}$, such that $\mathbf{w}^t = \mathbf{w}^{t,+} - \mathbf{w}^{t,-}$. If we choose to retain the box constraints as well, then the bounds, $0 \le \mathbf{w}^{t,\pm} \le \overline{\mathbf{w}}$, must also hold. Since the inner-level problems are simply LPs, all the KKT conditions can be simplified to complementarity constraints,

$$\begin{aligned}
&\left.\begin{aligned}
&0 \le \alpha_j^t \perp y_j(\mathbf{x}_j'(\mathbf{w}^{t,+} - \mathbf{w}^{t,-}) - b_t) - 1 + \xi_j^t \ge 0, \\
&0 \le \xi_j^t \perp \frac{C}{|\mathcal{N}_t|} - \alpha_j^t \ge 0,
\end{aligned}\right\}, \quad \forall j \in \overline{\mathcal{N}}_t, \\
&0 \le \gamma^{t,+} \perp \overline{\mathbf{w}} - \mathbf{w}^{t,+} \ge 0, \\
&0 \le \gamma^{t,-} \perp \overline{\mathbf{w}} - \mathbf{w}^{t,-} \ge 0, \\
&0 \le \mathbf{w}^{t,+} \perp \mathbb{1} - \sum_{j \in \overline{\mathcal{N}}_t} y_j\alpha_j^t\mathbf{x}_j + \gamma^{t,+} \ge 0, \\
&0 \le \mathbf{w}^{t,-} \perp \mathbb{1} + \sum_{j \in \overline{\mathcal{N}}_t} y_j\alpha_j^t\mathbf{x}_j + \gamma^{t,-} \ge 0,
\end{aligned} \tag{4.13}$$

except for the equality constraints that are dual to b_t,

$$\sum_{j \in \overline{\mathcal{N}}_t} y_j\alpha_j^t = 0. \tag{4.14}$$

If the box constraint in the inner-level problems was to be dropped, then their corresponding dual variables, $\gamma^{\pm}$, would disappear from the complementarities above. Many other regularization schemes could also be incorporated into (3.1)) such as ones which use hinge loss combined with the so-called lasso penalty, $\mu_1\|\mathbf{w}\|_2^2 + \mu_2\|\mathbf{w}\|_1$ (elastic nets, [56]), ℓ_∞-norm regularization [2, 57] or the modified support vector machine which uses $\frac{1}{2}(\|\mathbf{w}\|_2^2 + b^2)$ for regularization [44]. The last regularization operator is nearly identical to classical SVMs except that the equality constraint, (4.14), vanishes from the dual (and subsequently from the first order conditions of the inner-level problems). This is useful if it is desired that all equality constraints be removed in order to apply solution techniques such as successive over-relaxation or linear programming chunking [8].

4.4. ℓ_0-norm Regularization. Let us assume that it was known *a priori* for the given n-dimensional data set, that at most $n_{\max}$ features are desired. This can be incorporated into the model by introducing the constraint, $\|\overline{\mathbf{w}}\|_0 \leq n_{\max}$, into the outer-level problem, where $\|\cdot\|_0$ is the so-called zero-norm[1], or the cardinality of a vector, i.e., it counts the number of non-zero elements in its argument. This constraint bounds the number of allowable features and causes the smallest feature weights to be dropped from the model. Since $\|\overline{\mathbf{w}}\|_0 = \mathbb{1}'\overline{\mathbf{w}}_\star$, the constraint can be reformulated using the $(\ )_\star$ function and its LPEC conditions, (4.6). Thus, the following complementarity constraints are added to (4.9), with $\boldsymbol{\delta}$ counting the selected features of $\overline{\mathbf{w}}$ and $\mathbf{d}$ being the multipliers to the constraints $\mathbb{1} - \boldsymbol{\delta} \geq 0$

$$\begin{aligned} \sum_{m=1}^{n} \delta_m &\leq n_{\max}, \\ 0 \leq \boldsymbol{\delta} \perp -\overline{\mathbf{w}} + \mathbf{d} &\geq 0, \\ 0 \leq \mathbf{d} \perp \mathbb{1} - \boldsymbol{\delta} &\geq 0. \end{aligned} \tag{4.15}$$

If the resulting LPEC can be solved efficiently, then the optimal value of $n_{\max}$ can be found using a line search or an scheme similar to recursive feature elimination [30], i.e., by successively decreasing $n_{\max}$ as long as the error rate continues to decrease. Alternately, the $\|\overline{\mathbf{w}}\|_0$ term can be moved into the inner-level objective as the regularization operator [7],

$$\underset{\substack{-\overline{\mathbf{w}} \leq \mathbf{w}^t \leq \overline{\mathbf{w}} \\ b_t \in \mathbb{R}}}{\arg\min} \left\{ \lambda \|\overline{\mathbf{w}}\|_0 + \frac{1-\lambda}{\ell} \sum_{j \in \overline{\mathcal{N}}_t} \max(1 - y_j(\mathbf{x}_j'\mathbf{w}^t - b_t), 0) \right\}. \tag{4.16}$$

This, however, converts the overall cross-validation problem into a *trilevel* problem, since the $(\ )_\star$ function, which is used to rewrite $\|\cdot\|_0$, adds a nested inner-level problem of its own to (4.16).

5. Kernel Bilevel Cross Validation

The SVMs considered thus far have all been linear machines and as such are unable to handle non-linear data sets effectively; this severely limits their usefulness to real data sets. We now demonstrate how one of the most powerful features of SVMs—their ability to deal with high-dimensional, highly nonlinear data using the *kernel trick*—can be incorporated into the bilevel model.

We continue this discussion using the bilevel classification example, (4.1), though the results below can easily be generalized to other kernel methods. The classification model was formulated to perform parameter and feature selection, taking advantage of the ability of the bilevel framework to handle multiple parameters. However, a glance at the first-order conditions, (4.3), shows that $\mathbf{w}^t$ depends, not only on the training data, but also on the multipliers, $\gamma^{t,\pm}$, of the box constraints. In order to apply the kernel trick and construct RKHS spaces for the kernel methods to operate in, it is essential that the hyperplane, $\mathbf{w}^t$, be expressed solely as a linear combination of the training data. This is a fundamental assumption that is at the heart of all kernel methods through the representer theorem. In order to make this so, we temporarily set aside feature selection, drop the box constraints

[1]It should be noted that the zero-norm is not really a norm because, the positive homogeneity condition does not hold except for very special cases, i.e., $\|a\mathbf{w}\|_0 \neq |a|\|\mathbf{w}\|_0$, in general. However, the term has found widespread use in both the optimization and machine learning communities.

(effectively causing $\gamma^{t,\pm}$ to drop out of the program) and work with the classical SV classifier. The resulting first order conditions within each fold include the following constraints:

$$\mathbf{w}^t = \sum_{j \in \overline{\mathcal{N}}_t} y_j \alpha_j^t \mathbf{x}_j, \quad \forall t = 1, \ldots, T. \tag{5.1}$$

Now, we can eliminate $\mathbf{w}^t$ within each fold of (4.9) using (5.1) and then apply the kernel trick, i.e., the resulting linear inner-product terms, $\mathbf{x}_i'\mathbf{x}_j$, are replaced with symmetric, positive semi-definite kernel functions, $\kappa(\mathbf{x}_i, \mathbf{x}_j)$. The final bilevel cross-validation model for SV classification *when the kernel is fixed* can be computed if we

$$\begin{aligned}
&\min_{\substack{C, b_t, \mathbf{z}^t,\\ \boldsymbol{\zeta}^t, \boldsymbol{\alpha}^t, \boldsymbol{\xi}^t}} \frac{1}{T} \sum_{t=1}^{T} \frac{1}{|\mathcal{N}_t|} \sum_{i \in \mathcal{N}_t} \zeta_i^t \\
&\text{s. t. } C \geq 0, \\
&\qquad \text{and for } t = 1, \ldots T, \\
&\qquad \left.\begin{aligned}
&0 \leq \zeta_i^t \perp y_i \left[\sum_{k \in \overline{\mathcal{N}}_t} y_k \alpha_k^t \kappa(\mathbf{x}_i, \mathbf{x}_k) - b_t \right] + z_i^t \geq 0, \\
&0 \leq z_i^t \perp 1 - \zeta_i^t \geq 0,
\end{aligned}\right\}, \qquad \forall i \in \mathcal{N}_t, \\
&\qquad \left.\begin{aligned}
&0 \leq \alpha_j^t \perp y_j \left[\sum_{k \in \overline{\mathcal{N}}_t} y_k \alpha_k^t \kappa(\mathbf{x}_j, \mathbf{x}_k) - b_t \right] - 1 + \xi_j^t \geq 0, \\
&0 \leq \xi_j^t \perp \frac{C}{|\overline{\mathcal{N}}_t|} - \alpha_j^t \geq 0,
\end{aligned}\right\}, \quad \forall j \in \overline{\mathcal{N}}_t, \\
&\qquad \sum_{j \in \overline{\mathcal{N}}_t} y_j \alpha_j^t = 0.
\end{aligned} \tag{5.2}$$

While it may not appear so at first glance, the optimization problem above is still an instance of an LPEC. Unfortunately, it is usually unreasonable to expect ready-made kernels for most machine learning tasks; in fact, most kernel families are parametrized, and the kernel parameters are typically determined via cross validation. Also, unlike its linear counterpart, this model is not capable of performing feature selection.

The issues of parameter selection (for regularization and the kernel) and feature selection can be combined as in the linear model if we use a parametrized kernel of the form $\kappa(\mathbf{x}_i, \mathbf{x}_k; \mathbf{p}, \mathbf{q})$. The nonnegative vector, $\mathbf{p} \in \mathbb{R}_+^n$, is the feature selection or scaling vector, and $\mathbf{q} \geq 0$ is a vector of kernel parameters. Let $P = \text{diag}(\mathbf{p})$. The parametrized versions of some commonly used kernels are shown below.

$$\begin{aligned}
&\text{Linear kernel} && \kappa(\mathbf{x}_i, \mathbf{x}_k; \mathbf{p}) = \mathbf{x}_i' P \mathbf{x}_k, \\
&\text{Polynomial kernel} && \kappa(\mathbf{x}_i, \mathbf{x}_k; \mathbf{p}, c, d) = (\mathbf{x}_i' P \mathbf{x}_k + c)^d, \\
&\text{Gaussian kernel} && \kappa(\mathbf{x}_i, \mathbf{x}_k; \mathbf{p}) = \exp\big(-(\mathbf{x}_i - \mathbf{x}_k)' P (\mathbf{x}_i - \mathbf{x}_k)\big).
\end{aligned} \tag{5.3}$$

Other kernels can be similarly extended and used in the model. Consequently, the new kernel parameters, $\mathbf{p}$ and $\mathbf{q}$, enter the outer level of the kernel model as variables in the problem. The introduction of the parametrized kernel is a very powerful extension to the linear model (4.9) as it is capable of picking the regularization parameters, kernel parameters and features leaving only the choice of kernel family to the user. Problem (5.2), optimized with added variables $\mathbf{p}$, $\mathbf{q}$ in the parametrized kernel, is an MPEC with non-linear complementarity constraints and in general is a very difficult problem to solve.

Alternately, taking advantage of the ability of the bilevel model to deal with a large number of hyper-parameters, the following kernel can be used:

$$\kappa(\mathbf{x}_i, \mathbf{x}_k; \beta_{i,k}, \mathbf{p}_k) = \exp\left((\mathbf{x}_i - \mathbf{x}_k)' P_k (\mathbf{x}_i - \mathbf{x}_k) big\right). \tag{5.4}$$

This is a generalization of the commonly used Gaussian kernel in that it contains a different width parameter, β_k, for each support vector which has to be determined. The resulting function becomes a Radial Basis Function Network. Kernels of this type are useful for time-series analysis and function approximations owing to the excellent fitting properties of the Gaussian.

6. Semi-supervised Learning and Transduction

We have, thus far, focused on model selection for *supervised* learning tasks such as classification and regression, with the label information available for all training points. Frequently, however, in applications like text classification, drug design, medical diagnosis, and graph and network search, the training set consists of a large number of unlabelled data points and a relatively small number of labelled training points. This necessitates *semi-supervised* learning, where training is performed using both the labelled and unlabelled data. If all the training data is unlabelled, the problem becomes one of *unsupervised* learning, e.g., clustering.

The concept of semi-supervised learning is closely related to that of *transductive learning*, which can be contrasted with the more typically performed *inductive learning*. In induction, the given labelled data is used to construct a robust decision rule that is valid everywhere. This rule is fixed after training and can subsequently be applied to the future test data. In transduction, the labelled training data and the unlabelled test data are both given. All available data is used to construct the decision rule in order to avoid over-fitting. The learning task in transduction is to only predict the labels for the specific test points and not for all future data. Performing transductive learning may result in improvement in generalization error bounds [54], thus reducing the number of labelled data required for good generalization.

Some additional notation is now introduced. As before, $\Omega = \{\mathbf{x}_i, y_i\}_{i=1}^{\ell}$ represents the set of labelled data, with $\ell = |\Omega|$. Let $\Psi = \{\mathbf{x}_i\}_{i=1}^{u}$ represent the unlabelled training data, with the corresponding labels (to be determined) being z_i, and $u = |\Psi|$. The sets, Ω and Ψ, are indexed by $\mathcal{N}$ and $\mathcal{M}$ respectively.

6.1. Semi-supervised regression. In bilevel semi-supervised regression, the labels of the unlabelled training data are treated as control variables, $\mathbf{z}$. The general bilevel model for semi-supervised machine learning problems can be formulated as

$$\begin{array}{l}
\underset{f,\mathbf{z},\boldsymbol{\lambda}}{\text{minimize}} \; \Theta(f, \mathbf{z}; \Omega, \Psi, \boldsymbol{\lambda}) \\
\text{subject to } \boldsymbol{\lambda} \in \Lambda, \\
\qquad f \in \underset{f \in \mathcal{F}}{\arg\min} \left\{ \mathcal{P}(f, \boldsymbol{\lambda}) + \sum_{j \in \mathcal{N}} \mathcal{L}_l(y_j, f(\mathbf{x}_j), \boldsymbol{\lambda}) + \sum_{j \in \mathcal{M}} \mathcal{L}_u(z_j, f(\mathbf{x}_j), \boldsymbol{\lambda}) \right\}.
\end{array} \tag{6.1}$$

In the model above, the loss functions, $\mathcal{L}_l$ and $\mathcal{L}_u$, are applied to the labelled and unlabelled data respectively, while $\mathcal{P}$ performs regularization. All the appropriate parameters, $\boldsymbol{\lambda}$, are optimized in the outer level; these parameters can include the regularization constant, tube width (for regression) and feature selection vectors among others. Optimizing the unknown labels, $\mathbf{z}$ in the outer level corresponds to inductive learning, while optimizing them in the inner level corresponds to transductive learning. A more robust solution can be found by combining both types of learning by optimizing $\mathbf{z}$ in both levels.

For semi-supervised support vector regression, we can choose both loss functions to be ε-insensitive and ℓ_2-norm regularization. For the case of one labeled training set, one unlabeled training set, and one test set, this yields the following bilevel program [34]:

$$
\begin{aligned}
&\underset{C,D,\varepsilon,\mathbf{w},b,\mathbf{z}}{\text{minimize}} \sum_{i\in\mathcal{N}} |\mathbf{x}_i'\mathbf{w} - b - y_i| \\
&\text{subject to } \varepsilon, C, D \geq 0, \\
&(\mathbf{w}, b) \in \underset{(\mathbf{w},b)\in\mathbb{R}^{n+1}}{\arg\min} \left\{ \begin{array}{l} \frac{1}{2}\|\mathbf{w}\|_2^2 + \frac{C}{|\mathcal{N}|}\sum_{j\in\mathcal{N}} \max(|\mathbf{x}_j'\mathbf{w} - b - y_j| - \varepsilon, 0) \\ \qquad + \frac{D}{|\mathcal{M}|}\sum_{j\in\mathcal{M}} \max(|\mathbf{x}_j'\mathbf{w} - b - z_j| - \varepsilon, 0) \end{array} \right\}.
\end{aligned}
\tag{6.2}
$$

The outer-level objective is simply the mean average deviation (MAD) on all the labelled data. The inner-level objective uses both the labelled and unlabelled data sets making this an instance of transductive learning. The labels, $\mathbf{z}$, are used in the inner-level loss function but are optimized as outer-level variables along with the hyper-parameters ε, C, and D. Additional upper and lower bounds can be imposed on these parameters if desired. This program can be converted to an LPEC as before. It should be noted that in typical semi-supervised learning problems, the number of unlabelled examples, u is far greater than the number of labelled examples, ℓ. This means that (6.2) will have a large number of outer-level variables ($\mathbf{z}$) and complementarity constraints arising from the unlabelled data points.

The model (6.2) performs simultaneous transductive learning and parameter selection. The quality of the "optimal" parameters can potentially be improved by combining semi-supervised learning with T-fold cross validation. This can be achieved if we

$$
\begin{aligned}
&\underset{C,D,\varepsilon,\mathbf{w}^t,b_t,\mathbf{z}}{\text{minimize}} \frac{1}{T}\sum_{t=1}^{T} \frac{1}{|\mathcal{N}_t|}\sum_{i\in\mathcal{N}_t} |\mathbf{x}_i'\mathbf{w} - b - y_i| \\
&\text{subject to } \varepsilon, C, D \geq 0, \\
&\quad \text{and for } t = 1, \ldots, T, \\
&(\mathbf{w}^t, b_t) \in \underset{(\mathbf{w},b)\in\mathbb{R}^{n+1}}{\arg\min} \left\{ \begin{array}{l} \frac{1}{2}\|\mathbf{w}\|_2^2 + \frac{C}{|\mathcal{N}_t|}\sum_{j\in\mathcal{N}_t} \max(|\mathbf{x}_j'\mathbf{w} - b - y_j| - \varepsilon, 0) \\ \qquad + \frac{D}{|\mathcal{M}|}\sum_{j\in\mathcal{M}} \max(|\mathbf{x}_j'\mathbf{w} - b - z_j| - \varepsilon, 0) \end{array} \right\},
\end{aligned}
\tag{6.3}
$$

so that the resultant program is again a novel combination of inductive and transductive learning. Here, the unlabelled data is used to train the decision rule *for each fold.* As there are T inner level problems, the complementarity conditions containing the unlabelled data will occur T times, though each time with a different $(\mathbf{w}^t, b_t)$ in the constraints.

6.2. Semi-supervised classification. Turning our attention to classification problems, we encounter several choices for both the inner- and outer-level loss functions. As always, we use the hinge loss for the labelled points. We look at three loss functions that were introduced in [16] for the unlabelled points in the inner level. The first is the so-called *hard-margin* loss,

$$\mathcal{L}_u(\mathbf{w}, b) = \begin{cases} \infty, & \text{for } -1 < \mathbf{x}'\mathbf{w} - b < 1, \\ 0, & \text{otherwise.} \end{cases} \tag{6.4}$$

This can be introduced into the inner level through the very hard constraint $\max(1 - |\mathbf{x}'\mathbf{w} - b|, 0) = 0$, resulting in the following inner-level optimization problem:

$$\begin{aligned} \min_{\mathbf{w}, b, \boldsymbol{\xi}, \mathbf{z}^+, \mathbf{z}^-} \quad & \frac{1}{2}\|\mathbf{w}\|_2^2 + C \sum_{j \in \mathcal{N}} \xi_j^t \\ \text{s. t.} \quad & y_i(\mathbf{x}_i'\mathbf{w} - b) \geq 1 - \xi_i, \xi_i \geq 0, \qquad \forall i \in \mathcal{N} \\ & \left. \begin{array}{l} -(\mathbf{x}_j'\mathbf{w} - b) \geq 1 - z_j^+, z_j^+ \geq 0, \\ (\mathbf{x}_j'\mathbf{w} - b) \geq 1 - z_j^-, z_j^- \geq 0, \\ z_j^+ z_j^- = 0 \end{array} \right\}, \quad \forall j \in \mathcal{M}. \end{aligned} \tag{6.5}$$

This results in a non-convex, quadratically-constrained quadratic program (QCQP) which is challenging to solve in general. Furthermore, the hard-margin condition might be too strong to allow for feasible solutions, leading us to consider soft-margin variants: the quadratic-margin penalty,

$$\mathcal{L}_u(\mathbf{w}, b) = \max(1 - (\mathbf{x}'\mathbf{w} - b)^2, 0), \tag{6.6}$$

and the non-convex hat-loss function,

$$\mathcal{L}_u(\mathbf{w}, b) = \max(1 - |\mathbf{x}'\mathbf{w} - b|, 0). \tag{6.7}$$

These loss functions arise from the relaxing the hard constraint $z_j^+ z_j^- = 0$ in (6.5) by moving it into the inner-level objective; if the product, $z_j^+ z_j^-$, is used directly, a quadratic penalty function, (6.6), results, and if the minimum error, $\min(z_j^+, z_j^-)$ is used, the hat loss function results. Using the quadratic penalty function for the unlabelled data is precisely the transductive idea proposed by Vapnik [54]. The "optimal" labels on the unlabelled data can be calculated as $\text{sign}(z_j^+ - z_j^-)$.

Finally, we can use the step function to formulate loss functions that use the number of misclassifications for both the labelled and unlabelled data sets if we solve

$$
\begin{aligned}
&\underset{C,D,\mathbf{w},b,\boldsymbol{\zeta},\mathbf{z}}{\text{minimize}} \ \frac{1}{|\mathcal{N}|}\sum_{i\in\mathcal{N}}\zeta_i \\
&\text{subject to } C, D \geq 0, \\
&\quad \boldsymbol{\zeta} \in \underset{0\leq\boldsymbol{\zeta}\leq\mathbb{1}}{\arg\min}\left\{\sum_{i\in\mathcal{N}}\zeta_i y_i(\mathbf{x}_i'\mathbf{w}-b)\right\}, \\
&\quad \mathbf{z} \in \underset{0\leq\mathbf{z}\leq\mathbb{1}}{\arg\min}\left\{\sum_{i\in\mathcal{M}}-\mathbf{z}_i(\mathbf{x}_i'\mathbf{w}-b)\right\}, \\
&\quad (\mathbf{w},b) \in \underset{(\mathbf{w},b)\in\mathbb{R}^{n+1}}{\arg\min}\left\{\begin{array}{l}\frac{1}{2}\|\mathbf{w}\|_2^2+\frac{C}{|\mathcal{N}|}\sum_{j\in\mathcal{N}}\max(1-y_j(\mathbf{x}_j'\mathbf{w}-b),0) \\ \qquad +\frac{D}{|\mathcal{M}|}\sum_{j\in\mathcal{M}}\max(1-z_j(\mathbf{x}_j'\mathbf{w}-b),0)\end{array}\right\}.
\end{aligned}
\tag{6.8}
$$

The outer-level objective performs misclassification minimization on the labelled data, with the first inner-level problem counting the number of misclassifications. The second inner-level problem computes the labels on the unlabelled data which are used to perform learning in the third inner-level problem. As in the regression case, the problem (6.8) and its variants that use the various loss functions above can be combined with cross validation to perform more effective parameter selection. Feature selection can also be incorporated into these models by adding extra constraints on $\mathbf{w}$ or by changing the regularization as discussed in Section 3. It is also relatively straightforward to kernelize the models discussed above as per the discussion in Section 5, as long as care is taken in dealing with the labelled and unlabelled kernels.

7. Incorporating Multitask Learning

We return to the problem of cross validation to demonstrate that *multitask learning* concepts can easily be incorporated in the bilevel setting. Multitask learning [9] is defined as learning multiple related tasks simultaneously. This type of learning is an instance of *inductive transfer*, otherwise called *transfer learning*, where the knowledge learned from some tasks may be applied to learning a related task more efficiently.

In the T-fold bilevel cross validation setting, each of the T inner-level problems attempts to construct a decision rule on subsets of the same training sample, which, by statistical learning theory assumptions, are drawn i.i.d. from the same distribution. Thus, the tasks of training within each fold are related and amenable to incorporating multitask principles. We do this by introducing new variables, $(\mathbf{w}^0, b_0)$, into the inner-level problems. For example, consider the following SV regression inner level, (2.1) with added multi-task terms (and including the bias term):

$$
(\mathbf{w}^t, b_t) \in \underset{(\mathbf{w},b)\in\mathbb{R}^{n+1}}{\arg\min}\left\{\begin{array}{l}\frac{1}{2}\|\mathbf{w}\|_2^2+\frac{C}{|\overline{\mathcal{N}}_t|}\sum_{j\in\overline{\mathcal{N}}_t}\max(|\mathbf{x}_j'\mathbf{w}-y_j|-\varepsilon,0) \\ \qquad +\frac{\lambda_{\mathbf{w}}}{2}\|\mathbf{w}-\mathbf{w}^0\|_2^2+\frac{\lambda_b}{2}(b-b_0)^2\end{array}\right\}.
\tag{7.1}
$$

The variables $(\mathbf{w}^0, b_0)$ enter the bilevel model as outer-level variables as do the parameters $\lambda_\mathbf{w}$ and λ_b. The multitask terms provide variance control by making each of the individual hyperplanes less susceptible to variations within their respective training sets. They also provide additional regularization. Finally, they ensure fold consistency because of the enforced task relatedness. We can replace (7.1) with its corresponding KKT conditions:

$$0 = (1+\lambda_\mathbf{w})\mathbf{w}^t - \lambda_w \mathbf{w}^0 + \sum_{i \in \overline{\mathcal{N}}_t} (\alpha_i^{t,+} - \alpha_i^{t,-})\mathbf{x}_i,$$

$$0 = \lambda_b(b_t - b_0) + \sum_{i \in \overline{\mathcal{N}}_t} (\alpha_i^{t,+} - \alpha_i^{t,-}),$$

$$\left.\begin{aligned} 0 \le \xi_i^t \perp \frac{C}{|\mathcal{N}_t|} - \alpha_i^{t,+} - \alpha_i^{t,-} \ge 0, \\ 0 \le \alpha_i^{t,+} \perp \xi_i^t + \varepsilon - \mathbf{x}_i'\mathbf{w}^t + b_t + y_i \ge 0, \\ 0 \le \alpha_i^{t,-} \perp \xi_i^t + \varepsilon + \mathbf{x}_i'\mathbf{w}^t - b_t - y_i \ge 0, \end{aligned}\right\}, \quad \forall i \in \overline{\mathcal{N}}_t. \tag{7.2}$$

From (7), we deduce

$$\begin{aligned} \mathbf{w}^t &= \frac{1}{1+\lambda_\mathbf{w}} \left[\lambda_\mathbf{w} \mathbf{w}^0 - \sum_{i \in \overline{\mathcal{N}}_t} (\alpha_i^{t,+} - \alpha_i^{t,-})\mathbf{x}_i \right], \\ b_t &= b_0 - \frac{1}{\lambda_b} \sum_{i \in \overline{\mathcal{N}}_t} (\alpha_i^{t,+} - \alpha_i^{t,-}), \end{aligned} \tag{7.3}$$

where it is understood that if $\lambda_b = 0$, then the latter expression for b_t reduces to

$$\sum_{i \in \overline{\mathcal{N}}_t} (\alpha_i^{t,+} - \alpha_i^{t,-}) = 0, \tag{7.4}$$

which does not involve b_t. In the interest of kernelizing (7), we postulate that

$$\mathbf{w}^0 \equiv \sum_{j \in \mathcal{N}} \beta_j \mathbf{x}^j, \tag{7.5}$$

for some scalars, β_j, to be determined. We obtain

$$\mathbf{w}^t \equiv \frac{1}{1+\lambda_\mathbf{w}} \left[\lambda_\mathbf{w} \sum_{j \in \mathcal{N}} \beta_j \mathbf{x}^j - \sum_{j \in \overline{\mathcal{N}}_t} (\alpha_j^{t,+} - \alpha_j^{t,-})\mathbf{x}^j \right]. \tag{7.6}$$

This last expression can be substituted into the complementarities in (7) to give

$$\begin{aligned} & 0 \le \xi_i^t \perp \frac{C}{|\mathcal{N}_t|} - \alpha_i^{t,+} - \alpha_i^{t,-} \ge 0, \\ & 0 \le \alpha_i^{t,+} \perp \xi_i^t + \varepsilon - \frac{1}{1+\lambda_\mathbf{w}} \left[\lambda_\mathbf{w} \sum_{j \in \mathcal{N}} \beta_j \mathbf{x}_i'\mathbf{x}_j - \sum_{j \in \overline{\mathcal{N}}_t} (\alpha_j^{t,+} - \alpha_j^{t,-})\mathbf{x}_i'\mathbf{x}_j \right] \\ & \qquad\qquad + b_t + y_i \ge 0, \\ & 0 \le \alpha_i^{t,-} \perp \xi_i^t + \varepsilon + \frac{1}{1+\lambda_\mathbf{w}} \left[\lambda_\mathbf{w} \sum_{j \in \mathcal{N}} \beta_j \mathbf{x}_i'\mathbf{x}_j - \sum_{j \in \overline{\mathcal{N}}_t} (\alpha_j^{t,+} - \alpha_j^{t,-})\mathbf{x}_i'\mathbf{x}_j \right] \\ & \qquad\qquad - b_t - y_i \ge 0. \end{aligned} \tag{7.7}$$

The "kernel trick" can now be applied to (7.7); see Section 5 for details.

8. Optimization Methods for Bilevel Models

The bilevel and multilevel model selection models proposed here require the solutions of LPECs/MPECs. There exist several approaches that can deal with the complementarity constraints that arise in and LPECs/MPECs. Some of these are: penalty methods, which allow for the violation of the complementarity constraints, but penalize them through a penalty term in the outer-level objective; smoothing methods, that construct smooth approximations of the complementarity constraints; and relaxation methods, that relax the complementarity constraints while retaining the relaxations in the constraints.

LPECs (or MPECs) are difficult to solve since they contain linear (or nonlinear) complementarity constraints; it is known that linear complementarity problems belong to the class of NP-complete problems [11]. Furthermore, the complementarity constraints cause the feasible region of a bilevel program to lack closedness and convexity or, even possibly, be disjoint [40]. Aside from these obvious sources of intractability, stationary points for MPECs *always* fail to satisfy linear independence constraint qualification (LICQ) or Mangasarian-Fromovitz constraint qualification (MFCQ) in the nonlinear programming sense. There is yet another consideration, that of local optimal points, which is particularly important in the machine learning context. Machine learning problems lead to well-posed complementarity problems, in general, that have multiple local minima [41] which can be useful, especially if it is hard to construct globally optimal solutions

8.1. Stationarity and constraint qualification for MPECs. In this subsection, we introduce some standard assumptions and definitions from MPEC theory in order to better understand the properties of MPECs at optimality. We consider bilevel programs of the type shown below, which is slightly different from the Bracken and McGill formulation, (1.4),

$$\begin{aligned} &\min_{x,y} F(x,y) \\ &\text{s. t. } G(x,y) \geq 0, \\ &\qquad y \in \left\{ \begin{array}{ll} \arg\min_{y} f(x,y) & \\ \text{s. t. } g_i(x,y) \geq 0, & \forall i = 1,\ldots,m \end{array} \right\}. \end{aligned} \tag{8.1}$$

Introducing Lagrange multipliers, $\lambda_i \geq 0$, for the inner-level constraints, (8.1) can be rewritten using either the first-order KKT conditions or a variational inequality as follows:

$$
\begin{aligned}
&\min_{x,y} F(x,y) \\
&\text{s. t. } G(x,y) \geq 0, \\
&\qquad \nabla f(x,y) - \sum_{i=1}^{m} \lambda_i \nabla g_i(x,y) = 0, \\
&\qquad 0 \leq \lambda_i \perp g_i(x,y) \geq 0, \ \forall i = 1,\ldots,m
\end{aligned}
\tag{8.2}
$$

$$
\Longleftrightarrow \quad
\begin{aligned}
&\min_{x,y} F(x,y) \\
&\text{s. t. } G(x,y) \geq 0, \\
&\qquad (u-y)'\nabla f(x,y) \geq 0, \ \text{for some } y, \\
&\qquad u \in \{y \mid g_i(x,y) \geq 0, \ \forall i = 1,\ldots,m\}\,.
\end{aligned}
$$

The two formulations above are equivalent nonlinear programs; we shall use the one with the inner-level KKT conditions. As noted above, LICQ or MFCQ, which are necessary to guarantee the existence of the multipliers, λ_i, at stationarity, fail to hold for (8.2) because the gradients of the complementarity constraints, $\lambda_i g_i(x,y) = 0$, are never linearly independent. Denoting the feasible region of the LPEC/MPEC (including the complementarities) is S_0, and the set of multipliers that satisfies the first-order KKT conditions of the inner-level problem is $\Lambda(x,y)$, we can define a key regularity assumption called the sequentially bounded constraint qualification (SBCQ).

Definition 1 (SBCQ). For any convergent subsequence $\{(x^k, y^k)\} \subseteq S_0$, there exists, for each k, a multiplier vector, $\lambda^k \in \Lambda(x^k, y^k)$, and $\{\lambda^k\}_{k=1}^{\infty}$ is bounded.

If SBCQ is satisfied, then it guarantees the non-emptiness of the set of multipliers, $\Lambda(x,y)$, and the existence of bounds on the multipliers on bounded sets. More importantly, it also guarantees the equivalence of (8.1) and (8.2) with regard to global optima; equivalence with regard to local optima can also be guaranteed if the functions $g_i(x,y)$ are convex in y. The SBCQ condition is weak and is easily satisfied under (implied by) other stronger constraint qualifications for the inner-level problem such as MFCQ.

In order to derive stationarity conditions for the MPEC, (8.2), we can relate it to the tightened and relaxed non-linear programs, where the first-order equality constraints have been collected into $H(x,y,\lambda)$,

$$
\begin{aligned}
&\min_{x,y} F(x,y) \ \text{(tightened)} \\
&\text{s. t. } G(x,y) \geq 0, \ H(x,y,\lambda) = 0, \\
&\qquad \lambda_i = 0, \qquad \forall i \in \mathcal{I}_\alpha, \\
&\qquad g_i(x,y) = 0, \qquad \forall i \in \mathcal{I}_\gamma, \\
&\qquad \lambda_i = 0, \ g_i(x,y) = 0, \qquad \forall i \in \mathcal{I}_\beta.
\end{aligned}
\qquad
\begin{aligned}
&\min_{x,y} F(x,y) \ \text{(relaxed)} \\
&\text{s. t. } G(x,y) \geq 0, \ H(x,y,\lambda) = 0, \\
&\qquad \lambda_i = 0, \qquad \forall i \in \mathcal{I}_\alpha, \\
&\qquad g_i(x,y) = 0, \qquad \forall i \in \mathcal{I}_\gamma, \\
&\qquad \lambda_i \geq 0, \ g_i(x,y) \geq 0, \qquad \forall i \in \mathcal{I}_\beta.
\end{aligned}
\tag{8.3}
$$

and with the Lagrangian function,

$$
\begin{aligned}
&\mathcal{L}(x,y,\lambda_i,\mu,\nu,u,v) \\
&\qquad = F(x,y) - \mu G(x,y) - \nu H(x,y,\lambda) - \sum_{i=1}^{m} u_i \lambda_i - \sum_{i=1}^{m} v_i g_i(x,y),
\end{aligned}
\tag{8.4}
$$

where

$$\begin{aligned} \mathcal{I}_\alpha &:= \{i \mid \lambda_i = 0, g_i(x,y) > 0\}, \\ \mathcal{I}_\beta &:= \{i \mid \lambda_i = 0, g_i(x,y) = 0\}, \\ \mathcal{I}_\gamma &:= \{i \mid \lambda_i > 0, g_i(x,y) = 0\}. \end{aligned} \tag{8.5}$$

If the index set, $\mathcal{I}_\beta$, is empty, then *strict complementarity* is said to hold and if not, the complementarity constraints in $\mathcal{I}_\beta$ are said to be *degenerate*. We can now define some stationarity concepts.

Definition 2 (B-stationarity). A feasible point (x^*, y^*, λ^*) is said to be *Bouligand or B-stationary* if it is a local minimizer of an LPEC obtained by linearizing all the MPEC functions about the point (x^*, y^*, λ^*) i.e., $\nabla F(x,y)'z \geq 0$, $\forall z \in \mathcal{T}_{\text{lin}}(x^*, y^*, \lambda^*)$, where $\mathcal{T}_{\text{lin}}$ denotes the tangent cone.

This is a primal stationarity condition and is very general. However, as a certificate, it is not very useful as verifying it is combinatorially expensive due to the difficulty in characterizing the tangent cone. Alternately, we can look at various dual stationarity conditions.

Definition 3 (W-stationarity). A feasible point (x^*, y^*, λ^*) is said to be *weakly or W-stationary* if there exist multipliers μ, ν, u and $v \geq 0$ such that

$$\begin{gathered} \nabla \mathcal{L}(x, y, \lambda_i, \mu, \nu, u, v) = 0, \\ \mu \geq 0, \qquad u_i = 0, \quad \forall i \in \mathcal{I}_\gamma, \qquad v_i = 0, \quad \forall i \in \mathcal{I}_\alpha. \end{gathered} \tag{8.6}$$

The conditions above are simply the non-trivial first-order KKT conditions of the tightened nonlinear program. W-stationarity is a very important concept for computational purposes as it can help identify points that are feasible but not stationary[2].

Definition 4 (S-stationarity). A feasible point (x^*, y^*, λ^*) is said to be *strongly or S-stationary* if the W-stationarity conditions, (8.6), and the condition: $\forall i \in \mathcal{I}_\beta$, $u_i, v_i \geq 0$, hold.

As in the weak case, the conditions for S-stationarity are simply the first-order KKT conditions for the relaxed nonlinear program. Finally, it can be shown that if "LICQ for MPECs" holds, then B-stationarity is equivalent to S-stationarity [50]. This discussion can be easily extended to the case where the outer-level problem may have equality constraints.

We now discuss some approaches to solving MPECs.

[2]W-stationarity concepts can be strengthened by enforcing additional constraints on the multipliers in (8.4). For example, replacing $\lambda_i g_i(x,y) = 0$ with $\min(\lambda_i, g_i(x,y)) = 0$ in (8.2) yields a non-smooth nonlinear program. The first-order KKT conditions for the latter can be written using the Clarke generalized gradient, and are precisely the conditions for Clarke or C-stationarity. See [55] for more details.

8.2. Nonlinear programming approaches. In machine learning, since the inner level problems are typically linear or quadratic, the reformulated bilevel program, yields an LPEC of the following general form

$$
\begin{aligned}
&\min_{x,y} c'x + d'y \\
&\text{s. t. } 0 \le y \perp w = Nx + My + q \ge 0, \\
&\qquad Ax + By + p \ge 0, \\
&\qquad Gx + Hy + f = 0.
\end{aligned}
\tag{8.7}
$$

where some subset of variables of y are the multipliers λ_i. The complementarity condition can also be expressed using $\min(y, w) = 0$. This equality condition is equivalent to $y - (y - w)_+ = 0$. Here, $\mathbf{r}_+ = \max(\mathbf{r}, 0)$, the component wise plus function applied to some vector $\mathbf{r} \ge 0$.

8.2.1. *Inexact solutions.* This solution approach can be thought of as similar to the well-known machine learning technique of *early stopping.* As mentioned before, inexact and approximate solutions as well as local minima yield fairly good optimal points in the machine learning context. We take advantage of this fact and use the relaxation approach to solve MPECs. This method simply involves replacing all instances of "hard" complementarity constraints of the form

$$0 \le y \perp w \ge 0 \equiv y \ge 0, \quad w \ge 0, \quad y'w = 0$$

with relaxed, "soft" complementarity constraints of the form

$$0 \le y \perp_{\text{tol}} w \ge 0 \equiv y \ge 0, \quad w \ge 0, \quad y'w \le \text{tol}$$

where tol > 0 is some prescribed tolerance of the complementarity conditions. If the machine learning problem yields an LPEC, the resulting inexact formulation will be a quadratically constrained quadratic program. For general MPECs, the relaxation will be a nonlinearly constrained optimization problem which can be solved using off-the-shelf NLP solvers such as FILTER [24, 25] or SNOPT [27], which are freely available on the NEOS server [13]. Both these solvers implement the sequential quadratic programming (SQP) method; FILTER uses trust-region based SQP while SNOPT uses line search based SQP.

Inexact cross validation was used to solve bilevel cross validation for support vector regression, (2.5), in [3] and support vector classification, (4.1), in [36] using FILTER. In spite of the fact that FILTER provides no guarantee of global optimality and generally converges to locally optimal solutions, this method performed well with regard to generalization error, indicating that local optimal solutions can be practically satisfactory. The reported results also compared favourably with grid search techniques with regard to parameter and feature selection and objective values. However, they were more efficient than grid search, especially with regard to feature selection.

8.2.2. *Smooth approximations.* The condition, $\min(y, Nx + My + q) = 0$, can be replaced by a function $\phi(y, w)$, possibly non-smooth, such that $\phi(y, w) = 0 \equiv 0 \le y \perp w \ge 0$. The Fischer-Burmeister function [23], $\phi(y, w) = y + w - \sqrt{y^2 + w^2}$, is a non-smooth example of such a function. This function is smoothed using a parameter ε to give the smoothed Fischer-Burmeister function, $\phi(y, w) = y + w - \sqrt{y^2 + w^2 + \varepsilon^2}$. The smoothed function is everywhere differentiable and yields the following approximation of (8.7):

$$
\begin{aligned}
&\min_{x,y,w} c'x + d'y \\
&\text{s. t. } w = Nx + My + q \geq 0,\ y \geq 0, \\
&\qquad Ax + By + p \geq 0, \\
&\qquad Gx + Hy + f = 0 \\
&\qquad y_i + w_i - \sqrt{y_i^2 + w_i^2 + \varepsilon_k^2} = 0, \quad \forall i = 1, \ldots, m.
\end{aligned}
\tag{8.8}
$$

Pang and Fukushima [26] showed that for decreasing values of ε_k, the sequence of stationary points to the nonlinear program (8.8), (x^k, y^k, w^k), converges to a B-stationary point, (x^*, y^*, w^*), if weak second order necessary conditions hold at each (x^k, y^k, w^k), and LICQ for MPECs holds at (x^*, y^*, w^*). Various methods can be used to solve the sequence of problems (8.8); for example, the sequential quadratic programming (SQP) algorithm [35].

Another approach that was proposed for nonlinear and mixed complementarity problems involves solving the non-smooth equation, $y = (y - w)_+$; the right hand side of the equation, $\max(y - w, 0)$, is not differentiable at zero, and can be replaced by an everywhere differentiable smooth approximation. Chen and Mangasarian [15] propose several different smooth approximations to the max function generated from different parametrized probability density functions that satisfy certain consistency properties. One approximation generated from the smoothed Dirac delta function that is commonly used in neural network literature is

$$
p(z, \alpha) = z + \frac{1}{\alpha} \log\left(1 + e^{-\alpha z}\right), \quad \alpha > 0,
\tag{8.9}
$$

where α is some smoothing parameter. Now, the smoothed non-linear equation representing the complementarity system is $\phi(y, w) = y - p(y - w, \alpha) = 0$.

8.2.3. *Exact penalty methods.* Penalty and augmented Lagrangian methods have been widely applied to solving LPECs and MPECs [33]. These methods typically require solving an unconstrained optimization problem. In contrast, exact penalty methods penalize only the complementarity constraints in the objective:

$$
\begin{aligned}
&\min_{x,y,w} c'x + d'y + \mu\phi(y, w) \\
&\text{s. t. } w = Nx + My + q \geq 0,\ y \geq 0, \\
&\qquad Ax + By + p \geq 0, \\
&\qquad Gx + Hy + f = 0.
\end{aligned}
\tag{8.10}
$$

One approach to solving exact penalty formulations like (8.10) is the successive linearization algorithm, where a sequence of problems with a linearized objective,

$$
c'(x - x^k) + d'(y - y^k) + \mu\big(\partial_x \phi(y^k, w^k)(x - x^k) + \partial_y \phi(y^k, w^k)(y - y^k)\big)
\tag{8.11}
$$

is solved to generate the next iterate. The algorithm requires concavity of the objective (to guarantee the existence of vertex solutions at each iteration) and lower-boundedness of the objective. An example of a differentiable penalty function is $\phi(y, w) = y'w$. The resulting quadratic program can be solved using the Frank-Wolfe method [41].

Alternately, the concave penalty function, $\min(y, w)$, has also been proposed. Various approaches can be used to handle the non-smoothness of the penalized objective function arising from this choice of $\phi(y, w)$. The most straight-forward approach is to use successive linearization with the gradients in the linearized objective being replaced by the supergradients [43],

$$\partial_x \phi = \sum_{j=1}^{m} \begin{cases} 0, & \text{if } y_j < w_j, \\ (1-\lambda_j)0 + \lambda_j N_j, & \text{if } y_j = w_j, , \\ N_j, & \text{if } y_j > w_j. \end{cases}$$

$$\partial_y \phi = \sum_{j=1}^{m} \begin{cases} I_j, & \text{if } y_j < w_j, \\ (1-\lambda_j)I_j + \lambda_j M_j, & \text{if } y_j = w_j, \\ M_j, & \text{if } y_j > w_j. \end{cases} \tag{8.12}$$

and $0 \leq \lambda \leq 1$. A second approach makes use of the fact that $\min(r, s)$, for any two scalars, r and s, can be computed as

$$\min(r, s) = \arg\min_{\rho,\sigma}\{\rho r + \sigma s \mid \rho, \sigma \geq 0, \rho + \sigma = 1\}. \tag{8.13}$$

This gives a separable bilinear program [42],

$$\begin{aligned} &\min_{x,y,w} c'x + d'y + \rho' r + \sigma' s \\ &\text{s. t.} w = Nx + My + q \geq 0, \ y \geq 0, \\ &\qquad Ax + By + p \geq 0, \\ &\qquad Gx + Hy + f = 0. \end{aligned} \tag{8.14}$$

which can be solved using a finite Frank-Wolfe method. A third approach requires replacing the non-smooth min with its smooth approximation, which can be defined analogous to the approximation for the max function shown in the previous subsection,

$$m(z, \alpha) = -\frac{1}{\alpha}\log(1 + e^{-\alpha z}), \quad \alpha > 0. \tag{8.15}$$

The application of these methods to the bilevel machine learning applications is presently under investigation.

8.3. Integer programming approaches. The connections between bilevel programs, MPECs and mixed integer programs (MIPs) are well known. It was shown in [1] that there exists a polynomial time reformulation to convert a mixed integer program to a bilevel program. Also demonstrated in [1] was an implicit reformulation of a bilevel program as a mixed integer program via MPECs. Specifically, a program with equilibrium constraints, such as (8.7), can be converted to a MIP by splitting the complementarity constraints through the introduction of integer variables, z, and a large finite constant θ.

$$\begin{aligned} &\min_{x,y,z} c'x + d'y \\ &\text{s. t. } 0 \leq Nx + My + q \leq \theta(1 - z), \\ &\qquad 0 \leq y \leq \theta z, \quad z \in \{0,1\}^m, \\ &\qquad Ax + By + p \geq 0, \\ &\qquad Gx + Hy + f = 0. \end{aligned} \tag{8.16}$$

Care must be taken to compute the value of θ large enough so as not to cut off parts of the feasible region. This is done by solving several LPs to obtain bounds on all the variables and constraints of (8.16) and setting θ to be equal to the largest bound. Once θ is fixed, the MIP can now be solved by using standard techniques such as branch and bound.

The biggest drawback of this approach is that the computation of the bound, θ, requires solving a very large number of LPs. Other drawbacks are that the approach can only be applied to LPECs with bounded feasible regions (thus ensuring that the feasible region of the MIP is also bounded) and does not necessarily always converge to a global optimum. These latter limitations tend to be less of a concern for bilevel programs arising from machine learning applications. However, all of the drawbacks mentioned here are all satisfactorily dealt with in the method of [32], wherein a parameter-free dual program of (8.16) is derived, reformulated as a minimax problem, and solved using Bender's approach. The application of this method to the bilevel machine learning applications is presently under investigation.

8.4. Other approaches. The discussion of the solution approaches above is not meant to be exhaustive. There are several other approaches to solving MPECs and LPECs such as active set identification methods [38], interior point methods [40, 39], implicit programming [14, 40] and non-smooth methods [49].

9. Conclusions

We showed how various important machine learning problems can be cast as bilevel programs. This includes cross validation for support vector regression and classification in order to perform parameter and feature selection, kernel methods, semi-supervised learning and multi-task learning. This is certainly not an exhaustive list of machine learning problems to which bilevel programming can be applied. Noting that all the methods proposed here were nonparametric methods, an interesting avenue of further research with regard to modelling is the incorporation of parametric or generative methods based on probability models into the bilevel framework.

In our approach, the inner-level problems in the bilevel programs are replaced by their first-order KKT conditions. This yields mathematical programs with equilibrium constraints, a class of non-convex, nonlinear, and generally hard problems. Under some mild conditions, the equivalence of bilevel programs and their corresponding MPECs can be guaranteed. However, the presence of complementarity constraints in MPECs is a major theoretical and computational challenge, as the principles of nonlinear programming theory cannot be directly extended to MPECs. However, additional stationarity and constraint qualification concepts enable one to guarantee the existence of solutions for these problems. In the machine learning context, many of the theoretical considerations are not a source of difficulty as machine learning applications generally yield reasonably well-posed and well-behaved MPECs.

A major outstanding open question is the development of efficient algorithms for bilevel programs. It should also be noted that machine learning problems, particularly support vector machines, yield elegant, convex, sparse and highly structured problems; it should not be surprising that a lot of these desirable characteristics get carried over to their bilevel counterparts. While the MPECs resulting from the bilevel programs are non-convex, they are certainly very sparse and highly structured.

The structure inherent in SVMs and kernel methods makes them an attractive target for decomposition methods. Well-known machine learning methods such as sequential minimal optimization (SMO) take this to the extreme by decomposing the problem to consider pairs of points. More recently, the method [32] has taken an important step in this direction of algorithmic development. Mangasarian showed that variables in MPECs such as misclassification minimization problems [41] are typically uncoupled, allowing the problem to be decomposed into smaller linear programs. Many of the algorithms presented in this monograph have similar structures. It has already been demonstrated that bilevel machine learning problems perform fairly well with regards to generalization error [3, 36] when solved using SQP-based methods such as FILTER. It should be noted that since the ultimate goal is to produce good generalization, the solutions found need not necessarily be highly accurate or global optimal. However, as the number of complementarities grows rapidly with increasing data set sizes, problem size becomes all but intractable for these general purpose solvers. An urgent need is to specifically exploit the structure and properties of machine learning problems to yield algorithms that are efficient and scalable. We present these models as challenges to mathematical programming researchers.

References

1. C. Audet, P. Hansen, B. Jaumard, and G. Savard, *Links between linear bilevel and mixed $0-1$ programming problems*, J. Optim. Theory Appl. **93** (1997), no. 2, 273–300.
2. K. Bennett and E. Bredensteiner, *Duality and geometry in* SVM *classifiers*, Proc. 7th Internat. Conf. on Machine Learning (P. Langley, ed.), Morgan Kaufmann, San Francisco, 2000, pp. 57–64.
3. K. Bennett, X. Ji, J. Hu, G. Kunapuli, and J.-S. Pang, *Model Selection via bilevel optimization*, International Joint Conference on Neural Networks (July 2006), pp. 1922–1929
4. K. Bennett and E. Parrado-Hernández *The interplay of optimization and machine learning research*, J. Mach. Learn. Res. **7** (2006), 1265–1281.
5. J. Bi, K. Bennett, M. Embrechts, C. Breneman, and M. Song, *Dimensionality reduction via sparse support vector machines*, J. Mach. Learn. Res. **3** (2003), 1229–1243.
6. J. Bracken and J. McGill, *Mathematical programs with optimization problems in the constraints*, Oper. Res. **21** (1973), 37–44.
7. P. Bradley and O. Mangasarian, *Feature selection via concave minimization and support vector machines*, Machine Learning Proc. 5th Internat. Conf. (ICML 1998) (J. Shavlik, ed.), Morgan Kaufmann, San Francisco, CA, 1998, pp. 82–90.
8. ______, *Optimization methods in massive data sets*, Handbook of Massive Datasets (J. Abello, P. M. Pardalos, and M. G. C. Resende, eds.), vol. 20, Kluwer Acad. Publ., Dordrecht, 1999.
9. R. Caruana, *Multitask learning*, Machine Learning **28** (1997), 41–75.
10. O. Chapelle, V. Vapnik, O. Bousquet, and S. Mukherjee, *Choosing multiple parameters for support vector machines*, Machine Learning **46** (2002), 131–159.
11. S. Chung, NP-*completeness of the linear complementarity problem*, J. Optim. Theory Appl. **60** (1989), 393–399.
12. C. Cortes and V. Vapnik, *Support-vector networks*, Machine Learning **20** (1995), 273–297.
13. J. Czyzyk, M. Mesnier, and J. More, *The* NEOS *server*, IEEE J. Comput. Sci. Engr. **5** (1998), 68–75.
14. X. Chen, and M. Fukushima, *A smoothing method for a mathematical program with P-matrix linear complementarity constraints*, Comput. Optim. Appl. **27** (2004), no. 3, 223–246.
15. C. Chen and O. Mangasarian, *A class of smoothing functions for nonlinear and mixed complementarity problems*, Comput. Optim. Appl. **5** (1996), 97–138.
16. A. Demiriz and K. Bennett, *Optimization approaches to semi-supervised learning*, Complementarity: Applications, Algorithms and extensions (Madison, 1999), Appl. Optim., vol. 50, Kluwer Acad. Publ., Dordrecht, 2001, pp. 121–141.
17. A. Demiriz, K. Bennett, C. Brenman, and M. Embrechts, *Support vector regression methods in cheminformatics*, Comput. Sci. Statist. **33** (2001), pp. 530–539; `http://www.galaxy.gmu.edu/interface/I01/I2001HTML`.

18. S. Dempe, *Foundations of bilevel programming*, Kluwer Acad. Publ., Dordrecht, 2002.
19. ———, *Annotated bibliography on bilevel programming and mathematical programs with equilibrium constraints*, Optimization **52** (2003), 333–359.
20. K. Duan, S. Keerthi, and A. Poo, *Evaluation of simple performance measures for tuning* SVM *hyperparameters*, Neurocomputing *51* (2003), 41–59.
21. T. Evgeniou and M. Pontil, *Regularized multi–task learning*, Proc. Internat. Conf. on Knowledge Discovery and Data Mining, Seattle, USA, 2004, pp. 109–117.
22. F. Facchinei and J.-S. Pang, *Finite-dimensional variational inequalities and complementarity problems*, Springer-Verlag, New York, 2003.
23. A. Fischer, *A special Newton-type optimization method*, Optimization **24** (1992), 269–282.
24. R. Fletcher and S. Leyffer, *Nonlinear programming without a penalty function*, Math. Programming **91** (2002), 239–270.
25. ———, *User manual for filter* SQP, University of Dundee, 1999; `http://www-unix.mcs.anl.gov/leyffer/papers/SQP_manual.pdf`.
26. M. Fukushima, Z. Luo, and J.-S. Pang, *A globally convergent sequential quadratic programming algorithm for mathematical programs with linear complementarity constraints*, Comput. Optim. Appl. **10** (1998), 5–34.
27. P. Gill, W. Murray, and M. Saunders, *User's guide for* SNOPT *version 6: A Fortran package for large-scale nonlinear programming*, Systems Optimization Laboratory, Stanford University, 2002; `http://www.cam.ucsd.edu/ peg/papers/sndoc6.pdf`.
28. G. Golub, M. Heath, and G. Wahba, *Generalised cross validation as a method for choosing a good ridge parameter*, Technometrics **21** (1979), 215–223.
29. I. Guyon and A. Elisseeff, *An introduction to variable and feature selection*, J. Mach. Learn. Res. **3** (2003), no. 3, 1157–1182.
30. I. Guyon, J. Weston, S. Barnhill, and V. Vapnik, *Gene selection for cancer classification using support vector machines*, Machine Learning **46** (2002), 389–422.
31. T. Hastie and S. Rossett, *The entire regularization path for the support vector machine*, J. Mach. Learn. Res. **5** (2004), 1391–1415.
32. J. Hu, J. Mitchell, J.-S. Pang, K. Bennett, and G. Kunapuli, *On the global solution of linear programs with complementarity constraints*, Manuscript, Rensselaer Polytechnic Institute, Troy, NY, 2007.
33. X. Huang, X. Yang, and K. Teo, *Partial augmented Lagrangian method and mathematical programs with complementarity constraints*, J. Global Optim. **35** (2006), no. 2, 235–254.
34. X. Ji, *A bilevel programming approach to semi-supervised learning*, Private Communication.
35. H. Jiang and D. Ralph, *Smooth* SQP *methods for mathematical programs with nonlinear complementarity constraints*, SIAM J. Optim. **10** (2000), no. 3, 779–808.
36. G. Kunapuli, K. Bennett, J. Hu, and J.-S. Pang, *Classification model selection via bilevel programming*, Optim. Methods Softw., 2007 (to appear).
37. G. Lanckriet, N. Cristianini, P. Bartlett, L. El Ghaoui, and M. Jordan, *Learning the kernel matrix with semidefinite programming*, J. Mach. Learn. Res. **5** (2004), 27–72.
38. G. Lin and M. Fukushima, *Hybrid algorithms with active set identification for mathematical programs with complementarity constraints*, Technical report 2002-008, Kyoto, Japan, 2002.
39. X. Liu and J. Sun, *Generalized stationary points and a robust interior point method for mathematical programs with equilibrium constraints*, Math. Programming **101** (2004), 231–261.
40. Z. Luo, J.-S. Pang, and D. Ralph, *Mathematical programs with equilibrium constraints*, Cambridge University Press, Cambridge, England, 1996.
41. O. Mangasarian, *Misclassification minimization*, J. Global Optim. **5** (1994), 309–323.
42. ———, *The linear complementarity problem as a separable bilinear program*, J. Global Optim. **6** (1995), 153–161.
43. ———, *Solution of general linear complementarity problems via nondifferentiable concave minimization*, Acta Math. Vietnam. **22** (1997), no. 1, 199–205.
44. ———, *Successive overrelaxation for support vector machines*, IEEE Trans. Neural Networks **10** (1999), 1032–1037.
45. O. Mangasarian and D. R. Musicant, *Robust linear and support vector regression*, IEEE Trans. Pattern Analysis and Machine Intelligence **22** (2000), no. 9, 950–956.
46. M. Momma and K. Bennett, *A pattern search method for model selection of support vector regression*, Proc. 2nd SIAM Internat. Conf. on Data Mining (R. Grossman, J. Han,

V. Kumar, H. Mannila, and R. Motwani, eds.), SIAM Publications, Philadelphia, PA, 2002; http://www.siam.org/meetings/sdm02/proceedings/sdm02-16.pdf.
47. B. Murdukhovich, *Metric approximation and necessary optimality conditions for general classes of nonsmooth extremal problems*, Soviet Math. Dokl. **22** (1980) 526–530.
48. C. Ong, A. Smola, and R. Williamson, *Learning the kernel with hyperkernels*, J. Mach. Learn. Res. **6** (2005) 1043–1071.
49. J. Outrata, M. Kocvara, and J. Zowe, *Nonsmooth approach to optimization problems with equilibrium constraints: theory, applications and numerical results*, Kluwer Acad. Publ., Dordrecht, 1998.
50. H. Scheel and S. Scholtes, *Mathematical programs with complementarity conditions: stationarity, optimality and sensitivity*, Math. Oper. Res. **25** (2000), no. 1, 1–22.
51. J. Shawe-Taylor and N. Cristianini, *Kernel methods for pattern analysis*, Cambridge University Press, Cambridge, 2004.
52. B. Schölkopf, R. Williamson, A. Smola, J. Shawe-Taylor, and J. Platt, *Support vector method for novelty detection*, Adv. Neural Inf. Process. Syst. **12** (2000), 582–588.
53. H. Van Stackelberg, *The theory of market economy*, Oxford University Press, Oxford, 1952.
54. V. Vapnik, *The nature of statistical learning theory*, 2nd ed., Springer-Verlag, New York, 2000.
55. J. Ye, *Necessary and sufficient optimality conditions for mathematical programs with equilibrium conditions*, J. Math. Anal. Appl. **307** (2005), 350–369.
56. H. Zou and T. Hastie, *Regularization and variable selection via the elastic net*, J. Roy. Statist. Soc. Ser. B **67**, no. 2 (2005), 301–320.
57. H. Zou and M. Yuan, *The F_∞-norm support vector machine*, Statist. Sinica (to appear); http://www2.isye.gatech.edu/ myuan/papers/fsvm.pdf.

Department of Mathematical Sciences, Rensselaer Polytechnic Institute, 110 8th Street, Troy NY 12180, USA.

E-mail address, G. Kunapuli: kunapg@rpi.edu

E-mail address, K. P. Bennett: bennek@rpi.edu

E-mail address, J. Hu: huj@rpi.edu

E-mail address, J.-S. Pang: pangj@rpi.edu

Centre de Recherches Mathématiques
CRM Proceedings and Lecture Notes
Volume **45**, 2008

Algorithms for Detecting Complete and Partial Horizontal Gene Transfers: Theory and Practice

Vladimir Makarenkov, Alix Boc, Alpha Boubacar Diallo, and Abdoulaye Baniré Diallo

Abstract. We describe two methods for detecting horizontal gene transfers in the framework of the complete and partial gene transfer models. In case of a complete gene transfer model a new fast backward selection algorithm for predicting horizontal gene transfer events is presented. The latter algorithm can rely either on the metric or on the topological optimization to identify horizontal gene transfers between branches of a given species phylogeny. In case of the topological optimization, we use the well-known Robison and Foulds (RF) topological distance, whereas in case of the metric optimization, the least-squares (LS) criterion is considered. We also formulate and prove the NP-hardness of the partial gene transfer problem. Second, an efficient algorithm for predicting partial transfers, using the Gauss and Seidel optimization, is discussed. We also show how to assess the reliability of a specific gene transfer or a whole gene transfer scenario. In the application section, we apply the new algorithm to detect possible gene transfers occurred during the evolution of the gene `rpl12e`.

1. Introduction

Horizontal gene transfer (HGT) is a direct transfer of genetic material from one lineage to another. The understanding that horizontal gene transfer might have played a key role in biological evolution is one of the most fundamental changes in our perception of general aspects of molecular biology in recent years [6, 18, 19]. Bacteria and Archaea have sophisticated mechanisms for the acquisition of new genes through HGT which may have been favoured by natural selection as a more rapid mechanism of adaptation than the alteration of gene functions through numerous point mutations. If the donor DNA and the recipient chromosome display some homologous sequences, the donor sequences can be stably incorporated into the recipient chromosome by homologous recombination. The three main mechanisms of HGT are the following: transformation, consisting of uptake of naked DNA from the environment; conjugation, which is mediated by conjugal plasmids or conjugal transposons; and transduction, consisting of DNA transfer by phage.

2000 *Mathematics Subject Classification.* Primary: 90C35; Secondary: 90C27.
The authors are grateful to Dr. Hervé Philippe for his help in the analysis of the `rpl2e` data.
This is the final form of the paper.

These transferring mechanisms can introduce sequences of DNA that display little similarity with the remaining DNA of the recipient cell [6].

There are a few ways to identify the genes that have been transferred horizontally. First, sequence analysis of the host genome may reveal areas with GC content or codon usage patterns atypical to it [17]. Second, if a sequence is found in only one organism and is absent from all other closely related organisms, it is more likely that it has been introduced horizontally into this organism rather than deleted from all the others. Third, the comparison of a morphology-based species tree or a molecular tree based on a molecule that is assumed to be refractory to horizontal gene transfer (e.g. 16S rRNA or 23S rRNA) against a phylogeny of an observed gene may reveal topological conflicts which can be explained by horizontal transfers.

Several attempts to use network-based models to depict horizontal gene transfers can be found (see for example: [35, 29, 4, 10], or [11]). A model of horizontal gene transfer that maps gene phylogenies into a species tree has been introduced by [10]. Mirkin *et al.* [25] and Hallett *et al.* [11] have developed algorithms allowing for simultaneous identification of gene duplications, gene losses, and horizontal gene transfers. The papers by Moret *et al.* [27, 28] give an overview of the network modeling in phylogenetics. In a recent paper published in the SFC2004 proceedings, [26] considered some approaches for biologically meaningful mapping of data of individual gene families into an evolutionary species tree. One approach first produces a gene tree, then maps it into the species tree, whereas the other approach first takes the gene phyletic profile, maps it into the species tree and then tunes it into a directed scenario based on the similarity data.

In this article we continue the work started in Ref. [3], where we described a HGT model based on least-squares, and in Ref. [22], where we showed the difference between complete and partial gene transfer models. First, we describe a polynomial-time HGT algorithm for the detection of complete transfers and test it with respect to the two optimization criteria: Least-squares (LS) and Robinson and Foulds (RF) topological distance. We also suggest how to assess the reliability of horizontal gene transfers identified by our algorithm. In the application section, we show how the new algorithm predicts transfers of the gene `rpl2e` for the group of 14 Archaea organisms which were originally examined in Ref. [24].

2. Algorithms for Predicting Horizontal Gene Transfers

2.1. Basic definitions. We start this section with some basic definitions about phylogenetic trees and tree metrics, generally following the terminology of Barthélemy and Guénoche [1, 2]. The distance $\delta(x, y)$ between two vertices x and y in a phylogenetic (i.e. additive) tree T is defined as the sum of the edge lengths in the unique path linking x and y in T. Such a path is denoted (x, y). A leaf is a vertex of degree one.

Definition 1. Let X be a finite set of n taxa. A *dissimilarity* d on X is a non-negative function on $(X \times X)$ such that for any x, y from X:

(1) $d(x, y) = d(y, x)$, and
(2) $d(x, y) = d(y, x) \geq d(x, x) = 0$.

	x2	x3	x4	x5
x1	6	6	4	2
x2		2	4	6
x3			4	6
x4				4

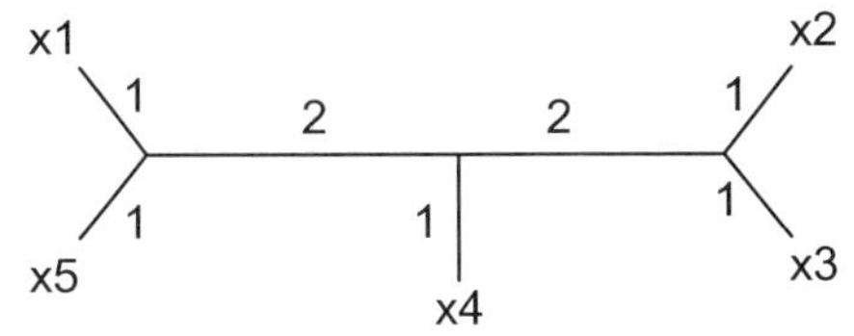

FIGURE 1. An example of a tree metric on the set X of 5 taxa.

Definition 2. A dissimilarity d on X satisfies the *four-point condition* if for any x, y, z, and w from X:

$$d(x,y) + d(z,w) \leq \text{Max}\{d(x,z) + d(y,w); d(x,w) + d(y,z)\}.$$

Definition 3. For a finite set X, a **phylogenetic tree** (i.e. an additive tree or a X-tree) is an ordered pair (T, ϕ) consisting of a tree T, with vertex set V, and a map ϕ :X$\rightarrow$ V with the property that, for all $x \in X$ with degree at most two, $x \in \phi(X)$. A phylogenetic tree is **binary** if ϕ is a bijection from X into the leaf set of T and every interior vertex has degree three.

The main theorem relating the four-point condition and dissimilarity representability by a phylogenetic tree (i.e., phylogeny) is as follows:

Theorem 2.1 (Zarestskii, Buneman, Patrinos & Hakimi, Dobson). *Any dissimilarity satisfying the four-point condition can be represented by a phylogenetic tree such that for any x, y from X, $d(x,y)$ is equal to the length of the path linking the leaves x and y in T. This dissimilarity is called a* tree metric. *Furthermore, this tree is unique.*

Figure 1 is an example of a tree metric on the set X of 5 taxa and the associated phylogenetic tree.

2.2. Optimization criteria. Here we present a fast greedy algorithm for predicting complete horizontal gene transfers. The algorithm for identifying HGTs proceeds by a progressive reconciliation of the given species and gene phylogenetic trees, denoted T and T' respectively. Usually, the species tree T is inferred from the genes that are refractory to horizontal gene transfer and genetic recombination (e.g., 16sRNA sequences). This tree represents the direct or tree-like evolution. The gene tree T' represents the evolution of a given gene which is supposed to undergo horizontal transfers.

At each step of the algorithm, all pairs of branches in T are tested against the hypothesis that a horizontal gene transfer has occurred between them. The considered HGT model assumes that the transferred gene supplants the entire homologous gene of the host or that the homologous gene is simply absent at the host genome. In such a model, the original species phylogenetic tree T is gradually transformed into the gene phylogenetic tree T' through a series of subtree moves (i.e., gene transfers or HGTs). The topology of the gene tree T' is kept fixed. The goal is to find the minimum possible sequence of trees $T, T_1, T_2, \ldots, T'$ that transforms T into T'. Obviously, a number of necessary biological rules should be taken into account. For instance, the transfers within the same lineage as well as some double-crossing transfers should be prohibited (for more detail, see [20, 30, 31, 10]).

We consider two optimization criteria which can be used at each algorithmic step to select the best HGT. The first optimization criterion that we consider is the *least-squares* (LS) *function* Q. It is computed as follows:

$$Q = \sum_{i} \sum_{j} \left(d(i,j) - \delta(i,j)\right)^2, \tag{2.1}$$

where $d(i,j)$ is the pairwise distance between the leaves i and j in the species tree T (or in the tree T_1 obtained from T after the first subtree move) and $\delta(i,j)$ the pairwise distance between i and j in the gene tree T'. The second criterion that can be useful for assessing discrepancy between the species and gene phylogenies is the Robinson and Foulds (RF) topological distance. The RF metric (Robinson and Foulds 1981) is an important and frequently used tool to compare the topologies of phylogenetic trees. This distance is equal to the minimum number of elementary operations, consisting of merging and splitting nodes, necessary to transform one tree into the other. This distance is also the number of bipartitions or Buneman's splits belonging to exactly one of the two trees. When the RF distance is considered, we can use it as an optimization criterion as follows: all possible transformations of the species tree, consisting of transferring one of its subtrees from one branch to another, are evaluated in a way that the RF distance between the transformed species tree T_1 and the gene tree T' is computed. The subtree transfer providing the minimum of the RF distance between T_1 and T' is retained. Note that the problem asking to find the minimum number of subtree transfer operations necessary to transform one tree into another (i.e. also known as *Subtree Transfer Problem*) *has been shown to be* NP-*hard* [12].

2.3. Greedy backward algorithm for predicting complete horizontal gene transfers. In this section we discuss the main features of our algorithm based on the backward selection of horizontal gene transfers. Consider a gene transfer in the species tree T going from b to a and transforming it into the tree T_1 (Fig. 2). The following timing constraint is considered (see also Ref. [22]): to allow the transfer between the branches (z,w) and (x,y) of the species tree T, the cluster combining the subtrees rooted by the vertices y and w must be present in the gene tree T'. Such a constraint enables us, first, to arrange the topological conflicts between T

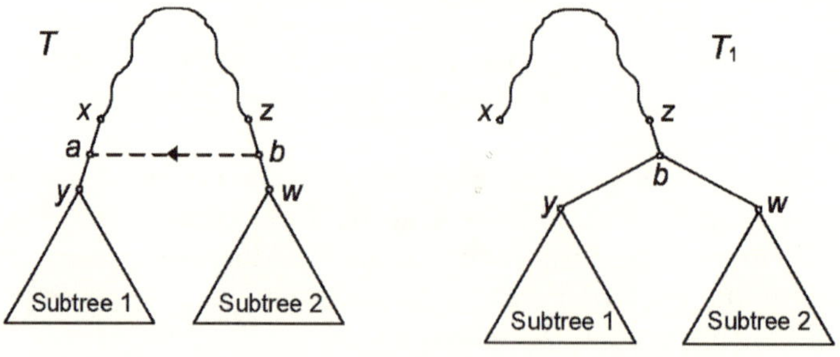

FIGURE 2. Subtree constraint: the transfer between the branches (z,w) and (x,y) of the species tree T can be allowed if and only if the cluster regrouping both affected subtrees is present in the gene tree; here, a single branch is depicted by a plane line and a path is depicted by a wavy line.

and T' that are due to the transfers between single species or their close ancestors and, second, to identify the transfers that have occurred deeper in the phylogeny (i.e., closer to the tree root). The usage of this constraint allows the method to follow the order that is opposite to the order of evolution and infer first the most recent HGTs which are easier to detect.

Proposition 2.1. *If all bipartitions corresponding to the branches of the path* (x, z) *in the transformed species tree* T_1 *(Fig. 2) can be found in the bipartition table of the gene tree* T'*, then the transfer from* b *to* a*, transforming the species tree* T *into* T_1*, is a part of a minimum cost* HGT *scenario transforming* T *into* T'*.*

This Proposition can be easily proved by induction on the number of branches of the path (x, z).

The main steps of the HGT detection algorithm are the following:

Preliminary step. Infer species and gene phylogenies, denoted respectively T and T', whose leaves are labeled by the same set of n taxa. Both species and gene trees must be rooted. If there exist identical subtrees with two or more leaves belonging to both T and T', reduce the size of the problem by replacing these subtrees with the same auxiliary taxa in both T and T'.

Step 1 $(\ldots k)$. Test all possible HGTs between pairs of branches in T_{k-1} ($T_{k-1} = T$ at Step 1) except the transfers between adjacent branches and those violating the evolutionary and subtree constraints. If no such a transfer exists, relax the subtree constraint. In our simulations described in the section Simulation study, this relaxation was necessary on average in 1.2% of cases. Search for the transfers satisfying the conditions of Proposition 2.1. If no such transfers exist, choose the best HGT with respect to the selected optimization criterion that can be in our case: the least-squares (LS) or the Robinson and Foulds (RF) metric. Reduce the size of the problem by contracting the newly-formed subtree in the transformed species tree T_k and the gene tree T'. In the list of the obtained HGTs, search for and eliminate the idle transfers using a backward procedure. An idle transfer is the transfer whose removal does not change the topology of the tree T_k.

Stopping condition and time complexity. The procedure stops when the LS or RF coefficient equals zero. Such a computation requires $O(kn^4)$ time to generate k transfers in a phylogenetic tree with n leaves. However, because of the progressive size reduction of the species and gene trees, the practical time complexity of this algorithm is rather $O(kn^3)$.

Proposition 2.2. *If the subtree constraint is not relaxed, the* HGT *detection algorithm requires at most* $n - 3$ *steps to transform a binary species tree with* n *leaves into a binary gene tree with the same set of* n *leaves.*

The proof of this proposition is based on the fact that the maximum value of the RF distance between two binary trees with n leaves is $2n - 6$ and that each subtree transfer satisfying the subtree constraint decreases the value of the RF distance by at least 2.

2.4. Partial gene transfer model. The partial gene transfer model is more general, but also more complex and challenging. It presumes that only a part of the transferred gene has been acquired by the host species through the process

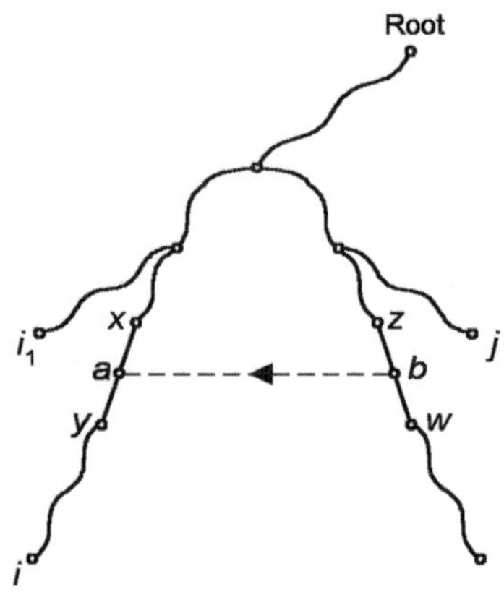

FIGURE 3. Evolutionary distance between the taxa i and j can be allowed to change after the addition of the branch (b, a) representing a partial HGT between the branches (z, w) and (x, y). Evolutionary distance between the taxa i_1 and j must not be affected by the addition of (b, a).

of homologous recombination [22]. This means that the traditional species phylogenetic tree is transformed into a directed phylogenetic network (i.e. a directed connected graph). For example, Denamur *et al.* [5] proposed a method to identify gene segments being transferred horizontally. This method was applied to detect partial HGTs of the mut$U$ and mut$S$ genes within E. coli evolutionary trees. Because many analyzes are now directed at understanding the evolution of complete genomes, the partial gene transfer model could be also useful if one wanted to model the transfer of a portion of a genome.

In a phylogenetic tree, there is always a unique path connecting a pair of nodes. Adding to it a HGT branch creates an extra path between certain nodes. Figure 3 illustrates the case where the evolutionary distance between the taxa i and j can be affected by the addition of the HGT branch (b, a) representing partial gene transfer from b to a. It is relevant to assume that the HGT from b to a can affect the evolutionary distance between the taxa i and j if and only if the destination point a is located on the path between i and the root of the tree; the position of j is fixed. Thus, in the reticulate phylogeny T in Fig. 3 the evolutionary distance $d_1(i, j)$ between the taxa i and j can be computed as follows:

$$d_1(i,j) = (1-\alpha)d(i,j) + \alpha\big(d(i,a) + d(j,b)\big), \tag{2.2}$$

where α indicates the fraction, unknown in advance, of the transferred gene and d is the internode distance in the species tree before the addition of the HGT branch (b, a).

On the contrary, the distance between the taxa i_1 and j (Fig. 3) must not be affected by the addition of (b, a). Figure 4 illustrates the other cases where the addition of a HGT branch must not affect the length of the evolutionary path between i and j.

The least-squares loss function Q to be minimized with the *unknown vector of edge lengths* ℓ in T and the *unknown fraction of the transferred gene* α is as follows:

(2.3) $$Q(L,\alpha)$$
$$= \sum_{ij \in S} \left((1-\dot{a}) \sum_{k \in \text{path}(ij)} \ell_{ij}^k + \alpha \left(\sum_{k \in \text{path}(ia)} \ell_{ia}^k + \sum_{k \in \text{path}(jb)} \ell_{jb}^k \right) - \delta(i,j) \right)^2$$
$$+ \sum_{ij \notin S} \left(\sum_{k \in \text{path}(ij)} \ell_{ij}^k - \delta(i,j) \right)^2 \longrightarrow \min,$$

where $\delta(i,j)$ is the given gene dissimilarity between i and j; ℓ_{ij}^k is the length of the branch k of the path (ij) in T; α is the fraction of the transferred gene $(0 \le \alpha \le 1)$; and S is the set of pairs of taxa $\{ij\}$ such that the transfer (ba) can affect the evolutionary distance between them.

To show the NP-hardness of the least-squares optimization in the context of the partial gene transfer the following problem can be stated:

Given: Species phylogenetic tree T (with the associated tree metric d on the set of taxa X), gene dissimilarity $\boldsymbol{\delta}$ on X, and a fixed non-negative value ε.

Problem. Find the minimum number of partial gene transfers k such that:

(2.4) $$Q = \sum_i \sum_j \left(d_k(i,j) - \delta(i,j)\right)^2 \le \varepsilon,$$

where $d_k(i,j)$ is the network distance between i and j, computed using Formulae 2.2 and 2.3, in the phylogenetic network T_k obtained from T after the addition of k partial gene transfers.

Theorem 2.2. *The minimum number of partial gene transfer problem (MNPGT problem) is* NP*-hard.*

The proof of this theorem is based on a polynomial-time reduction from the *Subtree Transfer Problem* (STR problem) that consists of finding the minimum number of complete gene transfers to transform a given species tree T into a given gene tree T'. The STR problem is identical to the problem of adding to T the minimum number of complete gene transfers such that $Q = \sum_i \sum_j \left(d_k(i,j) - \delta(i,j)\right)^2 \le 0$ (i.e., the case of $\varepsilon = 0$ is considered), where $d_k(i,j)$ is the pairwise distance between i and j in the phylogenetic tree (i.e., a particular case of a phylogenetic network). Here, the tree T_k is obtained from T after the addition of k complete gene transfers (i.e., a particular case of a partial transfer) and $\delta(i,j)$ is the given tree metric associated with T'.

Several important timing constraints have to be incorporated into this model, in addition to those taken into account in the complete HTS model, to identify the interactions between HGTs that are not intelligible from an evolutionary point of view. Some of these constraints, but not all of them, were initially pointed out by Page and Charleston [30, 31]. For instance, double-crossing transfers between two lineages (Figs. 5a and b) must be forbidden. In this case, the HGT events affect the ancestor of the species from the previous transfer. Making the source and destination lineages contemporaneous for one HGT makes the other transfer impossible (Fig. 5).

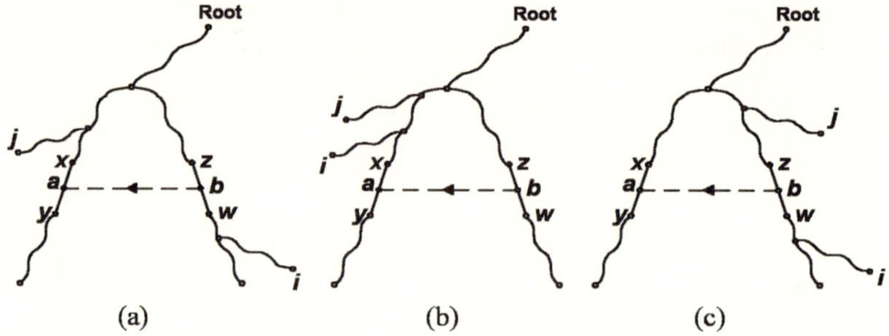

FIGURE 4. Three situations when the evolutionary distance between the taxa i and j must not be affected by the addition of the new branch (b, a) representing a partial HGT between the branches (z, w) and (x, y). Path between the taxa i and j cannot to go through the branch (b, a).

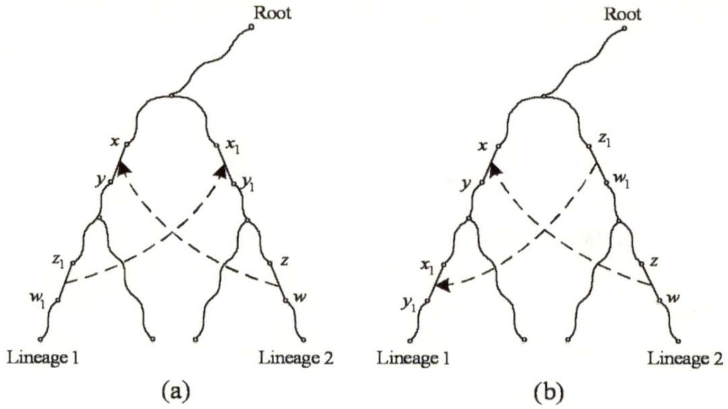

FIGURE 5. Transfers between two lineages crossing in such ways must be prohibited.

Note that the rule illustrated in Figure 5a is automatically taken into account in the complete gene transfer model, where its violation would be equivalent to the violation of the same lineage constraint (see Page and Charleston [30, 31]). For instance (Figure 5a), the HGT from (z, w) to (x, y) cannot be followed by the transfer from (z_1, w_1) to (x_1, y_1) because after the first HGT the branches (z_1, w_1) and (x_1, y_1) will be located on the same lineage (Lineage 2). We also identify two cases, where the evolutionary distance between the taxa i and j can be affected by multiple transfers (Figures 6a and b); and, two cases, where this distance must not be affected by them (Figures 6c and d). Failure to take these constraints into account can result in postulating transfers that are mutually incompatible.

Assume that a partial gene transfer between the branches (z, w) and (x, y) (i.e., from b to a in Fig. 3) of the species tree T has taken place. The lengths of all branches in T are reassessed in the least-squares sense after the addition of (b, a), whereas the length of (b, a) is assumed to be 0. To reassess the branch lengths of T, we have first to make an assumption about the value of the parameter α (eq. 2.2),

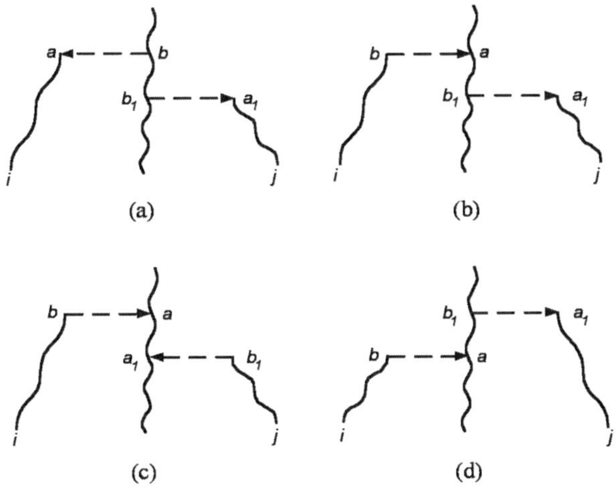

FIGURE 6. Cases (a) and (b): evolutionary path between the taxa i and j can go through both HGT branches (b, a) and (b_1, a_1). Cases (c) and (d): evolutionary path between the taxa i and j cannot go through both HGT branches (b,a) and (b_1, a_1).

indicating the gene fraction being transferred. This parameter can be estimated either by comparing sequence data corresponding to the subtrees rooted by the vertices y and w, or different values of α can be tested in the optimization problem.

Fixing the parameter α, we reduce to a linear system the system of equations establishing the correspondence between the experimental gene distances and the path-length distances in the HGT network. This system having generally more variables (i.e. branch lengths of T) than equations (i.e. pairwise distances in T; the number of equations is always $n(n-1)/2$ for n taxa) can be solved by approximation in the least-squares sense. Let us now show how the approximation problem can be stated and efficiently solved.

Let $\mathbf{A}_\alpha$ be the matrix of dimension $n(n-1)/2 \times m$, each row of which is associated with one pair of taxa of X, where n is the number of taxa and m is a number of edges in T. The value $a_{ij,e}$ of this matrix corresponding to the pair of taxa ij and the edge e is equal either to 1, or to α, or to $1-\alpha$ if the edge e is in the path (ij) in T, and is equal to 0 if not. Let ℓ be the vector of edge lengths of m elements and $\mathbf{d}$ be given vector of gene distances of $n(n-1)/2$ elements.

Fixing the value of α (e.g., values 0, 0.1, 0.2, ..., and 1.0 can be tested in turn), we obtain a linear system of $n(n-1)/2$ equations with m unknowns: $\mathbf{A}_\alpha \times \ell = \mathbf{d}$.

When $n \geq 4$, this system has more equations than unknowns. It can be solved by approximation in the least-squares sense:

$$(\mathbf{A} \times \ell - \mathbf{d})^2 \rightarrow \min . \tag{2.5}$$

After taking the gradient we have:

$$\mathbf{A}_\alpha^t \times (\mathbf{A}_\alpha \times l - \mathbf{d}) = 0. \tag{2.6}$$

Following algebraic manipulations, we obtain:

$$\mathbf{A}_\alpha^t \times \mathbf{A}_\alpha \times l = \mathbf{A}_\alpha^t \times \mathbf{d}. \tag{2.7}$$

Thus, we have: $\mathbf{B} \times \ell = \mathbf{c}$, where $\mathbf{B}$ is a $(m \times m)$ matrix, and $\mathbf{c}$ is a vector with m components.

Following Barthélemy and Guénoche [1] and Makarenkov and Leclerc [23], we apply a slightly modified Gauss-Seidel method to solve the above system. The method consists of decomposing $\mathbf{B}$ into its diagonal $(\mathbf{\Delta})$, its strictly upper triangular component $(-\mathbf{F})$, and its strictly lower triangular component $(-\mathbf{E})$:

$$\mathbf{B} = \begin{pmatrix} b_{11} & b_{12} & \dots & b_{1m} \\ b_{21} & b_{22} & \dots & b_{2m} \\ \dots & \dots & \dots & \dots \\ b_{m1} & b_{m2} & \dots & b_{mm} \end{pmatrix} = \begin{pmatrix} & & -\mathbf{F} \\ & \mathbf{\Delta} & \\ -\mathbf{E} & & \end{pmatrix} = \mathbf{\Delta} - \mathbf{E} - \mathbf{F}. \tag{2.8}$$

Then, we apply the iterative procedure:

$$\mathbf{\Delta} \times \ell^{k+1} = \mathbf{E} \times \ell^{k+1} + \mathbf{F} \times \ell^{(k)} + \mathbf{c}, \tag{2.9}$$

which allows us to compute gradually the components of the vector $\ell(j)^{(k+1)}$, corresponding to the edge lengths at the $k+1$th iteration, from those of $\ell(j)^k$. If the computed value of $\ell(j)^{(k+1)}$ is negative, it is replaced with the value 0. This operation is equivalent to the projection on the cone $\mathbf{L} \geq 0$, which ensures an appropriate solution.

The exact equation used in this method is the following for all $j = 1, 2, \dots, m$:

$$\ell(j)^{(k+1)} = \left(- \left(\sum_{j+1 \leq i \leq m} b_{ij}\ell(j)^{(k)} \right) - \left(\sum_{1 \leq i \leq j-1} b_{ij}\ell(j)^{(k+1)} \right) + c_j \right) \Big/ b_{jj}. \tag{2.10}$$

Thus, the main steps of the partial gene transfer algorithm can be stated as follows:

Preliminary step. This step corresponds to the preliminary step discussed in the context of the complete gene transfer model. It consists of inferring the species and gene phylogenies denoted respectively T and T' whose leaves are labeled by the same set X of n taxa. Because the classical Robinson and Foulds distance is defined only for tree topologies, we use the least-squares as a unique optimization criterion when modeling partial HGTs.

Step 2. Test all connections between pairs of branches in the species tree $\mathbf{T}$. For each HGT connexion satisfying evolutionary constraints, carry out the following optimization:

(a) Fix the value of the fraction of the gene being transferred α (e.g., one can try in turn the values of 0, 0.1, 0.2, ..., and 1.0). Compute using the Gauss–Seidel method the optimal lengths $\mathbf{l}$ of the edges in the species tree (or network, starting from Step 2) $\mathbf{T}$.

(b) Go back to the original equation system: $\mathbf{A}_\alpha \times \mathbf{l} = \mathbf{d}$. Fix the values of the vector $\mathbf{l}$ found using the Gauss-Seidel method and solve this problem by least-squares considering as unknown the parameter α.

(c) Then, fix the optimal value of α found and repeat the computation until both unknown parameters l and α converge to a certain solution.

All eligible pairs of branches in T can be processed in this way. The HGT connection providing the smallest value of the LS coefficient Q and satisfying the defined evolutionary constraints should be selected for the addition to the species tree T, transforming it into a phylogenetic network.

Step 3 $(2,\ldots,k)$. Run the algorithm until a fixed number k of partial gene transfers is found and added to T or the value of the LS criterion Q is lower than a pre-established threshold ε.

Time complexity of this algorithm is $O(kn^5)$ to add k partial horizontal gene transfers to the species tree with n leaves.

2.5. Bootstrap validation of horizontal gene transfers. Bootstrap analysis can be used to place confidence intervals on internal branches of evolutionary trees [7]. We designed a bootstrap validation procedure for computing the bootstrap scores either for a specific gene transfer or a whole gene transfer scenario. The following strategy was adopted to assess the reliability of obtained HGTs. Because we are mostly interested in the evolution of a given gene or a group of genes, the sequences used to build the species tree are not resampled. The species tree is taken as an *a priori* assumption of the method and held constant. The sequence data used to build the gene tree are drawn with replacement in order to create a series of pseudo-replicates. The HGT detection algorithm is then carried out on the bootstrapped pseudo-replicates. Thus, for all HGT branches appearing in the original scenario, we verify if they appear in the obtained transfer scenarios, using as input the original species tree and the gene tree inferred from the sets of pseudo-replicates. It is worth noting that among resampled datasets only those that give rise to a gene phylogenetic tree such that it contains the root branch separating this tree into exactly the same bipartition sets as the root branch of the original gene tree does, are eligible for the HGT bootstrap analysis.

Simulation study. A Monte Carlo study was conducted to test the ability of the new method to recover correct gene transfers. In the framework of *the complete* HGT *model only* we examined how the detection procedure performed depending on the model of sequence evolution, number of observed species, and sequence length. The results illustrated in Figs. 7 and 8, and reported in Tables 1 and 2 (see Appendix) were obtained from simulations carried out with random binary phylogenetic trees with 8, 16, 24, 32, 48, and 64 leaves, whereas the sequence length varied from 125 to 1000 sites. The simulation procedure consisted of the five basic steps described below:

1. A true tree topology, denoted T, was obtained using the random tree generation procedure proposed by Kuhner and Felsenstein [16]. The branch lengths of T were computed using an exponential distribution. Following the approach of Guindon and Gascuel [9], we added some noise to the branches of the true phylogenies to create a deviation from the molecular clock hypothesis. All the branch lengths of T were multiplied by $1+\alpha x$, where the variable x was obtained from a standard exponential distribution $\big(P(x>k)=\exp(-k)\big)$, where the constant a was a tuning factor for the deviation intensity. Following Guindon and Gascuel [9], a was fixed to 0.8. The random trees generated by this procedure are chosen to have the depth of $O\big(\log(n)\big)$, where n is the number of species (i.e. number of leaves in a binary phylogenetic tree).

2. Each random phylogeny was then submitted to the SeqGen program [32] to simulate sequence evolution along its branches according to the Jukes and Cantor [14], Kimura 2-parameter [15], and Jin–Nei Gamma[13] models.

3. To assess the quality of HGT detection by the new method, we developed a simulation program using the results of SeqGen. For each considered rooted tree,

viewed as an organismal phylogeny, our program created one random horizontal gene transfer that respected the evolutionary constraints discussed in the algorithmic section. During this operation, the program regenerated the DNA sequences for each tree node located in the subtree affected by the HGT. As the simulations were carried out for the complete gene transfer model, the HGT destination sequence was set identical to the source sequence and the new sequences were regenerated from it according to the selected evolutionary model.

4. The sequence to distance transformation corresponding to the considered model of evolution was then applied to the DNA sequences associated with the leaves of the phylogeny affected by the gene transfer. The NJ method [34] was used to infer the gene trees from the obtained distance matrix. The topology of the organismal phylogeny (i.e. true tree T) was supposed to be known.

5. The HGT detection method was then carried out to infer the transfer. The experiments were conducted using the procedures based on the RF and LS optimization. The simulations were carried out for 500 random rooted phylogenies with 8 and 16 leaves and 100 random rooted phylogenies with 24 to 64 leaves.

Figures 7 and 8 present the average simulation results obtained for random phylogenies with 8 to 64 leaves, using as optimization criteria the RF topological distance and LS function, respectively. These figures illustrate how the detection rate changes as the number of sites varies from 125 to 1000. As expected, the detection rate grows as the number of sites increases and the number of species decreases. Note that for the phylogenies with 8 to 32 leaves the best results were obtained under the Kumura and Jukes–Cantor models. For the phylogenies with 48 to 64 species the best performances were regularly obtained under the Kimura model, whereas the results found under the Jukes–Cantor model were the worst of the three evolutionary models.

This trend can be observed in the case of both optimization criteria. Obviously, with the short sequences we have a bigger phylogenetic error that can either appear like a HGT, when it does not occur, or disguise a real HGT. Tables 1 and 2 (see Appendix) report the false positive and false negative (indicated in parentheses) detection rates obtained using as optimization criteria the RF distance and LS function, respectively. A false positive HGT is an incorrect transfer found by the algorithm and a false negative HGT is the right transfer that has not been detected. A false positive HGT will always occur if the gene tree inferred by NJ (see Step 4 above) is different from the true gene tree (see Step 3 above), but it can also take place when both trees are identical but a transfer going to the direction opposite to the correct HGT disguises it, leading to the same gene tree (see [20]).

False negative HGTs are mostly due to the error of inferring the gene tree, but can also happen when a transfer going to the opposite direction disguises the correct HGT. As defined, the false positive detection rate is always bigger or equal to the negative one. The analysis of Tables 1 and 2 shows that the false negative rate is almost as big as the false positive rate when the tests were conducted with large phylogenies (48 and 64 species) and short sequences (125 and 250 sites). The false negative rate was noticeably lower than the false positive one in the case of the large phylogenies and long sequences. Furthermore, we have measured the recovery rates for the HGT source, destination, and source and destination combined (i.e. the latter parameter corresponds to the detection rate depicted in Figs. 7 and 8). These tests were carried out under the Jukes and Cantor model of sequence evolution and

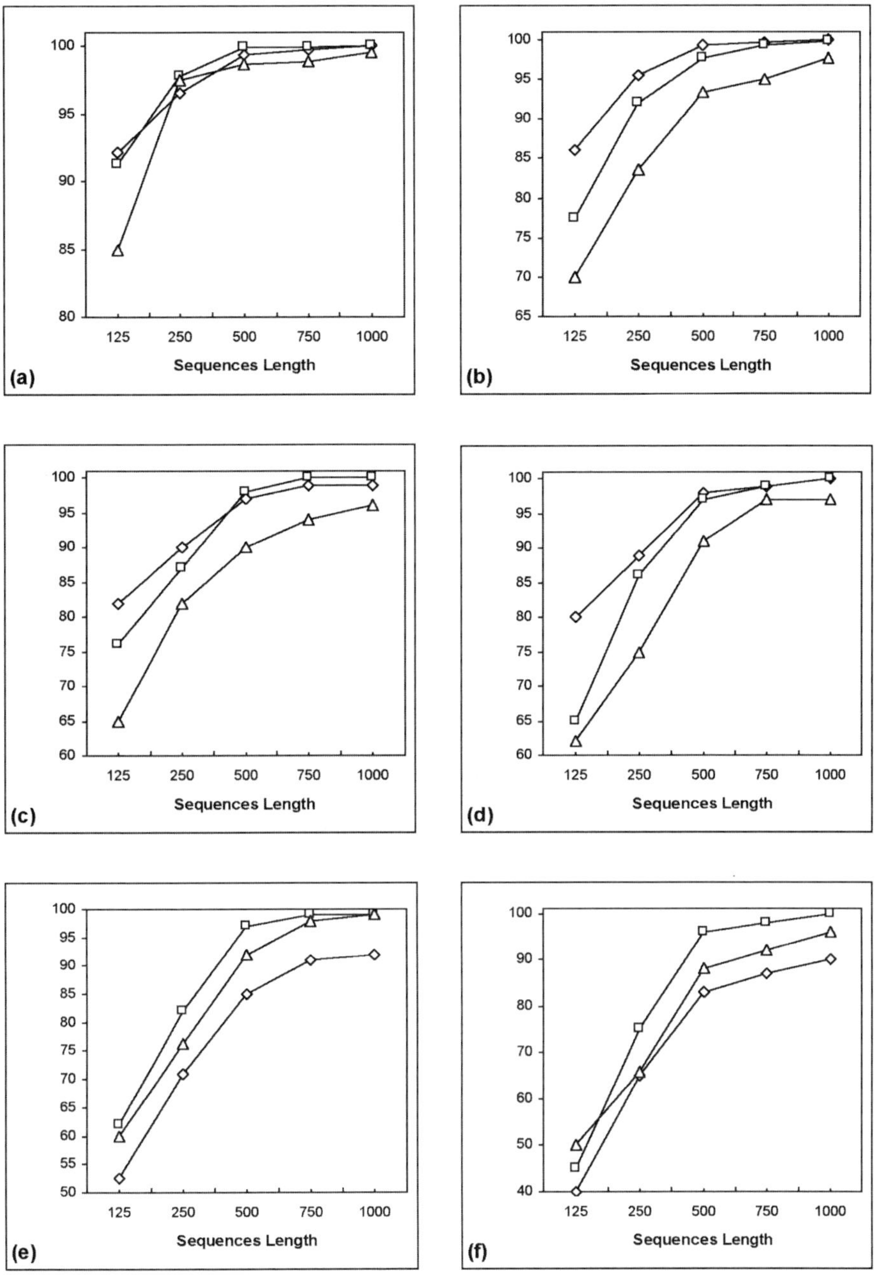

FIGURE 7. HGT detection rates obtained for random phylogenies with 8 to 64 leaves ($8-a$, $16-b$, $24-b$, $32-d$, $48-e$, $64-f$) using the RF topological distance for optimization. Jukes and Cantor ($\Diamond$), Kimura 2-parameter ($\Box$), and Jin–Nei Gamma (Δ) models were used for the tree generation.

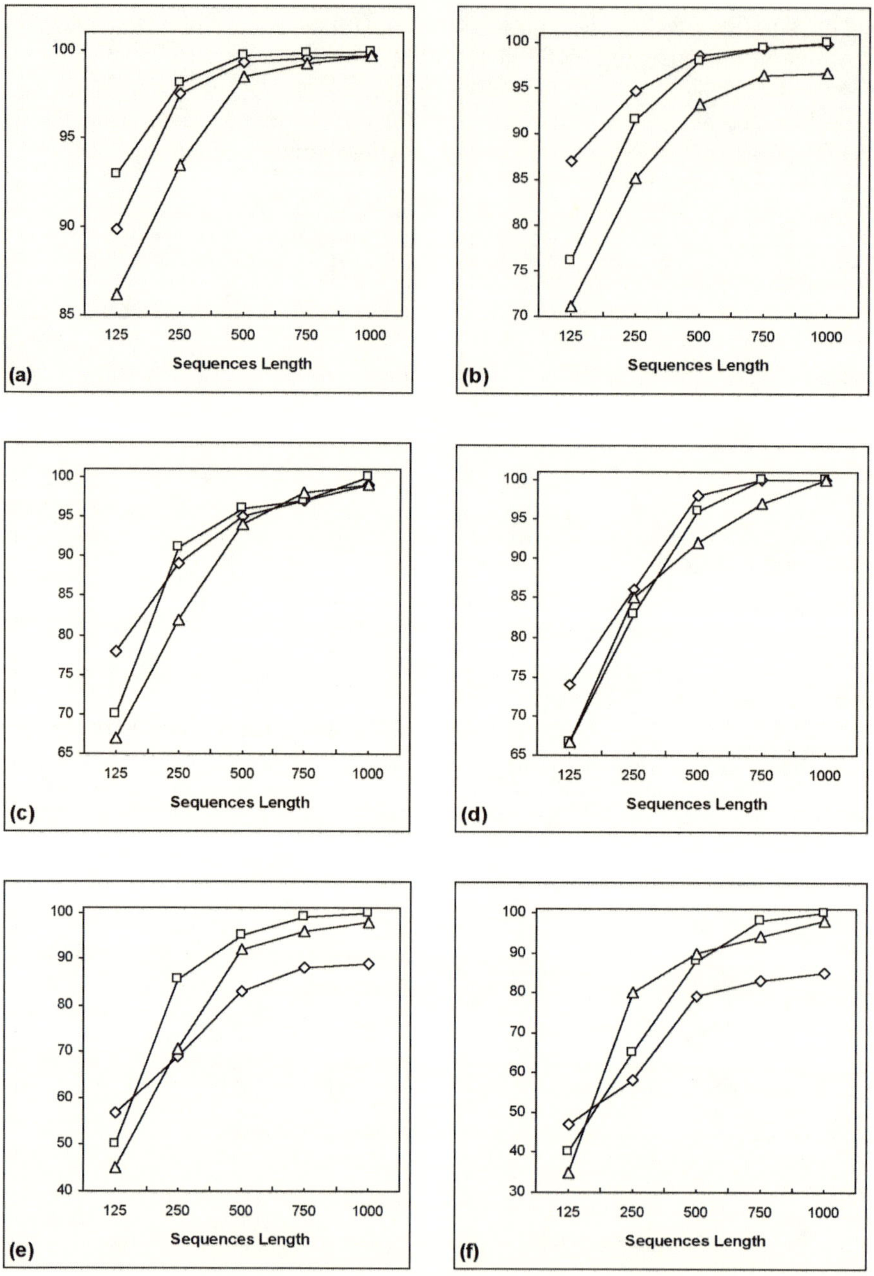

FIGURE 8. HGT detection rates obtained for random phylogenies with 8 to 64 leaves ($8-a$, $16-b$, $24-b$, $32-d$, $48-e$, $64-f$) using the LS function for optimization. Jukes and Cantor ($\Diamond$), Kimura 2-parameter ($\Box$), and Jin–Nei Gamma (Δ) models were used for the tree generation.

using the RF distance for the algorithmic optimization. Note that the transfer destinations were generally better detectable than their sources. The difference in the source-destination detection was more important for the short sequence. For example, for the sequences with 125 sites it varied, on average, from 6% (for 8 species) to 1% (for 64 species). However, for the longer sequences the source and destination rates were very similar.

Generally, the procedure based on the RF distance provided better results than that based on the LS function. Nevertheless, some noticeable exceptions (e.g. under the Kimura model for the phylogenies with 8 leaves or under the Jin–Nei model in the case of the short sequences) can be pointed out. The simulation study suggested that the accuracy of the transfer detection is highly dependable on the model of sequence evolution, number of considered species, and length of observed sequences.

Results and discussion. [Detecting horizontal transfers of the gene `rpl2e`] We first tested our algorithm on the phylogeny of 14 species of Archaea originally considered by Matte-Tailliez *et al* [24]. The latter authors discuss problems encountered when reconstructing some parts of the archaeal phylogeny, pointing out the evidence of HGT events perturbing the evolution of a number of considered genes. Matte-Tailliez *et al.* inferred the maximum likelihood tree (Figure 10, undirected lines) based on the concatenated 53 ribosomal proteins (7,175 positions) and compared it to the maximum likelihood phylogeny of the gene `rpl2e` (Figure 9) built for the same 14 organisms. The calculations of the best ML tree and its branch lengths for the 53 concatenated proteins were conducted using the PUZZLE program with Γ-law correction.

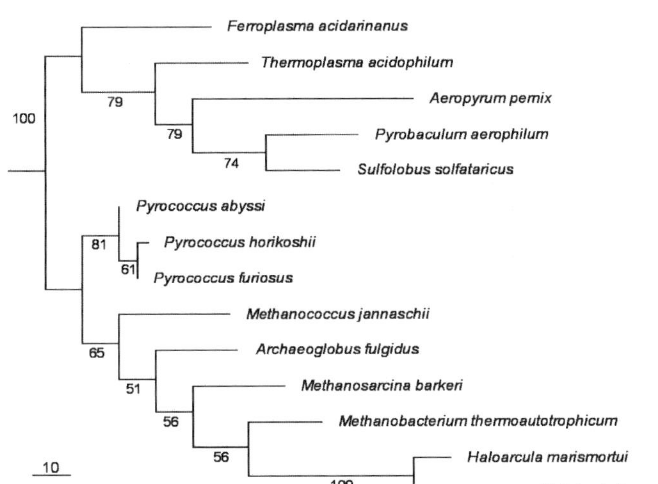

FIGURE 9. Maximum likelihood phylogenetic tree for the protein `rpl2e` (89 positions). Numbers close to branches are ML bootstrap scores obtained from the sampled protein sequences using the SeqBoot and Proml (JTT model) programs from the PHYLIP package (Felsenstein, 1989). Its topology is identical to the tree found by Matte-Taillez *et al* [24, Figure 3].

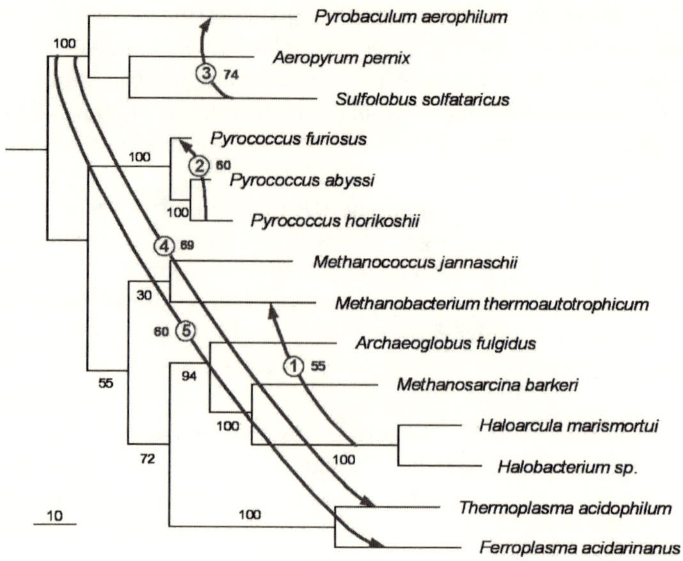

FIGURE 10. Species tree (Matte-Taillez *et al.* [24, Fig. 1a], with five reconciliation branches (denoted by arrows). Numbers close to branches are ML bootstrap scores computed by the RELL method upon 2,000 top-ranking trees using the *MOLPHY* program without correction for among-site variation. Numbers on HGT arrows indicate their order of appearance in the unique gene transfer scenario found by the HGT detection method. Bootstrap scores for transfers are indicated by numbers close to arrow circles. Arrows 4 and 5 depict the HGTs between the clades of *Thermoplasmatales* and *Crenarchaeota* also predicted by Matte-Taillez *et al* [24].

Given the topological incongruence of the obtained phylogenies, the authors hypothesized a few cases of lateral transfers of the gene `rpl2e`. More precisely, the case of the transfer between the clades of *Thermoplasmatales* (*Ferroplasma acidarmanus* and *Thermoplasma acidophilum*) and *Crenarchaeota* (*Aeropyrum pernix*, *Pyrobaculum aerophilum* and *Sulfolobus solfataricus*) was indicated as the most evident one.

In order to apply our method, we first reconstructed from the original sequences the topologies of the gene (Figure 9) and species trees (Figure 10, undirected lines). The computations were conducted in the framework of the complete gene transfer model, using the RF optimization and subtree constraint options (Figure 2). Five directed branches needed to reconcile the species and gene topologies have been found (Figure 10). The connection representing the transfer between the cluster of *Halobacterium sp.* and *Haloarcula marismortui* and the species *Methanobacterium thermoautotrophicum* was found in the first iteration. This transfer provided the biggest drop of the RF distance between the species and gene phylogenies; its bootstrap score is 55%.

In the second and third iterations, we found the reconciliation branches between the species *Pyrococcus horikoshii* and *Pyrococcus furiosus* and between *Sulfolobus solfataricus* and *Pyrobaculum aerophilum*. Both of these reconciliation branches link closely related species. Such kind of connections may be due to HGT as well as to local topological rearrangements necessary because of the tree reconstruction artifacts (e.g. attraction of long branches, unequal evolutionary rates, etc). The transfer branches 4 and 5 linking the cluster of *Crenarchaeota* to the species *Thermoplasma acidophilum* and *Ferroplasma acidarmanus* can be interpreted as HGT events that might have taken place between *Thermoplasmatales* and *Crenarchaeota*.

In the second and third iterations, we found the reconciliation branches between the species *Pyrococcus horikoshii* and *Pyrococcus furiosus* and between *Sulfolobus solfataricus* and *Pyrobaculum aerophilum*. Both of these reconciliation branches link closely related species. Such kind of connections may be due to HGT as well as to local topological rearrangements necessary because of the tree reconstruction artifacts (e.g. attraction of long branches, unequal evolutionary rates, etc). The transfer branches 4 and 5 linking the cluster of *Crenarchaeota* to the species *Thermoplasma acidophilum* and *Ferroplasma acidarmanus* can be interpreted as HGT events that might have taken place between *Thermoplasmatales* and *Crenarchaeota*.

Note, that HGT between these two groups was also predicted by Matte-Taillez *et al* [24]. In fact, the transfers 4 and 5 could consist of a unique transfer between the clades of *Thermoplasmatales* and *Crenarchaeota* that was separated into two transfers by our method due to the application of the subtree constraint (Figure 2) and the presence of the tree reconstruction artifacts. Figure 11 illustrates the evolution of the newly formed *Thermoplasmatales*-*Crenarchaeota* clade involving

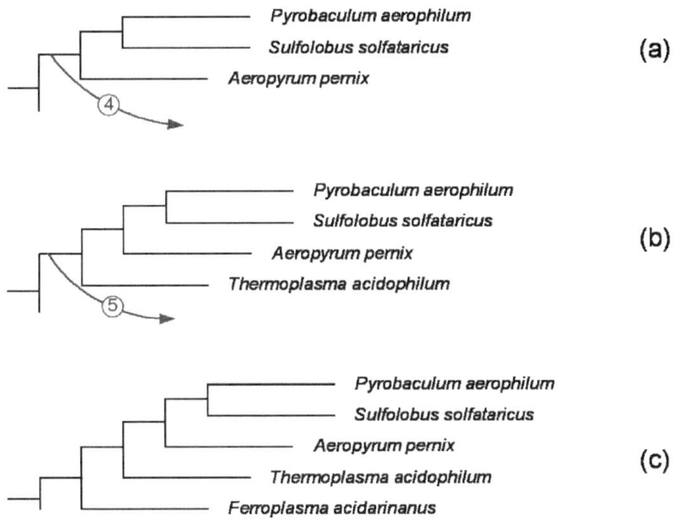

FIGURE 11. Changes in the *Crenarchaeota*-*Thermoplasmatales* cluster occurring after the addition of HGT branches 4 and 5. (a) This cluster after the transfer 3; the species *Thermoplasma acidophilum* joins the *Crenarchaeota* cluster. (b) This cluster after the transfer 4; the species *Ferroplasma acidarmanus* is added to the clade comprising three *Crenarchaeota* and *Thermoplasma acidophilum*. (c) This cluster after the transfer 5.

the HGTs 4 and 5. The usage of the LS criterion instead of RF leads to the solution consisting of 6 HGTs including all transfers from Figure 10 except the HGT number 2 that goes in the opposite direction. Note that a new reconciliation branch found with LS brings the species *Methanococcus jannaschii* to the cluster of 4 species including *Archaeoglobus fulgidus*. This reconciliation branch turns out to be useless and have a low bootstrap score of 14%.

3. Conlusion

We presented two polynomial-time algorithms for detecting horizontal gene transfer events. We considered the complete and partial gene transfer models, implying at each step, either the transformation of a species phylogeny into another tree or its transformation into a network structure. The algorithm for inferring complete gene transfers exploits the discrepancies between the species and gene phylogenies either to map the gene tree into the species tree by least-squares or to compute a topological distance between them and then estimate the possibility of a HGT event between each pair of branches of the species phylogeny. The models based on the optimization of the least-squares function and the Robinson and Foulds topological distance were introduced.

Inferred HGTs should be carefully analyzed using all available information about the data in hand in order to select the transfers that will be represented as a final solution. Each gene transfer branch added to the species phylogeny aids to resolve a conflict between it and the gene tree (i.e. helps to reconcile the species and gene phylogenies). A bootstrap validation procedure allowing one to assess the reliability of a specific gene transfer or whole gene transfer scenario was proposed. A comprehensive Monte Carlo study was carried out to test the ability of the new method to recover correct HGTs. It provided very encouraging results especially when the Robinson and Foulds distance was used as an optimization criterion. The example of the evolution of the gene `rpl2e` was considered in the application section. More simulation work is required to investigate the properties of the algorithm intended to infer partial gene transfers.

As any method of phylogenetic inferring, the new HGT detection method is subject to a number of artifacts which generally affect phylogenetic analysis; the main of them being: attraction of long branches, unequal evolutionary rates, and situations when the occurrence of some HGT events almost coincides with speciation events located closely to the recipient species. It is important to investigate in greater details the impact of these artifacts on the HGT detection technique introduced in this article. It would be also interesting to extend the presented model to the case, where the gene and species trees have different numbers of taxa; this situation can take place when some species have more than one copy of the gene under consideration.

The software implementing the new algorithms for detecting complete and partial horizontal gene transfers is freely available at the following URL address: `http://www.info2.uqam.ca/~boca05/software/` (this is a consol version running on the Unix and Windows platforms; it is distributed along with its `C++` source code). A graphical version of this program has been also implemented and included in the *T*-Rex web server [21] at the following URL: `http://www.trex.uqam.ca`.

References

1. J.-P. Barthélemy and A. Guénoche, *Les arbres et les représentations des proximités*, Paris, Masson, 1988.
2. ———, *Trees and proximity representations*, New York, Wiley, 1991.
3. A. Boc and V. Makarenkov, *New efficient algorithm for detection of horizontal gene transfer events*, 3rd Workshop on Algorithms in Bioinformatics (Budapest, 2003) (G. Benson and R. Page, eds.) WABI, Lecture Notes in Comput. Sci., Springer Verlag, 2003, pp. 190–201.
4. M. A. Charleston, *Jungle: a new solution to the host/parasite phylogeny reconciliation problem*, Math. Biosci. **149** (1998), 191–223.
5. E. Denamur, G. Lecointre, and P. Darlu *et al*, *Evolutionary implications of the frequent horizontal transfer of mismatch repair genes*, Cell. **103** (2000), 711–721.
6. W. F. Doolittle, *Phylogenetic classification and the universal tree*, Science **284** (1999), 2124–2129.
7. J. Felsenstein, *Confidence limits on phylogenies: an approach using the bootstrap*, Evolution **39** (1985), 738–791.
8. ———, PHYLIP—*Phylogeny inference package*, v. 3.2, Cladistics **5** (1989), 164–166.
9. S. Guindon and O. Gascuel, *Efficient biased estimation of evolutionary distances when substitution rates vary across sites*, Mol. Biol. Evol. **19** (2002), 534–543.
10. M. Hallett and J. Lagergren, *Efficient algorithms for lateral gene transfer problems*, RECOMB (N. El-Mabrouk, T. Lengauer, and D. Sankoff, eds.), ACM, New York, 2001, pp. 149–156.
11. M. Hallett, J. Lagergren, and A. Tofigh, *Simultaneous identification of duplications and lateral transfers*, RECOMB (P. E.Bourne and D. Gusfield, eds.), ACM, San Diego, 2004, pp. 347–356.
12. J. Hein, T. Jiang, L. Wang, and K. Zhang, *On the complexity of comparing evolutionary trees*, Discrete Appl. Math. **71** (1996), 153–169.
13. L. Jin and M. Nei, *Limitations of the evolutionary parsimony method of phylogenetic analysis*, Mol. Biol. Evol. **7** (1990), 82–102.
14. T. H. Jukes and C. Cantor, *Mammalian protein metabolism*, Evolution of Protein Molecules (H. N. Munro, ed.), New York Academic Press, 1969, pp. 21–132.
15. M. A Kimura, *Simple method for estimating evolutionary rate of base substitutions through comparative studies of nucleotide sequences*, J. Mol. Evol. **16** (1980), 111–120.
16. M. Kuhner and J. Felsenstein, *A simulation comparison of phylogeny algorithms under equal and unequal evolutionary rates*, Mol. Biol. Evol. **11** (1994), 459–468.
17. J. G. Lawrence and H. Ochman, *Amelioration of bacterial genomes: rates of change and exchange*, J. Mol. Evol. **44** (1997), 383–397.
18. P. Legendre (guest ed.), *Special section on reticulate evolution*, J. Classification **17** (2000), 153–195.
19. P. Legendre and V. Makarenkov, *Reconstruction of biogeographic and evolutionary networks using reticulograms*, Syst. Biol. **51** (2002), 199–216.
20. W. P. Maddison, *Gene trees in species trees*, Syst. Biol. **46** (1997), 523–536.
21. V. Makarenkov, T-Rex: *reconstructing and visualizing phylogenetic trees and reticulation networks*, Bioinformatics **17** (2001), 664–668.
22. V. Makarenkov, A. Boc, C. F. Delwiche, A. B. Diallo, and H. Philippe, *New efficient algorithm for modeling partial and complete gene transfer scenarios*, Data Science and Classification (V. Batagelj, H.-H. Bock, A. Ferligoj, and A. Ziberna, eds.), IFCS 2006, Studies in Classification, Data Analysis, and Knowledge Organization, part VIII, Springer Verlag, 2006, pp. 341–349.
23. V. Makarenkov and B. Leclerc, *An algorithm for the fitting of a phylogenetic tree according to a weighted least-squares criterion*, J. Classification **16** (1999), 3–26.
24. O. Matte-Tailliez, C. Brochier, P. Forterre, and H. Philippe, *Archaeal phylogeny based on ribosomal proteins*, Mol. Biol. Evol. **19** (2002), 631–639.
25. B. Mirkin, T. I. Fenner, M. Galperin, and E. Koonin, *Algorithms for computing parsimonious evolutionary scenarios for genome evolution, the last universal common ancestor and dominance of horizontal gene transfer in the evolution of prokaryotes*, BMC Evol. Biol. **3** (2003); 2.
26. B. Mirkin, *Mapping gene family data onto evolutionary trees*, Comptes rendus des 11es Rencontres de la Sociéte Francophone de Classification, (M. Chavent, O. Dordan, C. Lacomblez, M. Langlais, and B. Patouille, eds.), University of Bordeaux, pp. 61-68.

27. B. M. E. Moret, L. Nakhleh, T. Warnow, C. R. Linder, A. Tholse, A. Padolina, J. Sun, and R. Timme, *Phylogenetic networks: modeling, reconstructibility, and accuracy*, IEEE/ACM Trans. on Comput. Biol. and Bioinf. **1** (2004), 13–23.
28. L. Nakhleh, D. Ruths, and H. Innan,*Gene trees, species trees, and species networks*, Meta-analysis and Combining Information in Genetics (R. Guerra and D. Allison, eds.), 2005, pp. 1–27.
29. R. D. M. Page, *Maps between trees and cladistic analysis of historical associations among genes, organisms, and areas*, Systematic Biol. **43** (1994), 58–77.
30. R. D. M. Page and M. A. Charleston, *From gene to organismal phylogeny: reconciled trees*, Bioinformatics **14** (1998), 819–820.
31. ———, *Trees within trees: phylogeny and historical associations*, Trends Ecol. Evol. **13** (1998b), 356–359.
32. A. Rambaut and N. C. Grassly, *Seq-Gen: an application for the Monte Carlo simulation of* DNA *sequence evolution along phylogenetic trees*, Comput. Appl. Biosci. **13** (1996), 235–238.
33. D. R. Robinson and L. R. Foulds, *Comparison of phylogenetic trees*, Math. Biosciences **53** (1981), 131–147.
34. N. Saitou and M. Nei, *The neighbor-joining method: a new method for reconstructing phylogenetic trees*, Mol. Biol. Evol. **4** (1987), 406–425.
35. A. von Haeseler and G. A. Churchill, *Network models for sequence evolution*, J. Mol. Evol. **37** (1993), 77–85

Appendix A

This Appendix includes the results of the tests described in the section Simulation Study. The results reported in Tables 1 and 2 correspond to the graphics represented in Figures 7 (optimization using the RF distance) and 8 (optimization using the LS function). They were obtained from simulations carried out for random binary phylogenies with 8, 16, 24, 32, 48, and 64 leaves, whereas the sequence length varied from 125 to 1000 sites. Note that the sum of the HGT detection rate shown in Figures 7 and 8 and of the false negative detection rate reported in Tables 1 and 2 is always 100%.

Département d'Informatique, Université du Québec à Montréal, C.P. 8888, Succ. Centre-ville, Montréal (Québec), H3C 3P8, Canada.

E-mail address, V. Makarenkov: `makarenkov.vladimir@uqam.ca`

E-mail address, A. Boc: `boc.alix@courrier.uqam.ca`

E-mail address, A. B. Diallo: `diallo.alpha_boubacar@courrier.uqam.ca`

McGill Centre for Bioinformatics and School of Computer Science, McGill University, 3775 University Street, Montréal (Québec), H3A 2B4, Canada.

E-mail address: `banire@mcb.mcgill.ca`

TABLE 1. False positive and false negative (in parentheses) detection rates obtained for random phylogenies with 8 to 64 leaves using the RF distance as an optimization criterion. A false positive HGT is an incorrect transfer found by the algorithm and a false negative HGT is the right transfer that has not been found. For each sequence length, the simulations were carried out for 500 random phylogenies with 8 and 16 leaves and 100 random phylogenies with 24 to 64 leaves.

RF rates (in %)			Sequence length				
			125	250	500	750	1000
Species number	8	Jukes-Cantor	14.9(7.8)	5.9(3.5)	1.1(0.7)	0.3(0.3)	0.0(0.0)
		Kimura	12.9(8.7)	3.3(2.2)	0.2(0.1)	0.1(0.1)	0.0(0.0)
		Jin-Nei	20.1(15.0)	3.9(2.5)	1.6(1.3)	1.1(1.1)	0.5(0.5)
	16	Jukes-Cantor	25.7(14.0)	7.1(4.5)	1.2(0.7)	0.4(0.3)	0.0(0.0)
		Kimura	35.1(22.5)	11.9(7.9)	3.2(2.3)	0.6(0.6)	0.1(0.0)
		Jin-Nei	43.0(30.0)	22.5(16.5)	7.6(6.6)	5.3(4.9)	2.3(2.3)
	24	Jukes-Cantor	36(18)	15(10)	4(3)	1(1)	1(1)
		Kimura	43(24)	24(13)	4(2)	2(0)	0(0)
		Jin-Nei	55(35)	33(18)	19(10)	9(6)	5(4)
	32	Jukes-Cantor	37(20)	29(11)	4(2)	1(1)	1(0)
		Kimura	60(35)	31(14)	8(3)	3(1)	2(0)
		Jin-Nei	70(38)	47(25)	16(9)	8(3)	8(3)
	48	Jukes-Cantor	65(48)	49(29)	28(15)	1(1)	1(0)
		Kimura	55(38)	46(18)	9(3)	3(1)	2(0)
		Jin-Nei	70(40)	58(24)	19(8)	8(3)	8(3)
	64	Jukes-Cantor	70(60)	45(35)	27(17)	23(13)	20(10)
		Kimura	65(55)	35(25)	14(4)	12(2)	10(0)
		Jin-Nei	60(50)	44(34)	22(12)	18(8)	14(4)

TABLE 2. False positive and false negative (in parentheses) detection rates obtained for random phylogenies with 8 to 64 leaves using the LS function as an optimization criterion. A false positive HGT is an incorrect transfer found by the algorithm and a false negative HGT is the right transfer that has not been found. For each sequence length, the simulations were carried out for 500 random phylogenies with 8 and 16 leaves and 100 random phylogenies with 24 to 64 leaves.

RF rates (in %)			Sequence length				
			125	250	500	750	1000
Species number	8	Jukes-Cantor	17.2(10.1)	5.0(2.5)	0.8(0.7)	0.8(0.5)	0.3(0.3)
		Kimura	10.8(7.0)	2.8(1.9)	0.3(0.3)	0.2(0.2)	0.1(0.1)
		Jin-Nei	18.6(13.8)	7.8(6.5)	1.7(1.5)	0.9(0.8)	0.5(0.3)
	16	Jukes-Cantor	25.5(13.0)	7.6(5.3)	2.2(1.4)	0.8(0.5)	0.1(0.1)
		Kimura	37.6(23.8)	11.9(8.4)	2.3(2.0)	0.6(0.6)	0.0(0.0)
		Jin-Nei	40.9(28.8)	20.9(14.8)	8.1(6.7)	3.8(3.6)	3.3(3.3)
	24	Jukes-Cantor	43(22)	13(11)	5(5)	3(3)	1(1)
		Kimura	59(30)	26(9)	7(4)	4(3)	1(0)
		Jin-Nei	67(33)	26(18)	12(6)	6(2)	3(1)
	32	Jukes-Cantor	47(26)	21(14)	5(2)	0(0)	0(0)
		Kimura	56(33)	31(17)	9(4)	0(0)	0(0)
		Jin-Nei	50(33)	31(15)	12(8)	11(3)	4(0)
	48	Jukes-Cantor	53(43)	38(31)	33(7)	22(12)	19(11)
		Kimura	60(50)	34(14)	16(5)	5(1)	2(0)
		Jin-Nei	65(55)	50(29)	25(8)	12(4)	10(3)
	64	Jukes-Cantor	63(53)	52(42)	41(21)	27(17)	25(15)
		Kimura	70(60)	45(35)	22(12)	15(2)	10(0)
		Jin-Nei	75(65)	40(20)	20(10)	16(6)	12(2)

Centre de Recherches Mathématiques
CRM Proceedings and Lecture Notes
Volume **45**, 2008

Nonlinear Knowledge in Kernel Machines

Olvi L. Mangasarian and Edward W. Wild

ABSTRACT. We give a unified presentation of recent work in applying prior knowledge to nonlinear kernel approximation [26] and nonlinear kernel classification [25]. In both approaches, prior knowledge over general nonlinear sets is incorporated into nonlinear kernel approximation or classification problems as linear constraints in a linear program. The key tool in this incorporation is a theorem of the alternative for convex functions that converts nonlinear prior knowledge implications into linear inequalities without the need to kernelize these implications. Effectiveness of the proposed approximation formulation is demonstrated on two synthetic examples as well as an important lymph node metastasis prediction problem arising in breast cancer prognosis. Effectiveness of the proposed classification formulation is demonstrated on two publicly available datasets, including a breast cancer prognosis dataset. Upon the introduction of prior knowledge, all these problems exhibit marked improvements over nonlinear kernel approximation and classification approaches that do not utilize such knowledge.

1. Introduction

Prior knowledge has been used effectively in improving classification both for linear [7] and nonlinear [6] kernel classifiers as well as for nonlinear kernel approximation [23, 21]. In all these applications prior knowledge was converted to linear inequalities that were imposed on a linear program. The linear program generated a linear or nonlinear classifier, or a linear or nonlinear function approximation, all of which were more accurate than the corresponding results that did not utilize prior knowledge. However, whenever a nonlinear kernel was utilized in these applications, kernelization of the prior knowledge was not a transparent procedure that could be easily related to the original sets over which prior knowledge was given. In contrast, prior knowledge over arbitrary general sets has been recently incorporated *without* kernelization of the prior knowledge sets into nonlinear kernel approximation [26] and nonlinear kernel classification [25]. Here, we present a unified formulation of both approaches, which is made possible through the use of a fundamental theorem

2000 *Mathematics Subject Classification.* Primary: 62-07; Secondary: 90C90.
The research described in this Data Mining Institute Report 06-06, November 2006, was supported by National Science Foundation Grants CCR-0138308 and IIS-0511905, the Microsoft Corporation and ExxonMobil.
This is the final form of the paper.

of the alternative for convex functions that we describe in Section 2 of the paper. An interesting, novel approach to knowledge-based support vector machines that modifies the hypothesis space rather than the optimization problem is given in [15]. In another recent approach, prior knowledge is incorporated by adding additional points labeled based on the prior knowledge to the dataset [24].

In Section 2 we describe a general formulation for incorporating prior knowledge into kernel machines. Section 3 shows how this formulation can be specialized to incorporate prior knowledge into nonlinear kernel approximation, while Section 4 shows how to specialize the formulation to incorporate prior knowledge into nonlinear kernel classification. Numerical examples from show prior knowledge can improve both approximation and classification are presented in Section 5. Section 6 concludes the paper.

We describe our notation now. All vectors will be column vectors unless transposed to a row vector by a prime $'$. The scalar (inner) product of two vectors x and y in the n-dimensional real space R^n will be denoted by $x'y$. For $x \in R^n$, $\|x\|_1$ denotes the 1-norm: $\sum_{i=1}^n |x_i|)$ while $\|x\|$ denotes the 2-norm: $\sum_{i=1}^n (x_i)^2)^{\frac{1}{2}}$ and x_+ denotes the vector $\max\{x, 0\}$. The notation $A \in R^{m\times n}$ will signify a real $m \times n$ matrix. For such a matrix, A' will denote the transpose of A, A_i will denote the ith row of A and $A_{\cdot j}$ the jth column of A. A vector of ones in a real space of arbitrary dimension will be denoted by e. Thus for $e \in R^m$ and $y \in R^m$ the notation $e'y$ will denote the sum of the components of y. A vector of zeros in a real space of arbitrary dimension will be denoted by 0. For $A \in R^{m\times n}$ and $B \in R^{n\times k}$, a *kernel* $K(A, B)$ maps $R^{m\times n} \times R^{n\times k}$ into $R^{m\times k}$. In particular, if x and y are column vectors in R^n then, $K(x', y)$ is a real number, $K(x', A')$ is a row vector in R^m and $K(A, A')$ is an $m \times m$ matrix. We shall make no assumptions whatsoever on our kernels other than symmetry, that is $K(x', y)' = K(y', x)$, and in particular we shall not assume or make use of Mercer's positive definiteness condition [28, 27, 4]. The base of the natural logarithm will be denoted by ε. A frequently used kernel in nonlinear classification is the Gaussian kernel [28, 3, 18] whose ijth element, $i = 1, \ldots, m$, $j = 1, \ldots, k$, is given by: $\big(K(A, B)\big)_{ij} = \varepsilon^{-\mu\|A_i{}' - B_{\cdot j}\|^2}$, where $A \in R^{m\times n}$, $B \in R^{n\times k}$ and μ is a positive constant. The abbreviation "s.t." stands for "subject to".

2. General Formulation

We wish to impart knowledge to a function learned on a dataset in R^n represented by the m rows of the matrix $A \in R^{m\times n}$ with associated labels represented by the m-dimensional vector d. For approximation problems, d will be real valued, that is $d \in R^m$. For classification problems, d will have values $+1$ or -1 according to the class of each example, that is $d \in \{-1, +1\}^m$. The learned function f from R^n to R is defined as follows:

$$f(x) = K(x', B')u - \gamma \tag{2.1}$$

where $B \in R^{k\times n}$ is an arbitrary basis matrix and $K(x', B') : R^{1\times n} \times R^{n\times k} \longrightarrow R^{1\times k}$ is an arbitrary kernel function. In an approximation setting, we will use $f(x)$ directly to obtain a predicted value at $x \in R^n$, while in a classification setting we will use the value of the sign of $f(x)$ to obtain the predicted class of x. In general, the matrix B is set equal to A [18]. However, in reduced support vector machines [12, 10] $B = \bar{A}$, where $\bar{A}$ is a submatrix of A whose rows are a small subset of the rows of A. In fact, B can be an arbitrary matrix in $R^{k\times n}$. The variables $u \in R^k$ and $\gamma \in R$ are parameters determined by the following optimization problem:

$$\min_{u,\gamma} \nu L(u,\gamma,A,d) + \|u\|_1, \tag{2.2}$$

where L is an arbitrary loss function, and ν is a positive parameter that controls the weight in the optimization problem of data fitting versus complexity reduction. In approximation problems, the loss function L will measure the difference between the actual values of the given data d and the predicted values given by f. For classification problems, the loss function L will be related to misclassification rate. The 1-norm of u is chosen in order to promote solution sparsity [2, 30]. In Section 3 we will demonstrate that the use of the 1-norm of the error between the predicted and actual values for L in an approximation setting will result in a linear program, while Section 4 shows that use of the hinge loss for L results in a linear program for classification problems.

We now impose prior knowledge on the construction of our learned function $f(x) = K(x', B')u - \gamma$ through the following implication:

$$g(x) \le 0 \Longrightarrow K(x',B')u - \gamma \ge \phi(x), \quad \forall x \in \Gamma. \tag{2.3}$$

Here, $g(x) : \Gamma \subset R^n \longrightarrow R^p$ is a p-dimensional function defined on a subset Γ of R^n that determines the region in the input space where prior knowledge requires that $K(x', B')u - \gamma$ be larger than some known function $\phi(x) : \Gamma \subset R^n \longrightarrow R$. In previous work [23, 5, 6], prior knowledge implications such as (2.3) could not be handled directly as we shall do here by using Theorem 2.1 below. Instead, the inequality $g(x) \le 0$ was kernelized. This led to an inequality not easily related to the original constraint $g(x) \le 0$. In addition, in [23, 5, 6] could only handle linear $g(x)$ and $\phi(x)$. The implication (2.3) can be written in the following equivalent logical form:

$$g(x) \le 0,\ K(x',B')u - \gamma - \phi(x) < 0, \quad \text{has no solution } x \in \Gamma. \tag{2.4}$$

It is precisely implication (2.3) that we shall convert to a system which is linear in the parameters of f, (u, γ), by means of the following theorem of the alternative for convex functions. The alternatives here are that either the negation of (2.4) holds, or (2.5) below holds, but *never both.*

Theorem 2.1 (Prior knowledge as system of linear inequalities). *For a fixed $u \in R^k$, $\gamma \in R$, the following are equivalent:*

(i) *The implication* (2.3) *or equivalently* (2.4) *holds.*

(ii) *There exists $v \in R^p$, $v \ge 0$ such that:*

$$K(x',B')u - \gamma - \phi(x) + v'g(x) \ge 0, \quad \forall x \in \Gamma, \tag{2.5}$$

where it is assumed for the implication (i)$\Longrightarrow$(ii) ***only,*** *that $g(x)$ and $K(x', B')$ are convex on Γ, $\phi(x)$ is concave on Γ, Γ is a convex subset of R^n, $u \ge 0$ and that $g(x) < 0$ for some $x \in \Gamma$.*

PROOF. (i)$\Longrightarrow$(ii)]: This follows from [17, Corollary 4.2.2], the fact that the functions $g(x)$ and $K(x', B')u - \gamma - \phi(x)$ of (2.4) are convex on Γ and that $g(x) < 0$ for some $x \in \Gamma$.

(i)$\Longleftarrow$(ii): If (i) did not hold then there there exists an $x \in \Gamma$ such that $g(x) \le 0$, $K(x', B')u - \gamma - \phi(x) < 0$, which would result in the contradiction:

$$0 > K(x',B')u - \gamma - \phi(x) + v'g(x) \ge 0. \tag{2.6}$$

□

We note immediately that in the proposed application of converting prior knowledge to linear inequalities in the parameters (u, γ) all we need is the implication (i)$\Longleftarrow$(ii), which **requires no assumptions whatsoever** on the functions $g(x)$, $K(x', B')$, $\phi(x)$ or on the parameter u. We further note that the implication (2.3) can represent fairly complex knowledge such as $K(x', B')u - \gamma$ being equal to any desired function whenever $g(x) \leq 0$.

We note that Theorem 2.1 can also be invoked on the following prior knowledge implication, which is similar to (2.3):

$$h(x) \leq 0 \Longrightarrow K(x', B')u - \gamma \leq -\psi(x), \quad \forall x \in \Lambda. \tag{2.7}$$

We now incorporate the prior knowledge contained in implications (2.3) and (2.7) into the optimization problem (2.2) as follows:

$$\begin{aligned}
\min_{u,\gamma,z_1,\ldots,z_\ell,q_1,\ldots,q_t} \quad & \nu L(u, \gamma, A, d) + \|u\|_1 + \sigma \left(\sum_{i=1}^{\ell} z_i + \sum_{j=1}^{t} q_j \right) \\
\text{s.t.} \quad & K(x^{i'}, B')u - \gamma - \phi(x^i) + v'g(x^i) + z_i \geq 0, \\
& z_i \geq 0, \; i = 1, \ldots, \ell, \\
& v \geq 0, \\
& -K(x^{j'}, B')u + \gamma - \psi(x^j) + r'h(x^j) + q_j \geq 0, \\
& q_j \geq 0, \; j = 1, \ldots, t, \\
& r \geq 0.
\end{aligned} \tag{2.8}$$

We note that we have discretized the variable $x \in \Gamma$ and $x \in \Lambda$ in the constraints above to the finite meshes of points $\{x^1, x^2, \ldots, x^\ell\}$ and $\{x^1, x^2, \ldots, x^t\}$ in order to convert a semi-infinite program [8] with an infinite number of constraints into a finite mathematical program. We have also added nonnegative slack variables $z_1, z_2, \ldots, z_\ell$ and $q_1, q_2, \ldots, q_t$ to allow small deviations in the prior knowledge. The sum of these nonnegative slack variables for the prior knowledge inequalities is minimized with weight $\sigma > 0$ in the objective function in order to drive them to zero to the extent possible. Thus, the magnitude of the parameter σ enforces prior knowledge while the magnitude of ν enforces data fitting.

We turn now to specific formulations for knowledge-based kernel approximation and knowledge-based kernel classification.

3. Knowledge-Based Kernel Approximation

In the approximation setting, we wish to approximate a function, f, given exact or approximate function values of a dataset of points represented by the rows of the matrix $A \in R^{m \times n}$. Thus, for each point A_i we are given an exact or inexact value of f, denoted by a real number d_i, $i = 1, \ldots, m$. We therefore desire the parameters (u, γ) of (2.1) to be determined such that:

$$K(A, B')u - e\gamma - d \approx 0. \tag{3.1}$$

An appropriate choice of L to enforce the above condition is:

$$L(u, \gamma, A, d) = \|K(A, B')u - e\gamma - d\|_1. \tag{3.2}$$

This loss function is the sum of the absolute values of the differences between the predicted values $f(A_i)$ and the given values d_i, $i = 1, \ldots, m$. In order to incorporate this loss function into the optimization problem (2.8) we introduce a vector $s \in R^m$ defined by:

$$-s \leq K(A, B')u - e\gamma - d \leq s. \tag{3.3}$$

We can then incorporate the loss defined by (3.2) into (2.8) as follows:

$$\begin{aligned}
&\min_{u,\gamma,z_1,\ldots,z_\ell,q_1,\ldots,q_t,s} \nu\|s\|_1 + \|u\|_1 + \sigma\left(\sum_{i=1}^{\ell} z_i + \sum_{j=1}^{t} q_j\right) \\
&\text{s.t. } K(x^{i'}, B')u - \gamma - \phi(x^i) + v'g(x^i) + z_i \geq 0, \\
&\quad z_i \geq 0,\ i = 1, \ldots, \ell, \\
&\quad v \geq 0, \\
&\quad - K(x^{j'}, B')u + \gamma - \psi(x^j) + r'h(x^j) + q_j \geq 0, \\
&\quad q_j \geq 0,\ j = 1, \ldots, t, \\
&\quad r \geq 0, \\
&\quad - s \leq K(A, B')u - e\gamma - d \leq s.
\end{aligned} \tag{3.4}$$

Note that at the solution of (3.4), $\|s\|_1 = \|K(A, B')u - e\gamma - d\|_1$. Note further that (3.4) is the same as [26, Equation 10], except that here we have explicitly included implication (2.7) in addition to (2.3).

4. Knowledge-Based Kernel Classification

The classification problem consists of classifying the points represented by the rows of the matrix $A \in R^{m\times n}$ into the two classes $+1$ and -1 according to their labels given as $d \in \{-1, +1\}^m$. Thus, for each point A_i we are given its label $d_i \in \{-1, +1\}$, $i = 1, \ldots, m$ and we seek to find a function f of the form in (2.1) that satisfies, to the extent possible, the separation condition:

$$D(K(A, B')u - e\gamma) \geq 0, \tag{4.1}$$

where $D \in R^{m\times m}$ is the diagonal matrix with diagonal d. Note that this condition is satisfied if and only if d_i and $f(A_i)$ both have the same sign. Support vector machines attempt to impose a stronger condition, $D(K(A, B')u - e\gamma) \geq e$, using the hinge loss:

$$L(u, \gamma, A, d) = \left\|\left(e - D(K(A, B')u - e\gamma)\right)_+\right\|_1. \tag{4.2}$$

Note that the hinge loss involves the plus function, $(x)_+ = max\{x, 0\}$, introduced in Section 1. This loss function can be added to the optimization problem (2.8) by introducing a nonnegative slack variable $s \in R^m$ as follows:

$$D(K(A, B')u - e\gamma) + s \geq e, s \geq 0. \tag{4.3}$$

The loss (4.2) is incorporated into the optimization problem (2.8) as follows:

$$
\begin{aligned}
&\min_{u,\gamma,z_1,\ldots,z_\ell,q_1,\ldots,q_t,s} \nu\|s\|_1 + \|u\|_1 + \sigma\left(\sum_{i=1}^{\ell} z_i + \sum_{j=1}^{t} q_j\right)\\
&\text{s.t. } K(x^{i'}, B')u - \gamma - \phi(x^i) + v'g(x^i) + z_i \geq 0,\\
&\quad z_i \geq 0,\ i = 1,\ldots,\ell,\\
&\quad v \geq 0,\\
&\quad -K(x^{j'}, B')u + \gamma - \psi(x^j) + r'h(x^j) + q_j \geq 0,\\
&\quad q_j \geq 0,\ j = 1,\ldots,t,\\
&\quad r \geq 0,\\
&\quad D(K(A, B')u - e\gamma) + s \geq 0,\\
&\quad s \geq 0.
\end{aligned} \tag{4.4}
$$

We note that at the solution, $\|s\|_1 = \left\|\left(e - D(K(A,B')u - e\gamma)\right)_+\right\|_1$. Note further that (4.4) is the same as the optimization problem in [25] with $\phi(x) = \psi(x) = \alpha$, $\forall x$.

5. Numerical Experience

The effectiveness of our proposed formulation has been illustrated on three approximation tasks [26] and two classification tasks [25]. We describe these experiments and their results here. It is important to point out that the present formulation is very different in nature from that presented in [6] and [23]. Our primary concern here is to incorporate prior knowledge in an explicit and transparent manner without having to kernelize it as was done in [6] and [23]. In particular, we are able to directly incorporate general implications involving nonlinear inequalities in a linear program by utilizing Theorem 2.1. Synthetic examples were used to show how our approach uses nonlinear prior knowledge to obtain approximations or classifiers that are much better than those obtained without prior knowledge. Although the given prior knowledge for the synthetic examples is strong, the examples illustrate the simplicity and effectiveness of our approach to incorporate prior knowledge into nonlinear support vector classification and approximation. The Wisconsin Prognostic Breast Cancer (WPBC) dataset was used to demonstrate situations in which prior knowledge and data are combined to obtain a better approximation or classifier than by using only prior knowledge or data alone.

5.1. Approximation datasets. The effectiveness of our proposed approximation formulation (3.4) has been illustrated on two synthetic datasets and the Wisconsin Prognostic Breast Cancer (WPBC) database, available from [16].

5.1.1. *Two-dimensional hyperboloid function.* The first approximation example is the two-dimensional hyperboloid function:

$$\eta(x_1, x_2) = x_1 x_2. \tag{5.1}$$

This function was studied in [23]. The given data consists of eleven points along the line $x_1 = x_2$, $x_1 \in \{-5, -4, \ldots, 4, 5\}$. The given values at these points are the actual function values.

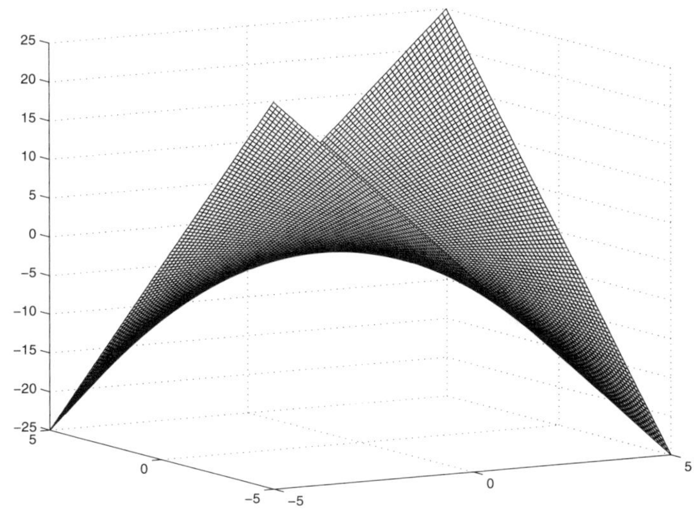

FIGURE 1. The exact hyperboloid function $\eta(x_1, x_2) = x_1x_2$.

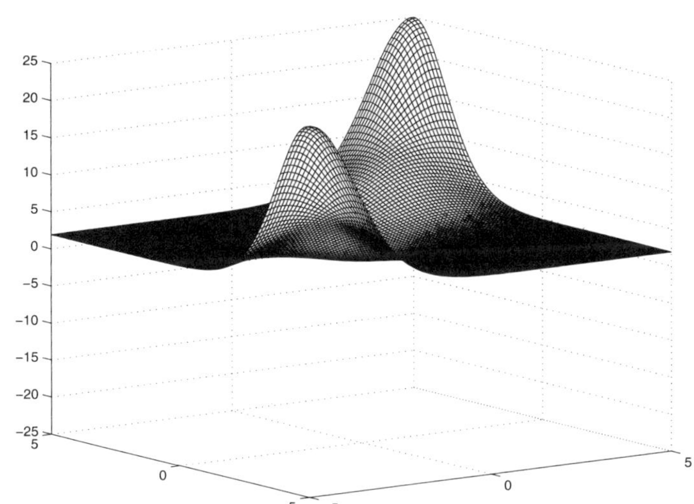

FIGURE 2. Approximation of the hyperboloid function $\eta(x_1, x_2) = x_1x_2$ based on eleven exact function values along the line $x_2 = x_1$, $x_1 \in \{-5, -4, \dots, 4, 5\}$, but *without* prior knowledge.

Figure 1 depicts the two-dimensional hyperboloid function of (5.1). Figure 2 depicts the approximation of the hyperboloid function by a surface based on the eleven points described above *without* prior knowledge. Figures 1 and 2 are taken from [23].

Figure 3 depicts a much better approximation than that of Fig. 2 of the hyperboloid function by a nonlinear surface based on the same eleven points above *plus*

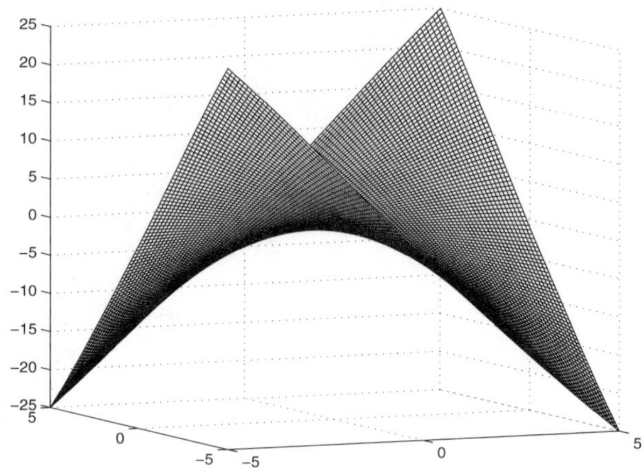

FIGURE 3. Approximation of the hyperboloid function $\eta(x_1, x_2) = x_1x_2$ based on the same eleven function values as Fig. 2 *plus* prior knowledge consisting of the implication (5.2).

prior knowledge. The prior knowledge consisted of the implication:

$$x_1x_2 \leq 1 \Longrightarrow f(x_1, x_2) \leq x_1x_2, \tag{5.2}$$

which, because of the nonlinearity of x_1x_2, cannot be handled by [23]. Note that even though the prior knowledge implication (5.2) provides only partial information regarding the hyperboloid (5.1) being approximated, applying it is sufficient to improve our kernel approximation substantially as depicted in Fig. 3. The prior knowledge implication (5.2) was applied in its equivalent inequality (2.5) form, at discrete points as stated in the inequality constraints of (3.4). In this example, the knowledge was applied at eleven points along the line $x_1 = -x_2$, $x_1 \in \{-5, -4, \ldots, 4, 5\}$.

It is instructive to compare (5.2) with the prior knowledge used in [23] to obtain a visually similar improvement. In that work, the following prior knowledge was used:

$$(x_1, x_2) \in \left\{(x_1, x_2) \,\middle|\, -\frac{1}{3}x_1 \leq x_2 \leq -\frac{2}{3}x_1\right\} \Longrightarrow f(x_1, x_2) \leq 10x_1, \tag{5.3}$$

and

$$(x_1, x_2) \in \left\{(x_1, x_2) \,\middle|\, -\frac{2}{3}x_1 \leq x_2 \leq -\frac{1}{3}x_1\right\} \Longrightarrow f(x_1, x_2) \leq 10x_2. \tag{5.4}$$

These implications were implemented by replacing $f(x_1, x_2)$ with its nonlinear kernel approximation (2.1) and by kernelizing the resulting prior knowledge [23, Equation 18]. The result can then be incorporated into a linear program with no discretization required [23, Proposition 3.1]. However, as is noted in [23], the implications (5.3) and (5.4) are not correct everywhere, but are merely intended to coarsely model the global shape of $\eta(x_1, x_2)$. This inexactness arises because of the limitation that knowledge be linear in the input space, and because the use of the nonlinear kernel to map knowledge in the input space to higher dimensions is difficult to interpret. In contrast, the prior knowledge of implication (5.2) is always correct and exactly captures the shape of the function. Thus, this example illustrates that there is a significant gain in usability due to the fact that the knowledge may be nonlinear in input space features.

5.1.2. *Two-dimensional tower function.* For the second approximation example, we considered the following function:

$$\tau(x_1, x_2) = \begin{cases} 4, & \text{when } \|(x_1, x_2)\| < 1, \\ 3, & \text{when } 1 \leq \|(x_1, x_2)\| < 2, \\ 2, & \text{when } 2 \leq \|(x_1, x_2)\| < 3, \\ 1, & \text{when } 3 \leq \|(x_1, x_2)\| < 4, \\ 0, & \text{otherwise} \end{cases} \tag{5.5}$$

which is shown in Fig. 4. Due to the visual appearance of this function, we refer to it as the *tower* function.

The data used to approximate the tower function of (5.5) consists of 400 equally spaced points on the grid $[-4,\ 4] \times [-4,\ 4]$, with given values defined using the following equation:

$$f(x_1, x_2) = \min\{\tau(x_1, x_2), 2\}, \tag{5.6}$$

where $\tau(x_1, x_2)$ is given by (5.5). This misleading data explains the chopped-off appearance that is shown by the approximation of Fig. 5 which is a poor approximation of the tower function based on this data *without* prior knowledge.

Figure 6 shows an approximation of the tower function using the data described above *plus* the following prior knowledge:

$$(x_1, x_2) \in [-4, 4] \times [-4, 4] \Longrightarrow f(x_1, x_2) = \tau(x_1, x_2), \tag{5.7}$$

where $\tau(x_1, x_2)$ is the exact value of the tower function of (5.5). This implication was enforced at 2500 equally spaced points on the grid $[-4, 4] \times [-4, 4]$. The approximation depicted in Fig. 6 was made by setting the parameters C and ν of

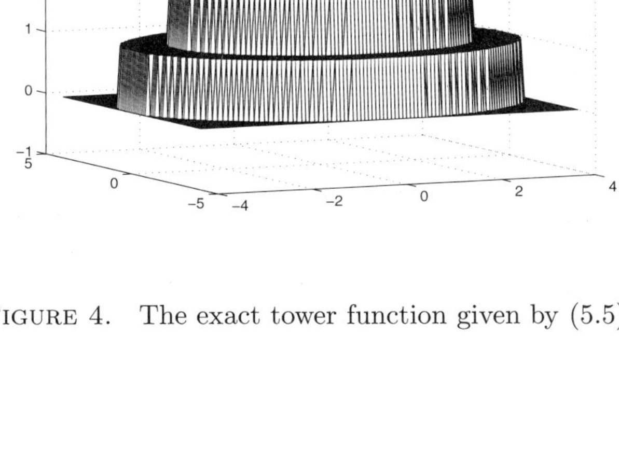

FIGURE 4. The exact tower function given by (5.5).

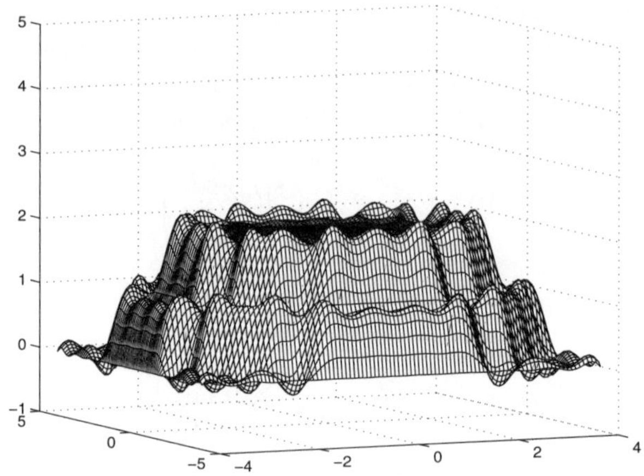

FIGURE 5. An approximation of the tower function in (5.5) using 400 equally spaced points on $[-4,4]\times[-4,4]$ given by (5.6) *without* prior knowledge.

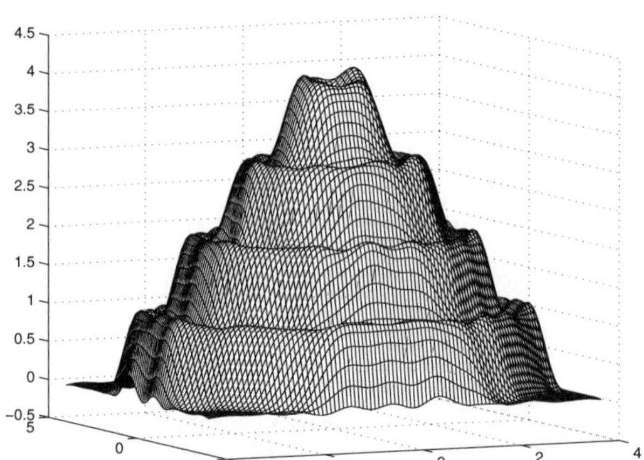

FIGURE 6. An approximation of the tower function in (5.5) using 400 equally spaced points on $[-4,4]\times[-4,4]$ given by (5.6) *with* the prior knowledge described in (5.7).

(3.4) to 10^1 and 10^{20} respectively. Thus, this example illustrates that despite poor initial data, a substantially improved approximation using prior knowledge can be made by incorporating prior knowledge in the form of an implication such as (5.7).

5.1.3. *Predicting lymph node metastasis.* To demonstrate the effectiveness of our approximation formulation (3.4) on a real-world dataset, a potentially useful application of knowledge-based approximation to breast cancer prognosis [22, 29, 14] was considered. An important prognostic indicator for breast cancer recurrence is the number of metastasized lymph nodes under a patient's armpit which could be as many as 30. To obtain this number, a patient must optionally undergo a potentially debilitating surgery in addition to the removal of the breast tumor. Thus, it is useful to approximate the number of metastasized lymph nodes using available information. The Wisconsin Prognostic Breast Cancer (WPBC) data, in which the primary task is to determine time to recurrence [16], contains information on the number of metastasized lymph nodes for 194 breast cancer patients, as well as thirty cytological features obtained by a fine needle aspirate and one histological feature, tumor size, obtained during surgery. Mangasarian et al. demonstrated in [23] that a function that approximated the number of metastasized lymph nodes using four of these features could be improved using prior knowledge. The formulation developed here has been used to approximate the number of metastasized lymph nodes using only the tumor size.

In order to simulate the situation where an expert provides prior knowledge regarding the number of metastasized lymph nodes based on tumor size, the following procedure was used. First, 20% of the data was randomly selected as "past data." This past data was used to develop prior knowledge, while the remaining 80% of the data, the "present data," was used for evaluation. The goal is to simulate the situation in which an expert can provide prior knowledge, but no more data is available. To generate such prior knowledge, kernel approximation was used to find a function $\phi(x) = K(x', B')u - \gamma$, where B is the matrix containing the past data and K is the Gaussian kernel defined in Section 1. This function was then used as the basis for our prior knowledge. Since this function was not believed to be accurate for areas where there was little data in the past data set, this knowledge was imposed only on the region $p(x) \geq 0.1$, where $p(x)$ was the density function for the tumor sizes in B estimated with the `ksdensity` routine, available in the MATLAB statistics toolbox [19]. The following prior knowledge implication was considered:

$$p(x) \geq 0.1 \Longrightarrow f(x) \geq \phi(x) - 0.01. \tag{5.8}$$

That is, the number of metastasized lymph nodes was greater than the predicted value on the past data, with a tolerance of 0.01. This implication incorporates a typical oncological surgeon's advice that the number of metastasized lymph nodes increases with tumor size. In order to accurately simulate the desired conditions, this knowledge was formed by observing only the past data. No aspect of the prior knowledge was changed after testing began on the present data.

TABLE 1. Leave-one-out root-mean-squared-error (RMSE) of approximations with and without knowledge on the present WPBC data. Best result is in bold.

Approximation	RMSE
Without knowledge	5.92
With knowledge	**5.04**
Improvement due to knowledge	14.8%

Table 1 illustrates the improvement resulting from the use of prior knowledge. The first two entries compare the leave-one-out error of function approximations without and with prior knowledge. When training functions on each training set, ten points of the training set were selected as a tuning set. This set was used to choose the value of C from the set $\{2^i | i = -7, \ldots, 7\}$. The kernel parameter was set to 2^{-7}, which gave a smooth curve on the past data set. This value was fixed before testing on the present data. For the approximation *with* knowledge, the parameter ν was set to 10^6, which ensured that the prior knowledge would be taken into account by the approximation. Implication (5.8) was imposed as prior knowledge, and the discretization for the prior knowledge was 400 equally spaced points on the interval $[1, 5]$. This interval approximately covered the region on which $p(x) \geq 0.1$. We note that the use of prior knowledge led to a 14.8% improvement. In our experience, such an improvement is difficult to obtain in medical tasks, and indicates that the approximation with prior knowledge is more potentially useful than the approximation without prior knowledge.

In order to further illustrate the effectiveness of using prior knowledge, two other experiments were performed. First, the root-mean-squared-error (RMSE) of the function ϕ was calculated on the present data, which was not used to create ϕ. The resulting RMSE was 6.12, which indicates that this function does not, by itself, do a good job predicting the present data. The leave-one-out error on the present data of an approximation that included the present data *and* the past data, but *without* prior knowledge was also calculated. This approach led to less than one percent improvement over the approximation without knowledge shown in Table 1, which indicates that the prior knowledge *in the form of the implication* (5.8) contains more useful information than the *raw* past data alone. These results indicate that the inclusion of the prior knowledge with the present data is responsible for the 14.8% improvement.

5.2. Classification datasets. The effectiveness of our proposed classification formulation (4.4) has been illustrated on two publicly available datasets: the Checkerboard dataset [9], and the Wisconsin Prognostic Breast Cancer (WPBC) dataset [16].

5.2.1. *Checkerboard Problem.* The first classification example was based on the frequently utilized checkerboard dataset [9, 11, 20, 12, 6]. This synthetic dataset contains two-dimensional points in $[-1, 1] \times [-1, 1]$ labeled so that they form a checkerboard. For this example, a dataset consisting of only the sixteen points at the center of each square in the checkerboard was used to generate a classifier without knowledge. The rows of both matrices A and B of (4.4) were set equal to the coordinates of the sixteen points, which are the standard values. Figure 7 shows a classifier trained on these sixteen points *without* any additional prior knowledge.

Figure 8 shows a much more accurate classifier trained on the same sixteen points as used in Figure 7, *plus* prior knowledge representing only the leftmost two squares in the bottom row of the checkerboard. This knowledge was imposed via the following implications:

$$
\begin{aligned}
-1 \leq x_1 \leq -0.5 \wedge -1 \leq x_2 \leq -0.5 &\Longrightarrow f(x_1, x_2) \geq 0, \\
-0.5 \leq x_1 \leq 0 \wedge -1 \leq x_2 \leq -0.5 &\Longrightarrow f(x_1, x_2) \leq 0.
\end{aligned}
\tag{5.9}
$$

The implication on the first line was imposed at 100 uniformly spaced points in $[-1, -0.5] \times [-1, -0.5]$, and the implication on the second line were imposed at 100

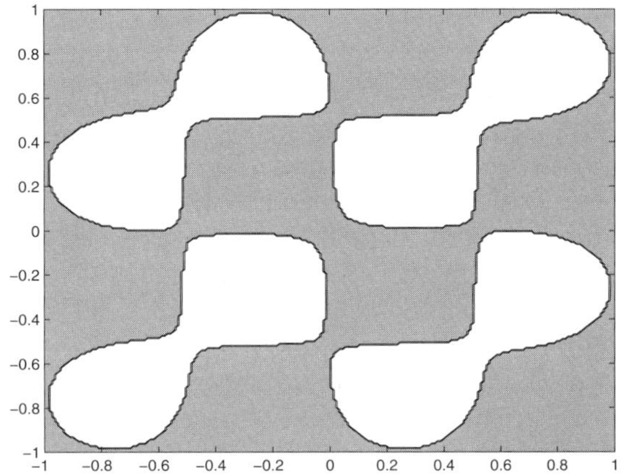

FIGURE 7. A classifier for the checkerboard dataset trained using only the sixteen points at the center of each square *without* prior knowledge. The white regions denote areas where the classifier returns a value greater than zero, and the gray regions denote areas where the classifier returns a value less than zero. A uniform grid consisting of $40,000$ points was used to create the plot utilizing the obtained classifier.

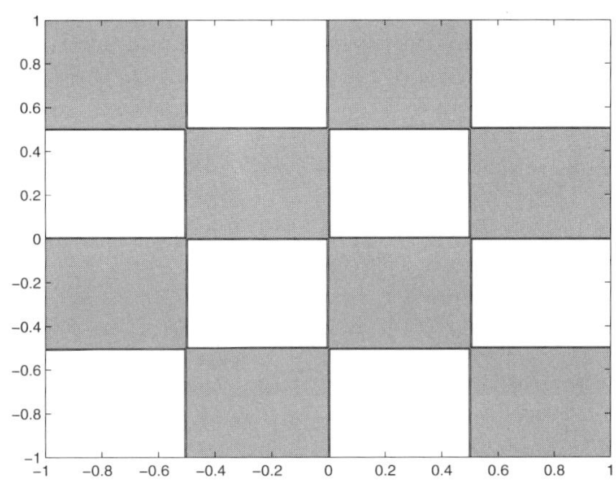

FIGURE 8. A classifier for the checkerboard dataset trained using the sixteen points at the center of each square *with* prior knowledge representing the two leftmost squares in the bottom row given in (5.9). The white regions denote areas where the classifier returns a value greater than zero, and the gray regions denote areas where the classifier returns a value less than zero. A uniform grid consisting of $40,000$ points was used to create the plot utilizing the knowledge-based classifier.

uniformly spaced points in $[-0.5, 0] \times [-1, -0.5]$. No prior knowledge was given for the remaining squares of the checkerboard. We note that this knowledge is very similar to that used in [6], although our classifier is more accurate here. This example demonstrates that knowledge of the form used in [6] can be easily applied with our proposed approach without kernelizing the prior knowledge.

5.2.2. *Predicting breast cancer survival time.* We conclude our experimental results with a potentially useful application of the Wisconsin Prognostic Breast Cancer (WPBC) dataset [16, 13]. This dataset contains thirty cytological features obtained from a fine needle aspirate and two histological features, tumor size and the number of metastasized lymph nodes, obtained during surgery for breast cancer patients. The dataset also contains the amount of time before each patient experienced a recurrence of the cancer, if any. Here, the task of predicting whether a patient will remain cancer free for at least 24 months is considered. Past experience with this dataset has shown that an accurate classifier for this task is difficult to obtain. In this dataset, 81.9% of patients are cancer free after 24 months. To our knowledge, the best result on this dataset prior to [25] is 86.3% correctness obtained by Bennett in [1]. It is possible that incorporating expert information about this task is necessary to obtain higher accuracy on this dataset. In [25] it was demonstrated that with sufficient prior knowledge, our approach can achieve 91.0% correctness.

To obtain prior knowledge for this dataset, the number of metastasized lymph nodes was plotted against the tumor size, along with the class label, for each patient. An oncological surgeon's advice was then simulated by selecting regions containing patients who experienced a recurrence withing 24 months. In a typical machine learning task, not all of the class labels would be available. However, the purpose here is to demonstrate that if an expert is able to provide useful prior knowledge, our approach can effectively apply that knowledge to learn a more accurate classifier. We leave studies on this dataset in which an expert provides knowledge without all of the labels available to future work. In such studies, the expert would be given information regarding the class of only data points in a training set that is a subset of all the data, and then give advice on the class of points in the entire dataset. The prior knowledge constructed for this dataset is depicted in Fig. 9 and consists of the following three implications:

$$\left\|\begin{pmatrix}(5.5)x_1\\ x_2\end{pmatrix} - \begin{pmatrix}(5.5)7\\ 9\end{pmatrix}\right\| + \left\|\begin{pmatrix}(5.5)x_1\\ x_2\end{pmatrix} - \begin{pmatrix}(5.5)4.5\\ 27\end{pmatrix}\right\| - 23.0509 \leq 0 \implies f(x) \geq 1,$$

$$\begin{pmatrix}-x_2 + 5.7143x_1 - 5.75\\ x_2 - 2.8571x_1 - 4.25\\ -x_2 + 6.75\end{pmatrix} \leq 0 \implies f(x) \geq 1, \tag{5.10}$$

$$\frac{1}{2}(x_1 - 3.35)^2 + (x_2 - 4)^2 - 1 \leq 0 \implies f(x) \geq 1.$$

The class $+1$ represents patients who experienced a recurrence in less than 24 months. Here, x_1 is the tumor size, and x_2 is the number of metastasized lymph nodes. Each implication is enforced at the points in the dataset for which the left-hand side of the implication is true. These regions are shown in Fig. 9. The first implication corresponds to the region closest to the upper right-hand corner. The

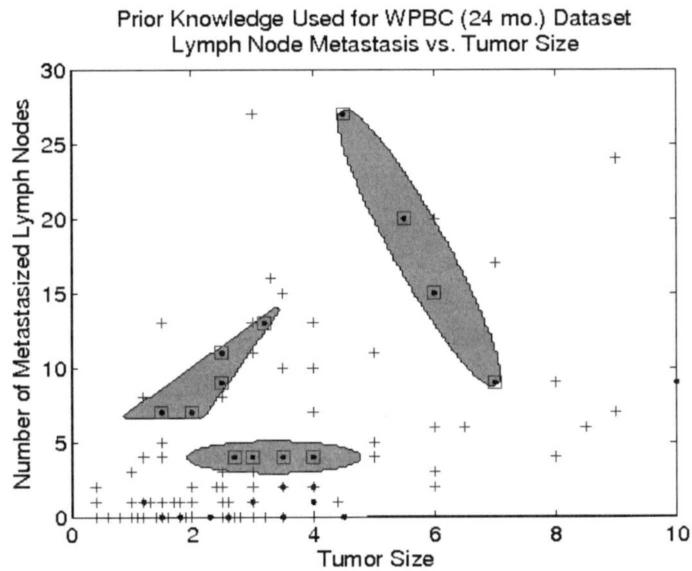

FIGURE 9. Number of metastasized lymph nodes versus tumor size for the WPBC (24 mo.) dataset. The solid dots represent patients who experienced a recurrence within 24 months of surgery, while the crosses represent the cancer free patients. The shaded regions which correspond to the areas in which the left-hand side of one of the three implications in Equation (5.10) is true simulate an oncological surgeon's prior knowledge regarding patients that are likely to have a recurrence. Prior knowledge was enforced at the points enclosed in squares.

triangular region corresponds to the second implication, and the small elliptical region closest to the x_1 axis corresponds to the third implication. Although these implications seem complicated, it would not be difficult to construct a more intuitive interface similar to standard graphics programs to allow a user to create arbitrary regions. Applying these regions with our approach would be straightforward.

In order to evaluate our proposed approach, the misclassification rates of two classifiers on this dataset were compared. One classifier is learned without prior knowledge, while the second classifier is learned using the prior knowledge given in (5.10). For both cases the rows of the matrices A and B of (4.4) were set to the usual values, that is to the coordinates of the points of the training set. The misclassification rates are computed using leave-one-out cross validation. For each fold, the parameter ν and the kernel parameter μ were chosen from the set $\{2^i \mid i \in \{-7, \ldots, 7\}\}$ by using ten-fold cross validation on the training set of the fold. In the classifier with prior knowledge, the parameter σ was set to 10^6, which corresponds to very strict adherence to the prior knowledge. The results are summarized in Table 2. The reduction in misclassification rate indicates that our classification approach can use appropriate prior knowledge to obtain a classifier on this difficult dataset with 50% improvement.

TABLE 2. Leave-one-out misclassification rate of classifiers with and without knowledge on the WPBC (24 mo.) dataset. Best result is in bold.

Classifier	Misclassification Rate
Without knowledge	0.1806
With knowledge	**0.0903**
Improvement due to knowledge	50.0%

6. Conclusion and Outlook

We have proposed a unified, computationally effective framework for handling general nonlinear prior knowledge in kernel approximation and kernel classification problems. We have reduced such prior knowledge to easily implemented linear constraints in a linear programming formulation. We have demonstrated the effectiveness of our approach on three synthetic problems and two important real world problems arising in breast cancer prognosis. Possible future extensions are to even more general prior knowledge, such as that where the right hand side of the implications (2.3) and (2.7) are replaced by very general nonlinear inequalities involving the kernel function (2.1). Another important avenue of future work is to construct an interface which allows users to easily specify arbitrary regions to be used as prior knowledge.

References

1. K. P. Bennett, *Decision tree construction via linear programming*, Proc. 4th Midwest Artificial Intelligence and Cognitive Sci. Soc. Conf. (Illinois) (M. Evans, ed.), Midwest Artificial Intelligence and Cognitive Science Society, Utica, IL, 1992, pp. 97-101.
2. P. S. Bradley and O. L. Mangasarian, *Feature selection via concave minimization and support vector machines*, Proc. 15th Internat. Conference on Machine Learning (California) (J. Shavlik, ed.), Morgan Kaufmann Publishers Inc., San Francisco, CA, pp. 82–90; `ftp://ftp.cs.wisc.edu/math-prog/tech-reports/98-03.ps`.
3. V. Cherkassky and F. Mulier, *Learning from data—concepts, theory and methods*, John Wiley & Sons, New York, 1998.
4. N. Cristianini and J. Shawe-Taylor, *An introduction to support vector machines*, Cambridge University Press, Cambridge, MA, 2000.
5. G. Fung, O. L. Mangasarian, and A. Smola, *Minimal kernel classifiers*, J. Machine Learning Research **3** (2002), 303–321; `ftp://ftp.cs.wisc.edu/pub/dmi/tech-reports/00-08.ps`.
6. G. Fung, O. L. Mangasarian, and J. Shavlik, *Knowledge-based nonlinear kernel classifiers*, 16th Conf. on Learning Theory (COLT 03); Workshop on Kernel Machines (Berlin) (M. Warmuth and B. Schölkopf, eds.), Springer–Verlag, 2003, pp. 102–113; `ftp://ftp.cs.wisc.edu/pub/dmi/tech-reports/03-02.ps`.
7. ———, *Knowledge-based support vector machine classifiers*, Advances in Neural Information Processing Systems, vol. 15 (Cambridge, 2003) (S. Becker, S. Thrun, and K. Obermayer, eds.), MIT Press, Cambridge, MA, pp. 521–528; `ftp://ftp.cs.wisc.edu/pub/dmi/tech-reports/01-09.ps`.
8. M. A. Goberna and M. A. López, *Linear semi-infinite optimization*, John Wiley, New York, 1998.
9. T. K. Ho and E. M. Kleinberg, *Checkerboard dataset*, 1996; `http://www.cs.wisc.edu/math-prog/mpml.html`.
10. S.Y. Huang and Y.-J. Lee, *Theoretical study on reduced support vector machines*, Technical report, National Taiwan University of Science and Technology, 2004; `yuh-jye@mail.ntust.edu.tw`.

11. L. Kaufman, *Solving the quadratic programming problem arising in support vector classification*, Advances in Kernel Methods—Support Vector Learning (B. Schölkopf, C. J. C. Burges, and A. J. Smola, eds.), MIT Press, Cambridge, MA, 1999, pp. 147–167.
12. Y.-J. Lee and O. L. Mangasarian, RSVM*: Reduced support vector machines*, Proc. 1st SIAM International Conference on Data Mining, Chicago (2001), CD-ROM, 2001; `ftp://ftp.cs.wisc.edu/pub/dmi/tech-reports/00-07.ps`.
13. Y.-J. Lee, O. L. Mangasarian, and W. H. Wolberg, *Breast cancer survival and chemotherapy: a support vector machine analysis*, DIMACS Ser. Discrete Math. Theoret. Comput. Sci., vol. 55, Amer. Math. Soc., Providence, R.I., 2000, pp. 1–10; `ftp://ftp.cs.wisc.edu/pub/dmi/tech-reports/99-10.ps`.
14. ———, *Survival-time classification of breast cancer patients*, Comput. Optim. Appl. **25** (2003), 151–166; `ftp://ftp.cs.wisc.edu/pub/dmi/tech-reports/01-03.ps`.
15. Q. V. Le, A. J. Smola, and T. Gärtner, *Simpler knowledge-based support vector machines*, Proc. 23rd Internat. Conf. on Machine Learning, Pittsburgh, PA, 2006 `http://www.icml2006.org/icml2006/technical/accepted.html`.
16. P. M. Murphy and D. W. Aha, UCI *machine learning repository*, 1992; `www.ics.uci.edu/~mlearn/MLRepository.html`.
17. O. L. Mangasarian, *Nonlinear programming*, McGraw–Hill, New York, 1969; SIAM Classic in Applied Mathematics, vol. 10, 1994, Philadelphia (reprint).
18. ———, *Generalized support vector machines*, Advances in Large Margin Classifiers (Cambridge, MA) (A. Smola, P. Bartlett, B. Schölkopf, and D. Schuurmans, eds.), MIT Press, Cambridge, MA, 2000, pp. 135–146; `ftp://ftp.cs.wisc.edu/math-prog/tech-reports/98-14.ps`.
19. MATLAB, *User's guide*, The MathWorks, Inc., Natick, MA 01760, 1994-2006; `http://www.mathworks.com`.
20. O. L. Mangasarian and D. R. Musicant, *Lagrangian support vector machines*, Jc**1** (2001), 161–177; `ftp://ftp.cs.wisc.edu/pub/dmi/tech-reports/00-06.ps`.
21. R. Maclin, J. Shavlik, L. Torrey, T. Walker, and E. Wild, *Giving advice about preferred actions to reinforcement learners via knowledge-based kernel regression*, Proc. 20th National Conference on Artificial Intelligence, 2005, pp. 819–824.
22. O. L. Mangasarian, W. N. Street, and W. H. Wolberg, *Breast cancer diagnosis and prognosis via linear programming*, Oper. Res. **43** (1995), no. 4, 570–577.
23. O. L. Mangasarian, J. W. Shavlik, and E. W. Wild, *Knowledge-based kernel approximation*, J. Machine Learning Research **5** (2004), 1127–1141; `ftp://ftp.cs.wisc.edu/pub/dmi/tech-reports/03-05.ps`.
24. R. Maclin, J. Shavlik, T. Walker, and L. Torrey, *A simple and effective method for incorporating advice into kernel methods*, Proc. 21st National Conference on Artificial In telligence (Boston, MA), 2006.
25. O. L. Mangasarian and E. W. Wild, *Nonlinear knowledge-based classification*, Technical report 06-04, Data Mining Institute, University of Wisconsin, 2006; `ftp://ftp.cs.wisc.edu/pub/dmi/tech-reports/06-04.pdf`.
26. ———, *Nonlinear knowledge in kernel approximation*, IEEE Trans. Neural Networks **18** (2007), 300–306, ftp://ftp.cs.wisc.edu/pub/dmi/tech-reports/05-05.pdf.
27. B. Schölkopf and A. Smola, *Learning with kernels*, MIT Press, Cambridge, MA, 2002.
28. V. N. Vapnik, *The nature of statistical learning theory*, 2nd ed., Springer, New York, 2000.
29. W. H. Wolberg, W. N. Street, D. N. Heisey, and O. L. Mangasarian, *Computerized breast cancer diagnosis and prognosis from fine-needle aspirates*, Archives of Surgery **130** (1995), 511–516.
30. J. Zhu, S. Rosset, T. Hastie, and R. Tibshirani, 1*-norm support vector machines*, Advances in Neural Informmation Processing Systems 16–NIPS2003 (S. Thrun, S. K. Saul, and B. Schölkopf, eds.), MIT Press, Cambridge, MA, 2004.

Computer Sciences Department, University of Wisconsin, Madison, WI 53706, USA and Department of Mathematics, University of California at San Diego, La Jolla, CA 92093, USA.

E-mail address: olvi@cs.wisc.edu

Computer Sciences Department, University of Wisconsin, Madison, WI 53706, USA.

E-mail address: wildt@cs.wisc.edu

Centre de Recherches Mathématiques
CRM Proceedings and Lecture Notes
Volume **45**, 2008

Ultrametric Embedding: Application to Data Fingerprinting and to Fast Data Clustering

Fionn Murtagh

Abstract. We begin with pervasive ultrametricity due to high-dimensionality and/or spatial sparsity. How extent or degree of ultrametricity can be quantified leads us to the discussion of varied practical cases when ultrametricity can be partially or locally present in data. We show how the ultrametricity can be assessed in text or document collections, and in time series signals. An aspect of importance here is that to draw benefit from this perspective the data may need to be recoded. Such data recoding can also be powerful in proximity searching, as we will show, where the data is embedded globally and not locally in an ultrametric space.

1. Introduction

The topology or inherent shape and form of an object is important. In data analysis, the inherent form and structure of data clouds are important. Quite a few models of data form and structure are used in data analysis. One of them is a hierarchically embedded set of clusters,—a hierarchy. It is traditional (since at least the 1960s) to impose such a form on data, and if useful to assess the goodness of fit. Rather than fitting a hierarchical structure to data, our recent work has taken a different orientation: we seek to find (partial or global) inherent hierarchical structure in data. As we will describe in this article, there are interesting findings that result from this, and some very interesting perspectives are opened up for data analysis.

A formal definition of hierarchical structure is provided by ultrametric topology (in turn, related closely to p-adic number theory). We will return to this in Section 2 below. First, though, we will summarize some of our findings.

Ultrametricity is a pervasive property of observational data. It arises as a limit case when data dimensionality or sparsity grows. More strictly such a limit case is a regular lattice structure and ultrametricity is one possible representation for it. Notwithstanding alternative representations, ultrametricity offers computational efficiency (related to tree depth/height being logarithmic in number of terminal

2000 *Mathematics Subject Classification.* Primary: 62H30; Secondary: 68T10.
This is the final form of the paper.

nodes), linkage with dynamical or related functional properties (phylogenetic interpretation), and processing tools based on well known p-adic or ultrametric theory (examples: deriving a partition, or applying an ultrametric wavelet transform).

Local ultrametricity is also of importance. Practical data sets (derived from, or observed in, databases and data spaces) present some but not exclusively ultrametric characteristics. This can be used for forensic data exploration (fingerprinting data sets, as we discuss below). Local ultrametricity has been used to expedite search and discovery in information spaces (in [5] as discussed by us in [13], which we will not pursue further here). Such proximity searching and matching has traditionally been addressed ultrametrically by fitting a hierarchy to data. Below, we show a different way to embed the data (in a computationally highly efficient way) in an ultrametric space, using a principle employed in our local ultrametric work: namely, data recoding.

Our ultimate aim in this work is to proceed a lot further, and gain new insights into data (and observed phenomena and events) through ultrametric (topology) or equivalently p-adic (algebra) representation theory.

2. Quantifying Degree of Ultrametricity

Summarizing a full description in Murtagh [13] we explored two measures quantifying how ultrametric a data set is,—Lerman's and a new approach based on triangle invariance (respectively, the second and third approaches described in this section).

The triangular inequality holds for a metric space: $d(x,z) \leq d(x,y) + d(y,z)$ for any triplet of points x, y, z. In addition the properties of symmetry and positive definiteness are respected. The "strong triangular inequality" or ultrametric inequality is: $d(x,z) \leq \max\{d(x,y), d(y,z)\}$ for any triplet x, y, z. An ultrametric space implies respect for a range of stringent properties. For example, the triangle formed by any triplet is necessarily isosceles, with the two large sides equal; or is equilateral.

- Firstly, Rammal *et al.* [22] used discrepancy between each pairwise distance and the corresponding subdominant ultrametric. Now, the subdominant ultrametric is also known as the ultrametric distance resulting from the single linkage agglomerative hierarchical clustering method. Closely related graph structures include the minimal spanning tree, and graph (connected) components. While the subdominant provides a good fit to the given distance (or indeed dissimilarity), it suffers from the "friends of friends" or chaining effect.
- Secondly, Lerman [11] developed a measure of ultrametricity, termed H-classifiability, using ranks of all pairwise given distances (or dissimilarities). The isosceles (with small base) or equilateral requirements of the ultrametric inequality impose constraints on the ranks. The interval between median and maximum rank of every set of triplets must be empty for ultrametricity. We have used extensively Lerman's measure of degree of ultrametricity in a data set. Taking ranks provides scale invariance. But the limitation of Lerman's approach, we find, is that it is not reasonable to study ranks of real-valued (values in non-negative reals) distances defined on a large set of points.

- Thirdly, our own measure of extent of ultrametricity [13] can be described algorithmically. We examine triplets of points (exhaustively if possible, or otherwise through sampling), and determine the three angles formed by the associated triangle. We select the smallest angle formed by the triplet points. Then we check if the other two remaining angles are with small base, or equilateral (when all angles are equal). The approximation to equality is given by 2 degrees (0.0349 radians). Our motivation for the approximate ("fuzzy") equality is that it makes our approach robust and independent of measurement precision.

A supposition for use of our measure of ultrametricity is that we can define angles (and hence triangle properties). This in turn presupposes a scalar product. Thus we presuppose a normed vector space with a scalar product—a Hilbert space—to provide our needed environment. Quite a general way to embed data, to be analyzed, in a Euclidean space, is to use correspondence analysis [16]. This explains our interest in using correspondence analysis quite often in this work: it provides a convenient and versatile way to take input data in many varied formats (e.g., ranks or scores, presence/absence, frequency of occurrence, and many other forms of data) and map them into a Euclidean, factor space.

3. Ultrametricity and Dimensionality

3.1. Distance properties in very sparse spaces. Murtagh [13], and earlier work by Rammal *et al.* [21, 22], has demonstrated the pervasiveness of ultrametricity, by focusing on the fact that sparse high-dimensional data tend to be ultrametric. In Murtagh [13] it is shown how numbers of points in our clouds of data points are irrelevant; but what counts is the ambient spatial dimensionality. Among cases looked at are statistically uniformly (hence "unclustered", or without structure in a certain sense) distributed points, and statistically uniformly distributed hypercube vertices (so the latter are random 0/1 valued vectors). Using our ultrametricity measure, there is a clear tendency to ultrametricity as the spatial dimensionality (hence spatial sparseness) increases.

As [9] also show, Gaussian data behave in the same way and a demonstration of this is seen in Table 1. To provide an idea of consensus of these results, the 200,000-dimensional Gaussian was repeated and yielded on successive runs values of the ultrametricity measure of: 0.96, 0.98, 0.96.

In the following, we explain why high-dimensional and/or sparsely populated spaces are ultrametric.

As dimensionality grows, so too do distances (or indeed dissimilarities, if they do not satisfy the triangular inequality). The least change possible for dissimilarities to become distances has been formulated in terms of the smallest additive constant needed, to be added to all dissimilarities [23, 3, 4, 20]. Adding a sufficiently large constant to all dissimilarities transforms them into a set of distances. Through addition of a larger constant, it follows that distances become approximately equal, thus verifying a trivial case of the ultrametric or "strong triangular" inequality. Adding to dissimilarities or distances may be a direct consequence of increased dimensionality.

For a close fit or good approximation, the situation is not as simple for taking dissimilarities, or distances, into ultrametric distances. A best fit solution is given by [6] (and software is available in R [10]). If we want a close fit to the given

TABLE 1. Typical results, based on 300 sampled triangles from triplets of points. For uniform, the data are generated on $[0,1]$; hypercube vertices are in $\{0,1\}^{\text{Dimen}}$, and for Gaussian, the data are of mean 0, and variance 1. Dimen. is the ambient dimensionality. Isosc. is the number of isosceles triangles with small base, as a proportion of all triangles sampled. Equil. is the number of equilateral triangles as a proportion of triangles sampled. UM is the proportion of ultrametricity-respecting triangles (= 1 for all ultrametric).

No. points	Dimen.	Isosc.	Equil.	UM
Uniform				
100	20	0.10	0.03	0.13
100	200	0.16	0.20	0.36
100	2000	0.01	0.83	0.84
100	20000	0	0.94	0.94
100	200000	0	0.97	0.97
Hypercube				
100	20	0.14	0.02	0.16
100	200	0.16	0.21	0.36
100	2000	0.01	0.86	0.87
100	20000	0	0.96	0.96
100	200000	0	0.97	0.97
Gaussian				
100	20	0.12	0.01	0.13
100	200	0.23	0.14	0.36
100	2000	0.04	0.77	0.80
100	20000	0	0.98	0.98
100	200000	0	0.96	0.96

dissimilarities then a good choice would avail either of the maximal inferior, or subdominant, ultrametric; or the minimal superior ultrametric. Stepwise algorithms for these are commonly known as, respectively, single linkage hierarchical clustering; and complete link hierarchical clustering. (See [2, 11, 12] and other texts on hierarchical clustering.)

3.2. No "curse of dimensionality" in very high-dimensions. Bellman's [1] "curse of dimensionality" relates to exponential growth of hypervolume as a function of dimensionality. Problems become tougher as dimensionality increases. In particular problems related to proximity search in high-dimensional spaces tend to become intractable.

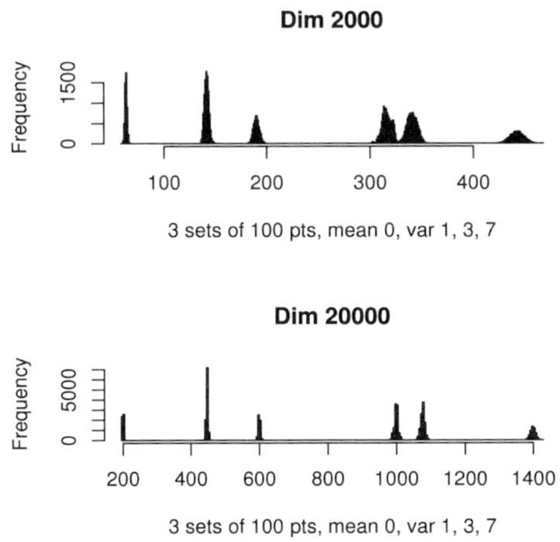

FIGURE 1. An illustration of how "symmetry" or "structure" can become increasingly pronounced as dimensionality increases. Shown are two simulations, each with 3 sub-populations of Gaussian-distributed data, in, respectively, ambient dimensions of 2000 and 20,000.

In a way, a "trivial limit" (Treves [24]) case is reached as dimensionality increases. This makes high -dimensional proximity search very different, and given an appropriate data structure—such as a binary hierarchical clustering tree—we can find nearest neighbours in worst case $O(1)$ or constant computational time [13]. The proof is simple: the tree data structure affords a constant number of edge traversals.

The fact that limit properties are "trivial" makes them no less interesting to study. Let us refer to such "trivial" properties as (structural or geometrical) regularity properties (e.g. all points lie on a regular lattice). First of all, the symmetries of regular structures in our data may be of importance. Secondly, "islands" or clusters in our data, where each "island" is of regular structure, may be exploitable. Thirdly, the mention of exploitability points to the application areas targeted: in this article, we focus on search and matching and show some ways in which ultrametric regularity can be exploited in practice. Fourthly, and finally, regularity by no means implies complete coverage (e.g., existence of all pairwise linkages) implying that interesting or revealing structure will be present in observed or recorded data sets.

Thus we see that in very high-dimensions, and/or in very (spatially) sparse data clouds, there is a simplification of structure, which can be used to mitigate any "curse of dimensionality". Figure 1 shows how the distances within and between clusters become tighter with increase in dimensionality.

4. Increasing Ultrametricity Through Data Recoding

4.1. Ultrametricity of text. In [15], words appearing in a text (in principle all, but in practice a set of the few hundred most frequent) are used to fingerprint the text. Rare words in a text corpus may be appropriate for querying the corpus for relevant texts, but such words are of little help for inter-text characterization and comparison. We also use entire words, with no stemming or other preprocessing. A full justification for such an approach to textual data analysis can be found in Murtagh [16].

So our methodology for studying a set of texts is to characterize each text with numbers of terms appearing in the text, for a set of terms. The χ^2 distance is an appropriate weighted Euclidean distance for use with such data [2, 15]. Consider texts i and i' crossed by words j. Let k_{ij} be the number of occurrences of word j in text i. Then, omitting a constant, the χ^2 distance between texts i and i' is given by $\sum_j 1/k_j(k_{ij}/k_i - k_{i'j}/k_{i'})^2$. The weighting term is $1/k_j$. The weighted Euclidean distance is between the *profile* of text i, viz. k_{ij}/k_i for all j, and the analogous *profile* of text i'. (Our discussion is to within a constant because we actually work on *frequencies* defined from the numbers of occurrences.)

Correspondence analysis allows us to project the space of documents (we could equally well explore the terms in the *same* projected space) into a Euclidean space. It maps the all-pairs χ^2 distance into the corresponding Euclidean distance. In the resulting factor space, we use our triangle-based approach for quantifying how ultrametric the data are.

We did this, [15], for a large number of texts (3 Jane Austen novels, James Joyce's *Ulysses*, technical reports—50 airline accident reports from the NTSB, National Transport Safety Board, fairy tales—209 fables of the Brothers Grimm, 214 dream reports from the DreamBank repository, Aristotle's *Categories*, etc.), finding consistent degree of ultrametricity results over texts of the same sort.

Some very intriguing ultrametricity characterizations were found in our work. For example, we found that the technical vocabulary of air accidents did not differ greatly in terms of inherent ultrametricity compared to the Brothers Grimm fairy tales. Secondly we found that novelist Austen's works were distinguishable from the Grimm fairy tales. Thirdly we found dream reports to be have higher ultrametricity level than the other text collections. Dream-like ultrametric characteristics of Joyce's *Ulysses* were also apparent.

Given that local ultrametricity on sets of terms or words implies locally hierarchical relationships between them, we have pursued this work in the direction of automating the task of ontology construction. See [19].

4.2. Data recoding in the correspondence analysis tradition. If the χ^2 distance (see above, Section 4.1) is used on data tables with constant marginal sums then it becomes a weighted Euclidean distance. This is important for us, because it means that we can directly influence the analysis by equi-weighting, say, the table rows in the following way: we double the row vector values by including an absence (0 value) whenever there is a presence (1 value) and vice versa. Or for a table of percentages, we take both the original value x and $100 - x$. In the correspondence analysis tradition [2, 14] this is known as *doubling* (*dédoublement*).

More generally, Booleanizing, or making qualitative, data in this way, for a varying (value-dependent) number of target value categories (or modalities) leads to the form of coding known as *complete disjunctive form.*

Such coding increases the embedding dimension, and data sparseness, and thus may encourage degree of ultrametricity. That it can do more we will now show.

The iris data has been very widely used as a toy data set since Fisher used it in 1936 ([8], taken from a 1935 article by Anderson) to exemplify discriminant analysis.

It consists of 150 iris flowers, each characterized by 4 petal and sepal, width and breadth, measurements. On the one hand, therefore, we have the 150 irises in $\mathbb{R}^4$. Next, each variable value was recoded to be a rank (all ranks of a given variable considered) and the rank was Boolean-coded (viz., for the top rank variable value, $1000\ldots$, for the second rank variable value, $0100\ldots$, etc.). Following removal of zero total columns, the second data set defined the 150 irises in $\mathbb{R}^{123}$. Actually, this definition of the 150 irises is in fact in $\{0,1\}^{123}$.

Our triangle-based measure of the degree of ultrametricity in a data set (here the set of irises), with 0 = no ultrametricity, and 1 = every triangle an ultrametric-respecting one, gave the following: for irises in $\mathbb{R}^4$, 0.017; and for irises in $\{0,1\}^{123}$: 0.948.

This provides a nice illustration of how recoding can dramatically change the picture provided by one's data. Furthermore it provides justification for data recoding if the ultrametricity can be instrumentalized by us in some way, e.g. to facilitate fast proximity search.

4.3. Ultrametricity of time series. In Murtagh [14] we use the following coding to show that chaotic time series are less ultrametric than, say, financial (futures, FTSE—Financial Times Stock Exchange index, stock price index), biomedical (EEG for normal and epileptic subjects, eyegaze trace), telecoms (web traffic) or meteorological (Mississippi water level, sunspots) time series; random generated (uniformly distributed) time series data are remarkably similar in their ultrametric properties; and ultrametricity can be used to distinguish various types of biomedical (EEG) signals.

A time series can be easily embedded in a space of dimensionality m, by taking successive intervals of length m, or a delay embedding of order m. Thus we define points

$$\mathbf{x}_r = (x_{r-m+1}, x_{r-m+2}, \ldots, x_{r-1}, x_r)^t \in \mathbb{R}^m$$

where t denotes vector transpose.

Given any $\mathbf{x}_r = (x_{r-m+1}, x_{r-m+2}, \ldots, x_{r-1}, x_r)^t \in \mathbb{R}^m$, let us consider the set of s such contiguous intervals determined from the time series of overall size n. For convenience we will take $s = \lfloor n/m \rfloor$ where $\lfloor . \rfloor$ is integer truncation. The contiguous intervals could be overlapping but for exhaustive or near-exhaustive coverage it is acceptable that they be non-overlapping. In our work, the intervals were non-overlapping. The quantification of the ultrametricity of the overall time series is provided by the aggregate over s time intervals of the ultrametricity of each $\mathbf{x}_r$, $1 \leq r \leq s$.

We seek to directly quantify the extent of ultrametricity in time series data. Earlier in this article we have seen how increase in ambient spatial dimensionality leads to greater ultrametricity. However it is not satisfactory from a practical point of view to simply increase the embedding dimensionality m insofar as short memory relationships are of greater practical relevance (especially for prediction). The greatest possible value of $m > 1$ is the total length of the time series, n. Instead we will look for an ultrametricity measurement approach for given and limited sized dimensionalities m. Our experimental results for real and for random data sets are for "window" lengths $m = 5, 10, \ldots, 105, 110$.

We seek local ultrametricity, i.e. hierarchical structure, by studying the following: Euclidean distance squared, $d_{jj'} = (x_{rj} - x_{rj'})^2$ for all $1 \leq j,\ j' \leq m$ in each time window, $\mathbf{x}_r$. It will be noted below in this section how this assumption of Euclidean distance squared has worked well but is not in itself important: in principle any dissimilarity can be used.

We enforce sparseness [21, 22, 13] on our given distance values, $\{d_{jj'}\}$. We do this by thresholding each unique value $d_{jj'}$, in the range $\max_{jj'} d_{jj'} - \min_{jj'} d_{jj'}$, by an integer in $\{1, 2\}$. Note that the range is chosen with reference to the currently considered time series window, $1 \leq j, j' \leq m$. Thus far, the recoded value, $d'_{jj'}$ is not necessarily a distance. With the extra requirement that $d'_{jj'} \longrightarrow 0$ whenever $j = j'$ it can be shown that $d'_{jj'}$ is a metric [14].

To summarize, in our coding, a small pairwise transition is mapped onto a value of 1; and a large pairwise transition is mapped onto a value of 2. A pairwise transition is defined not just for data values that are successive in time but for any pair of data values in the window considered.

This coding can be considered as (i) taking a local region, defined by the sliding window, and (ii) coding pairwise "change" $= 2$, versus "no change" $= 1$, relationships. Then, based on these new distances, we use the ultrametric triangle properties to assess conformity to ultrametricity. The average overall ultrametricity in the time series, quantified in this way, allows us to fingerprint our time series.

A wide range of window sizes (i.e., lengths), m, was investigated. Window size is not important: in relative terms the results found remain the same. Taking part of a time series and comparing the results to the full time series gave similar outcomes, thus indicating that the fingerprinting was an integral property of the data.

Our "change/no change" metric is crucial here, and not the input dissimilarity which is mapped onto it. Note too that generalization to multivariate time series is straightforward.

Eyegaze trace signals were found to be remarkably high in ultrametricity, which may be due to extreme values (truncated off-scale readings resulting from the subject's blinking) that were not subject to preprocessing. Web traffic was also very high in ultrametricity, due to to extreme values. All EEG data sets were close together, with clear separation between the normal sleep subject, and the epilepsy cases. The lowest ultrametricity was found for chaotic time series.

4.4. Fast clustering through Baire space embedding. The clustering of chemical compounds, based on chemical descriptors or representations, is important in the pharmaceutical and chemical sectors. It is used for screening and knowledge discovery in large databases of chemical compounds. A chemical compound is encoded (through various schemes that are not of relevance to us here) as a fixed length bit string (i.e. a set of Boolean or 0/1 values). We have started to look at a set of 1.2 million chemical compounds, each characterized (in a given descriptor or coding system, the Digital Chemistry bci1052 dictionary of fragments) by 1052 variables.

While attributes per chemical compound are roughly Gaussian in distribution, chemicals per attribute follow a power law. We found the probability of having more than p chemicals per attribute to be approximately $c/p^{1.23}$ for large p and for constant, c. This warrants normalization, which we do by dividing attribute/chemical presence values by the attribute marginal (i.e., attribute column sum). Any presence value is now a floating point value.

Consider now the very simplified example of two chemicals, x and y, with just one attribute, whose maximum precision of measurement is K. So let us consider

TABLE 2. Results for the three different data sets, each consisting of 7500 chemicals, are shown in immediate succession. The number of significant decimal digits is 4 (more precise, and hence more different clusters found), 3, 2, and 1 (lowest precision in terms of significant digits).

Sig. dig. k	No. clusters
4	6591
4	6507
4	5735
3	6481
3	6402
3	5360
2	2519
2	2576
2	2135
1	138
1	148
1	167

$x_K = 0.478$; and $y_K = 0.472$. In these cases, maximum precision, $|K| = 3$. For first decimal place $k = 1$, we find $x_k = y_k = 4$. For $k = 2$, $x_k = y_k$. But for $k = 3, x_k \neq y_k$. We now introduce the following distance:

$$d_B(x_K, y_K) = \begin{cases} 1 & \text{if} \quad x_1 \neq y_1 \\ \inf 2^{-n} & x_n \neq y_n, \quad 1 \leq n \leq |K| \end{cases}$$

So here $d_B(x_K, y_K) = 2^{-3}$. This distance is a greatest common prefix metric, and indeed ultrametric. Its maximum value is 1, i.e. it is a 1-bounded ultrametric. Our reason for use of d_B to denote this distance is due to it endowing a metric on the Baire space, the space of countably infinite sequences.

The case of multiple attributes is handled as follows. We have the set J of attributes. Hence we have $|J|$ values for each chemical structure. So the ith chemical structure, for each $j \in J$ value with precision $|K|$, is x_{iJK}. Collectively, all our data are expressed by x_{IJK}. As before, we normalize by column sums to work therefore on x^J_{IJK}. To find the Baire distance properties we work simultaneously on all J values, corresponding to a given chemical structure. Therefore the partition at level $k = 1$ has clusters defined as all those numbers indexed by i that share the same $k = 1$, or 1st, digit *in all J values.*

Table 2 demonstrates how this works. In Table 3 we look at k-means, using as input the cluster centers provided by the 1-significant digit Baire approach. Relatively very few changes were found. We note that the partitions in each case are dominated by a very large cluster. Further details on this work can be found in [18].

TABLE 3. Results of k-means for the same three data sets used heretofore, each relating to 7500 chemical structures, with 1052 descriptors. "Sig. dig.": number of significant digits used. "No. clusters": number of clusters in the data set of 7500 chemical structures, associated with the number of significant digits used in the Baire scheme. "Largest cluster": cardinality. "No. discrep.": number of discrepancies found in k-means clustering outcome. "No. discrep. cl.": number of clusters containing these discrepant assignments.

Sig. dig.	No. clusters	Largest cluster	No. discrep.	No. discrep. cl.
1	138	7037	3	3
1	148	7034	1	1
1	167	6923	9	7

5. Conclusions

We have been clear in this work in regard to where and when we used a Euclidean metric, or other dissimilarity, as input. We used correspondence analysis, for instance, for its property of "Euclideanizing" data in the form of counts or numbers of occurrences. Such treatment of the input data was to allow comparability of results, in a common framework, and in addition it was noted that very limited assumptions were made in regard to the input data.

It has been our aim in this work to link observed data with an ultrametric topology for such data. The traditional approach in data analysis, of course, is to impose structure on the data. This is done, for example, by using some agglomerative hierarchical clustering algorithm. We can always do this (modulo distance or other ties in the data). Then we can assess the degree of fit of such a (tree or other) structure to our data.

For our purposes, here, this is unsatisfactory.

Firstly, our aim was to show that ultrametricity can be naturally present in our data, globally or locally. We did not want any "measuring tool" such as an agglomerative hierarchical clustering algorithm to overly influence this finding. (Unfortunately [22] suffers from precisely this unhelpful influence of the "measuring tool" of the subdominant ultrametric. In other respects, [22] is a seminal paper.)

Secondly, let us assume that we did use hierarchical clustering, and then based our discussion around the goodness of fit. This again is a traditional approach used in data analysis, and in statistical data modeling. But such a discussion would have been unnecessary and futile. For, after all, if we have ultrametric properties in our data then many of the widely used hierarchical clustering algorithms will give precisely the same outcome, and furthermore the fit is by definition exact.

In linking data with an ultrametric embedding, whether local only, or global, we have, in this article, proceeded also in the direction of exploiting this achievement. While some applications, like discrimination between time series signals, or texts, have been covered here, other applications like bioinformatics database search and discovery, and analysis of large scale cosmological structures [17], have just been opened up. In [7] this methodology is applied to quantum statistics. There is a great deal of work to be accomplished.

References

[1] R. Bellman, *Adaptive control processes: a guided tour*, Princeton University Press, NJ, 1961.
[2] J. P. Benzécri, *L'analyse des données. I. Taxinomie*; II. *Correspondances*, 2nd ed., Dunod, Paris, 1979.
[3] F. Cailliez and J. P. Pagès, *Introduction à l'analyse de données*, Société de Mathématiques Appliquées et de Sciences Humaines (SMASH), Paris, 1976.
[4] F. Cailliez, *The analytical solution of the additive constant problem*, Psychometrika **48** (1983), 305–308.
[5] E. Chávez, G. Navarro, R. Baeza-Yates, and J. L. Marroquín, *Proximity searching in metric spaces*, ACM Comput. Surveys **33** (2001), 273–321.
[6] G. de Soete, *A least squares algorithm for fitting an ultrametric tree to a dissimilarity matrix*, Pattern Recognition Lett. **2** (1986), 133–137.
[7] A. A. Ezhov and A. Yu. Khrennikov, *On ultrametricity and a symmetry between Bose–Einstein and Fermi–Dirac systems*, p-Adic Mathematical Physics (A. Yu. Khrennikov, Z. Rakić and I. V. Volovich, eds.), 2nd Internat. AIP Conf. Proc., vol. 826, 2006, pp. 55–64.
[8] R. A. Fisher, *The use of multiple measurements in taxonomic problems*, Annals of Eugenics **7** (1936), 179–188.
[9] P. Hall, J. S. Marron, and A. Neeman, *Geometric representation of high-dimension low sample size data*, J. Roy. Statist. Soc. Ser. B **67** (2005), 427–444.
[10] K. Hornik, *A* CLUE *for* CLU*ster ensembles*, J. Statist. Softw. **14** (2005), (12).
[11] I. C. Lerman, *Classification et analyse ordinale des données*, Paris, Dunod, 1981.
[12] F. Murtagh, *Multidimensional clustering algorithms*, Physica-Verlag, 1985.
[13] ———, *On ultrametricity, data coding, and computation*, J. Classification **21** (2004), 167–184.
[14] ———, *Identifying the ultrametricity of time series*, Euro. Phys. J. Ser. B **43** (2005), 573–579.
[15] ———, *A note on local ultrametricity in text*, 2007; `http://arxiv.org/pdf/cs.CL/0701181`.
[16] ———, *Correspondence analysis and data coding with R and java*, Chapman & Hall/CRC, 2005.
[17] ———, *From data to the physics using ultrametrics: new results in high-dimensional data analysis*, p-Adic Mathematical Physics (A. Yu. Khrennikov, Z. Rakić and I. V. Volovich, eds.), 2nd Internat. AIP Conf. Proc., vol. 826, 2006, pp. 151–161.
[18] F. Murtagh, G. Downs, and P. Contreras, *Hierarchical clustering of massive, high-dimensional data sets by exploiting ultrametric embedding*, SIAM J. Sci. Comput., 2007 (in press).
[19] F. Murtagh, J. Mothe, and K. Englmeier, *Ontology from local hierarchical structure in text*, 2007; `http://arxiv.org/abs/cs.IR/0701180`.
[20] E. Neuwirth and L. Reisinger, *Dissimilarity and distance coefficients in automation-supported thesauri*, Inform. Syst. **7** (1982), 47–52.
[21] R. Rammal, J. C. Angles d'Auriac, and B. Doucot, *On the degree of ultrametricity*, J. Phys. Lett. **46** (1985), no. 20, 945–952.
[22] R. Rammal, G. Toulouse, and M. A. Virasoro, *Ultrametricity for physicists*, Rev. Modern Phys. **58** (1986), 765–788.
[23] W. S. Torgerson, *Theory and methods of scaling*, John Wiley & Sons, 1958.
[24] A. Treves, *On the perceptual structure of face space*, BioSystems **40** (1997), no. 1–2, 189–196.

Department of Computer Science, Royal Holloway, University of London, Egham, Surrey TW20 0EX, England.
E-mail address: `fmurtagh@acm.org`

Centre de Recherches Mathématiques
CRM Proceedings and Lecture Notes
Volume **45**, 2008

Selective Linear and Nonlinear Classification

Onur Seref, O. Erhun Kundakcioglu, and Panos M. Pardalos

ABSTRACT. We introduce a generalized support vector classification problem: Let $\mathbf{X}_i$, $i = 1, \dots, n$ be mutually exclusive sets of pattern vectors such that all pattern vectors $\mathbf{x}_{i,k}$, $k = 1, \dots, |\mathbf{X}_i|$ have the same class label $y_i \in \{-1, +1\}$. Select only one pattern vector $\mathbf{x}_{i,k^*}$ from each set $\mathbf{X}_i$ such that the margin between the set of selected positive and negative pattern vectors are maximized. This problem can be formulated as a quadratic mixed integer programming problem for linear and nonlinear selective classification. However, both formulations are highly combinatorial and intractable. Alternative approaches are developed to identify worse pattern vectors to be eliminated in order to obtain pattern vectors with large margin. The restricted free slack concept is introduced to identify such worse pattern vectors and eliminate them iteratively. Pooled free slack (PFS) and free slack per set (FSS) approaches are proposed for the distribution of the restricted free slack over the pattern vectors. These approaches involve convex quadratic formulations with linear constraints which can be solved efficiently. Primal and dual formulations of PFS and FSS are introduced for linear and nonlinear classification. Iterative elimination and direct selection methods are developed to select such pattern vectors using the alternative formulations. These methods are compared with a naïve method on simulated data. The iterative elimination method is also applied to neural data from a visuomotor categorical discrimination task to classify highly cognitive brain activities.

1. Introduction

Machine learning can simply be defined as the process by which a computer system improves its performance based on previous results. There have been very successful implementations of machine learning including language processing, medical diagnosis, financial analysis, search engines, and DNA sequence classification. Most of these applications rely on *pattern recognition* which is generally concerned with classifying objects based on their characteristics. Characteristics of an object are generally referred to as *features*, which are the qualitative and quantitative measures that distinguish that object from the other objects. Similarity between two objects can be evaluated as a function of the differences in a set of features they possess. Objects can be grouped into classes based on their similarity. These classes may be represented in different ways such as approximation functions or

2000 *Mathematics Subject Classification.* 90C20, 90C11.
This is the final form of the paper.

functions that define boundaries between the classes. Arranging objects into such classes based on their position relative to these functions is called *classification.*

Machine learning within the classification framework can be categorized into two main classes. *Supervised learning* refers to the capability of a system to learn from a set of examples, which is a set of input/output pairs. The input is usually a *vector of features* of an object, and the output is the *label* for the class this object belongs to. A set of objects each with a feature vector and a class label is called a *training set.* This set is used to derive classification functions. The trained system is capable of predicting the label of an object. The term *supervised* originates from the fact that the labels for the objects in the training set are provided as input, and therefore are determined by an outside source, which can be considered as the *supervisor.* On the contrary, *unsupervised learning* is the case where the objects are not labeled with any class information, and learning is about forming classes of objects based on similarities between their features.

Initially developed by Vladimir Vapnik [28], *Support Vector Machines* (SVMs) are the state-of-the-art supervised machine learning methods. SVM classifiers classify pattern vectors which are assumed to belong to two linearly separable sets from two different classes. The classification function is defined by a hyperplane that separates two classes. Although there are infinitely many hyperplanes that separate the two classes, the SVM classifier finds the hyperplane that maximizes the distance from the convex hulls of both classes by solving a quadratic convex optimization problem. The success and robustness of SVM classifiers rely on strong fundamentals from the statistical learning theory, from which generalization bounds for SVM classifiers are derived. SVMs can be extended to classification of nonlinear data by implicitly embedding the original data in a nonlinear space using *kernel functions* [25].

SVMs have a wide spectrum of application areas ranging from pattern recognition [15] and text categorization [12] to biomedicine [4, 5, 19], brain-computer interface [14, 8], and financial applications [10, 27]. The training is performed by optimizing a quadratic convex function that is subject to linear constraints. Quadratic programming (QP) is an extensively studied field of mathematics and there are many general purpose methods to solve QP problems such as quasi-newton, primal-dual and interior-point methods [3]. The general purpose methods are suitable for small size problems. For large problems faster methods are required. These faster methods usually involve chucking [20] and decomposition [21] techniques, which use subsets of points to find the optimal hyperplane. SVM Light [13] and LIBSVM [11] are among the most frequently used implementations that use chucking and decomposition methods efficiently. There are also alternative methods such as Generalized Proximal SVM (GEPSVM) [18] that approximate the two classes with two hyperplanes instead of a single hyperplane separating them.

In this study, we introduce a generalization of SVM classification. We consider n sets of pattern vectors with t pattern vectors in each set. A single pattern vector is chosen from each of the n sets such that the selected pattern vectors obtain the maximum margin with the standard SVM methods. All pattern vectors in a set share the same label. The standard SVM problem can be considered as a special case of selective classification where $t = 1$. In selective classification, an optimal pattern vector is chosen from each set in order to maximize the classification margin. The other pattern vectors do not have any effect on the classification.

Selective classification resembles the multiple instance (MI) classification problem, which consists of classifying positive and negative *bags* of pattern vectors, where each bag contains a number of pattern vectors with the same label. In the MI classification problem, classification is performed such that at least one pattern vector for each positive bag is classified as positive, and all the points for all negative bags are classified as negative. The MI classification problem was introduced in the context of drug activity prediction [7], then further developed [2, 16] and extended to protein family modeling [26]. In the drug activity prediction problem, a hypothesis class of axis-parallel rectangles are assumed where algorithms are developed to solve the problem [7]. Integer programming, expectation maximization, and kernel formulations are also proposed for MI classification problem [31, 32, 9, 1]. A benchmark of several MI classification algorithms is done with their non-multiple-instance counterparts [22]. Although the input data for the MI problem is similar to that of the selective classification problem, our objective is to select pattern vectors from both positive and negative classes to maximize the margin in between these classes, rather than classifying the positive bags from negative bags.

The main challenge in selective classification is that the selection process is a hard combinatorial optimization problem. We introduce efficient alternative approaches, which are successful on simulated data as well as on a real life neuroscience problem.

The remainder of the chapter is organized as follows. In Section 2, we give a general review of SVMs including the derivation of the fundamental formulations, which will be helpful for the subsequent sections. Selective classification formulations are introduced in Section 3, where the combinatorial selective classification problem and other alternative formulations are discussed. In Section 4, different algorithms based on the selective classification formulations are presented. In Section 5, computational results for the proposed alternative methods are reported. In Section 6, the developed methods are applied to neural data from a visuomotor task performed by macaque monkeys. Finally, we conclude our results in Section 7.

2. Support Vector Machines

We start the introductory review for SVMs with *maximal margin classifier* and further extend it to *soft margin classifiers*. Alternative formulations for different error norms are also given for SVMs.

In a typical *binary classification* problem, sets $\mathbf{S}^+$ and $\mathbf{S}^-$ are composed of pattern vectors $\mathbf{x}_i \in \mathbb{R}^d$, $i = 1, \dots, n$. If $\mathbf{x}_i \in \mathbf{S}^+$ then it is given the label $y_i = 1$; otherwise $\mathbf{x}_i \in \mathbf{S}^-$ and is given the label $y_i = -1$. The classification problem is to determine which class a new pattern vector $\mathbf{x}_i \notin \{\mathbf{S}^+ \cup \mathbf{S}^-\}$ belongs to. SVMs solve this problem by finding a hyperplane $(\mathbf{w}, b)$ that separates these two classes from each other with the maximum margin.

2.1. Maximal margin classifier. Maximal margin classifier is the simplest form of SVM classifiers that solves the problem of finding a separating hyperplane with the maximum margin. The underlying optimization problem for the maximal margin classifier is only feasible if the two classes of pattern vectors are linearly separable. However, most of the real life classification problems are not linearly separable. Nevertheless, the maximal margin classifier encompasses the fundamental methods used in standard SVM classifiers. The solution to the optimization problem in the maximal margin classifier minimizes the bound on the generalization error [29]. The basic premise of this method lies in the minimization of a convex optimization problem with linear inequality constraints, which can be solved efficiently by many alternative methods [3].

We start our review with the definition of a hyperplane, $\langle \mathbf{w} \cdot \mathbf{x} \rangle + b = 0$, which is represented as the normal vector $\mathbf{w}$ and the offset parameter b. There is inherent degree of freedom in specifying a hyperplane as $(\lambda \mathbf{w}, \lambda b)$. A *canonical hyperplane* is the one from which the closest pattern vector has a distance $1/\|\mathbf{w}\|$, i.e., $\min_{i=1,\dots,m} |\langle \mathbf{w} \cdot \mathbf{x}_i \rangle + b| = 1$.

Now, consider two pattern vectors $\mathbf{x}^+$ and $\mathbf{x}^-$, belonging to classes $\mathbf{S}^+$ and $\mathbf{S}^-$, respectively and they are the closest pattern vectors to a canonical hyperplane, such that $\langle \mathbf{w} \cdot \mathbf{x}^+ \rangle + b = 1$ and $\langle \mathbf{w} \cdot \mathbf{x}^- \rangle + b = -1$. It is easy to show that the margin between these pattern vectors and the hyperplane are both equal to $1/\|\mathbf{w}\|$.

Maximizing the margin $1/\|\mathbf{w}\|$ for the canonical hyperplane is equivalent to minimizing $\|\mathbf{w}\|$ or $\|\mathbf{w}\|^2$. In the following optimization problem, each point $\mathbf{x}_i$ with a label y_i satisfies the canonicality condition in the constraints (2.1b) while maximizing the margin by minimizing $\|\mathbf{w}\|$, as shown in Fig. 1.

$$\min \quad \frac{1}{2}\|\mathbf{w}\|^2 \tag{2.1a}$$

$$\text{subject to } y_i(\langle \mathbf{w} \cdot \mathbf{x}_i \rangle + b) \geq 1, \quad i = 1 \dots, n. \tag{2.1b}$$

From the solution to (2.1), a new pattern vector $\mathbf{x}^*$ can be classified as positive if $\langle \mathbf{w} \cdot \mathbf{x}^* \rangle + b > 0$, and negative otherwise.

2.2. Soft margin classifier. Most real life problems are composed of non separable data which is generally due to the noise. In this case *slack variables* ξ_i are introduced for each pattern vector $\mathbf{x}_i$ in the training set. The slack variables allow misclassifications for each pattern vector; however, they are subject to a penalty of $C/2$. In Fig. 1 soft margin is demonstrated that incurs penalty for misclassified pattern vectors. The maximum margin formulation can be augmented to soft margin formulation as follows.

$$\min \quad \frac{1}{2}\|\mathbf{w}\|^2 + \frac{C}{2}\sum_{i=1}^{n} \xi_i^2 \tag{2.2a}$$

$$\text{subject to } y_i(\langle \mathbf{w} \cdot \mathbf{x_i} \rangle + b) \geq 1 - \xi_i, \quad i = 1, \dots, n. \tag{2.2b}$$

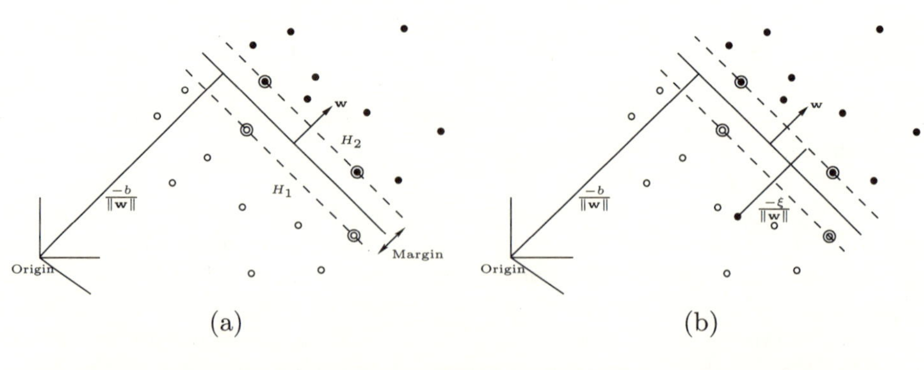

FIGURE 1. Maximal margin classifier (a) and soft margin classifier (b).

In (2.2), the nonnegativity of the slack variables are implicitly implied since the solution cannot be optimal when $\xi_i < 0$ for any pattern vector.

The 2-norm of the slack are penalized in the objective of (2.2). An alternative formulation (2.3) involves penalization of the 1-norm slack variables in the objective. However, we need to impose nonnegativity on the slack variables in order to be able to penalize them.

$$\text{(2.3a)} \qquad \min \quad \frac{1}{2}\|\mathbf{w}\|^2 + \frac{C}{2}\sum_{i=1}^{n}\xi_i$$

$$\text{(2.3b)} \qquad \text{subject to } y_i(\langle \mathbf{w}\cdot\mathbf{x}_i\rangle + b) \geq 1 - \xi_i, \quad i = 1,\dots,n$$

$$\text{(2.3c)} \qquad \xi_i \geq 0, \quad i = 1,\dots,n.$$

Both 1-norm and 2-norm SVM formulations are called the *primal* formulations and equivalent *dual* formulations can be obtained using the optimization theory. The significance of the dual formulations is that they do not involve inequality constraints, and they allow the *kernel trick* to be introduced for nonlinear classification. The standard method to obtain the dual formulation of the SVM problem consists of two parts. First we derive the *Lagrangian function* of the primal problem. This function provides a lower bound for the solution of the primal problem. Next we differentiate the Lagrangian function with respect to the primal variables and impose stationarity. We substitute the equivalent expressions for each primal variable back in the Lagrangian function and add them as constraints where appropriate. The dual problem is obtained by maximizing the resulting function with the new constraints. The dual problem is a concave maximization problem, which can also be solved efficiently. Here, we show how the dual is obtained for the 2-norm SVM formulation.

The Lagrangian function for the 2-norm SVM primal problem is given as follows.

$$\text{(2.4)} \qquad L(\mathbf{w}, b, \boldsymbol{\xi}, \boldsymbol{\alpha}) = \frac{1}{2}\|\mathbf{w}\|^2 + \frac{C}{2}\sum_{i=1}^{n}\xi_i^2 - \sum_{i=1}^{n}\alpha_i[y_i(\langle \mathbf{w}\cdot\mathbf{x_i}\rangle + b) - 1 + \xi_i].$$

Differentiating L with respect to the primal variables $\mathbf{w}$ and b, and assuming stationarity, we get the following expressions,

$$\text{(2.5)} \qquad \frac{\partial L}{\partial \mathbf{w}} = \mathbf{w} - \sum_{i=1}^{n} y_i\alpha_i\mathbf{x}_i = 0; \quad \frac{\partial L}{\partial b} = \sum_{i=1}^{n} y_i\alpha_i = 0; \quad \frac{\partial L}{\partial \xi_i} = \frac{\alpha_i}{C} = 0.$$

We can substitute the expressions in (2.5) back in the Lagrangian function to obtain the following dual formulation, which realizes the hyperplane $\mathbf{w}^* = \sum_{i=1}^{n} y_i\alpha_i^*\mathbf{x}_i$ with geometric margin $\gamma = 1/\|\mathbf{w}^*\|$.

$$\text{(2.6a)} \qquad \max \quad \sum_{i=1}^{n}\alpha_i - \frac{1}{2}\sum_{i=1}^{n}\sum_{j=1}^{n} y_i y_j \alpha_i\alpha_j\langle \mathbf{x}_i\cdot\mathbf{x}_j\rangle - \frac{1}{2C}\sum_{i=1}^{n}\alpha_i^2$$

$$\text{(2.6b)} \qquad \text{subject to } \sum_{i=1}^{n} y_i\alpha_i = 0$$

$$\text{(2.6c)} \qquad 0 \leq \alpha_i, \quad i = 1,\dots,n.$$

Note that from Karush–Kuhn–Tucker complementarity conditions, the constraints in the primal problem are binding for those with the corresponding dual variable $\alpha_i^* > 0$. Knowing $\mathbf{w}^*$, we can find b^* as follows:

$$b^* = \sum_{i:\alpha_i^*>0} y_i - \langle \mathbf{w}^* \cdot \mathbf{x}_i \rangle. \tag{2.7}$$

The decision rule $\text{sign}\big(f(\mathbf{x})\big)$ is equivalent to the hyperplane

$$f(\mathbf{x}) = \sum_{i=1}^{n} y_i \alpha_i^* \langle \mathbf{x} \cdot \mathbf{x}_i \rangle + b^*$$

and b^* is chosen such that $y_i f(\mathbf{x}_i) = 1 - \alpha_i^*/C$ for those pattern vectors with $\alpha_i^* > 0$.

The derivation for the 1-norm dual formulation is very similar to that of 2-norm, which is given as follows,

$$\max \quad \sum_{i=1}^{n} \alpha_i - \frac{1}{2} \sum_{i=1}^{n} \sum_{j=1}^{n} y_i y_j \alpha_i \alpha_j \langle \mathbf{x}_i \cdot \mathbf{x}_j \rangle \tag{2.8a}$$

$$\text{subject to} \quad \sum_{i=1}^{n} y_i \alpha_i = 0 \tag{2.8b}$$

$$0 \leq \alpha_i \leq C, \quad i = 1, \ldots, n. \tag{2.8c}$$

The decision rule $\text{sign}\big(f(\mathbf{x})\big)$ is equivalent to the hyperplane

$$f(\mathbf{x}) = \sum_{i=1}^{n} y_i \alpha_i^* \langle \mathbf{x} \cdot \mathbf{x}_i \rangle + b^*$$

and b^* is chosen such that $y_i f(\mathbf{x}_i) = 1$ for those pattern vectors with $0 < \alpha_i^* < C$.

2.3. Kernel trick. Kernels are introduced in classification to provide enhanced similarity measures between pattern vectors. They basically transform the, so called, *input space*, $\mathcal{X}$, in which the original pattern vectors reside, to a usually higher dimensional dot-product space $\mathcal{H}$ called the *feature space*, via a map $\mathbf{\Phi}\colon \mathcal{X} \to \mathcal{H}$, such that $\mathbf{K}(\mathbf{x}_i, \mathbf{x}_j) = \langle \mathbf{\Phi}(\mathbf{x}_i) \cdot \mathbf{\Phi}(\mathbf{x}_j) \rangle$. The main concept is focused on the dot product of two mapped pattern vectors. Mapping the pattern vectors may become computationally intractable, while implicitly finding their dot products in the feature space have the same complexity as in the input space, in general.

The kernel $\mathbf{K}$ is required to be positive semidefinite in order to define a dot product space and create a feature map. Here a positive semidefinite kernel is defined as a function on $\mathcal{X} \times \mathcal{X}$ for a nonempty set $\mathcal{X}$, which for all $\mathbf{x}_1, \ldots, \mathbf{x}_n \in \mathcal{X}$ gives rise to a positive semidefinite matrix $\mathbf{K}$ such that $\sum_{i,j} c_i c_j \mathbf{K}_{ij} \geq 0$ for all $c_i \in \mathbb{R}$. In the literature it was shown that any algorithm that works on dot products can be kernelized through the *kernel trick* [23]. Given an algorithm which is formulated in terms of a positive semidefinite kernel $\mathbf{K}$, one can construct an alternative algorithm by replacing kernel $\mathbf{K}$ by another positive semidefinite kernel $\widetilde{\mathbf{K}}$.

In the machine learning literature, the kernel trick is introduced by Mercer's theorem and explains the geometry of feature spaces [6]. It can be considered as the characterization of a kernel $\mathbf{K}(\mathbf{x}, \mathbf{x}^*)$. The conditions for Mercer's theorem are equivalent to the requirement that the corresponding matrix is positive semidefinite for any finite subset of $\mathcal{X}$.

The convenience of kernels in SVMs comes into picture in the dual formulation. The linear dot product $\langle \mathbf{x}_i \cdot \mathbf{x}_j \rangle$ can be replaced with an appropriate non-linear kernel $\mathbf{K}$.

In this study, we used *Gaussian kernel*, which is the most commonly used kernel both theoretically and practically. The Gaussian kernel can be represented as,

$$\mathbf{K}(\mathbf{x}_i, \mathbf{x}_j) = e^{-(\|\mathbf{x}_i - \mathbf{x}_j\|^2)/\sigma} \tag{2.9}$$

where, σ is referred to as the bandwidth. The smaller bandwidths are better in classifying intricate patterns, but worse in generalization, and larger bandwidths perform vice versa. Therefore, a fine-tuning for the bandwidth parameter is usually required. For the remainder of the chapter kernel representation is used wherever appropriate.

3. Selective Classification

The motivation for the selective classification comes from a real life problem, in which time series are to be aligned for improved classification results. The individual pattern vectors and the corresponding labels or values from the machine learning context are extended to sets of pattern vectors sharing the same label or value. First, we study the problem of choosing a single pattern vector from each set of possibilities in order to find an optimum separation between the selected positive and negative pattern vectors. This problem, which we call *hard selection* is highly combinatorial. Therefore, we introduce alternative formulations which we call *soft selection* problems. The soft selection problems, in contrast, slightly differ from the optimization problems introduced in Section 2. However, the general structure of the altered problems is quite similar to the original ones with a convex objective, and

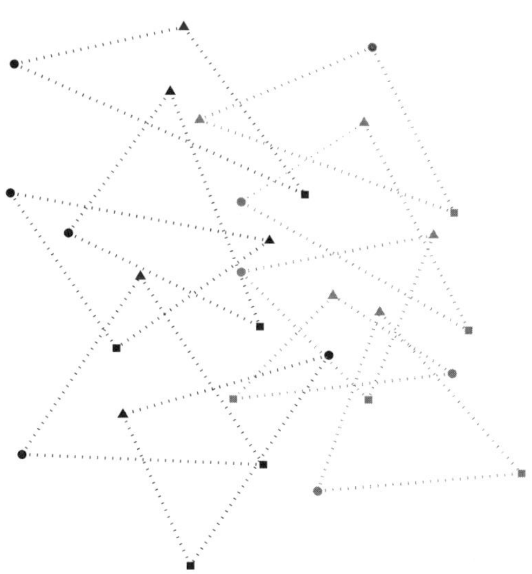

FIGURE 2. Sets of positive (black) and negative (gray) pattern vectors with 3 pattern vectors (circle, triangle and square) in each set.

linear constraints on a convex domain. This convenience allows us to derive dual formulations, in which we can apply the kernel trick to obtain nonlinear classifiers. This section includes formulations for the hard and soft selection problems.

In the standard SVM framework, the input for the classifier is the set of labeled pattern vectors in the training set $s = \{(\mathbf{x}_1, y_1), \ldots, (\mathbf{x}_n, y_n)\}$. In the case that we consider, instead of individual pattern vectors, there are sets $\mathbf{X}_i = \{\mathbf{x}_{i,1}, \ldots, \mathbf{x}_{i,t}\}$ with t pattern vectors sharing the same label y_i. Each set $\mathbf{X}_i$ can be considered as a $t \times d$ matrix, where each row is an individual pattern vector in $\mathbb{R}^d$. A two dimensional example is shown in Fig. 2 with 3 pattern vectors (circle, triangle and square) in each set of positive (black) and negative (gray) pattern vectors.

3.1. Hard selection. We start introducing the selective classification concept with the *hard selection* problem, in which a single pattern vector $\mathbf{x}_{i,k^*} \in \mathbf{X}_i$ is selected from each set $\mathbf{X}_i$ such that the margin between the selected positive and negative pattern vectors is maximized. The hard selection problem can be formulated as a mixed integer quadratic optimization problem as follows, by a modification over the standard SVM formulation,

$$\text{(3.1a)} \qquad \min \quad \frac{1}{2}\|\mathbf{w}\|^2 + \frac{C}{2}\sum_{i=1}^{n} \xi_i^2$$

$$\text{(3.1b)} \qquad \text{subject to } y_i(\langle \mathbf{w} \cdot \mathbf{x}_{i,k}\rangle + b) \geq 1 - \xi_i - M(1 - \widehat{v}_{i,k}),$$
$$i = 1, \ldots, n;\ k = 1, \ldots, t$$

$$\text{(3.1c)} \qquad \sum_{k=1}^{t} \widehat{v}_{i,k} = 1, \quad i = 1, \ldots, n$$

$$\text{(3.1d)} \qquad \widehat{v}_{i,k} \in \{0, 1\}, \quad i = 1, \ldots, n;\ k = 1, \ldots, t.$$

In (3.1), in the first set of constraints (3.1b), M is a sufficiently large positive number. If $\widehat{v}_{i,k} = 0$, then the corresponding constraint has no effect on the problem, and it is equivalent to removing the pattern vector $\mathbf{x}_{i,k}$ from the training set. If $\widehat{v}_{i,k} = 1$, the constraint becomes a regular constraint as in the standard SVM formulation. The second set of constraints (3.1c) ensure that only one pattern vector is chosen from each set. Note that since only one pattern vector is chosen from each set, a single slack variable ξ_i per set $\mathbf{X}_i$ is included in the formulation.

For any choice $\widehat{v}^* = \{\widehat{v}^*_{i,k} : i = 1, \ldots, n; k = 1, \ldots, t\}$ that satisfies constraints (3.1c) and (3.1d), the problem reduces to a standard SVM problem, for which the expression $\mathbf{w} = \sum_{i=1}^{n}\sum_{k=1}^{t} y_i \alpha_{i,k} \mathbf{x}_{\mathbf{i},\mathbf{k}}$ holds, as shown in Section 2. Substituting this expression back in the original combinatorial problem, we arrive at the following equivalent combinatorial formulation of the problem, where we can replace dot products with a nonlinear kernel function.

$$\text{(3.2a)} \quad \min \quad \frac{1}{2}\sum_{i=1}^{n}\sum_{k=1}^{t}\sum_{j=1}^{n}\sum_{l=1}^{t} y_i y_j \alpha_{i,k}\alpha_{j,l}\langle \mathbf{x}_{i,k} \cdot \mathbf{x}_{j,l}\rangle + \frac{C}{2}\sum_{i=1}^{n} \xi_i^2$$

$$\text{(3.2b)} \quad \text{subject to } y_i \left[\sum_{j=1}^{n}\sum_{l=1}^{t} y_j \alpha_{j,l}\langle \mathbf{x}_{i,k} \cdot \mathbf{x}_{j,l}\rangle + b\right] \geq 1 - \xi_i - M(1 - \widehat{v}_{i,k}),$$
$$i = 1, \ldots, n;\ k = 1, \ldots, t$$

$$\sum_{k=1}^{t} \widehat{\upsilon}_{i,k} = 1, \quad i = 1, \ldots, n \tag{3.2c}$$

$$\widehat{\upsilon}_{i,k} \in \{0, 1\}, \quad i = 1, \ldots, n; \; k = 1, \ldots, t. \tag{3.2d}$$

Formulations (3.1) and (3.2) are both highly combinatorial and intractable. Therefore we next introduce alternative formulations called soft selection problems that can be solved efficiently.

3.2. Soft selection. Our goal in *soft selection* is to replace the binary variables $\widehat{\upsilon}_{i,k}$ in the hard selection problem with continuous variables $\upsilon_{i,k}$ in order to obtain an easier formulation. The difficulty in hard selection is finding the best pattern vectors. We pursue a reverse approach for soft selection. Our primary goal is to diminish the influence of pattern vectors that are unlikely to be chosen by the hard selection problem, and consequently, obtain a larger margin. In order to achieve this goal, we propose a number of alternative formulations with additional continuous variables $\upsilon_{i,k}$ serving as free (unpenalized) slack variables for such pattern vectors.

In the standard soft margin classification, each pattern vector $\mathbf{x}_i$ has a slack variable ξ_i, but this slack is penalized in the objective function. We introduce the concept of *restricted free slack* for those pattern vectors which have high influence on the separating hyperplane's orientation and placement.

We also keep a single penalized slack ξ_i per set $\mathbf{X}_i$, which is used for each pattern vector $\mathbf{x}_{i,k} \in \mathbf{X}_i$ in order to keep the problem feasible for any input data. Our aim is to focus on those pattern vectors which require the most free slack to reduce their penalty. The single penalized slack concept also supports our focus.

There are two problems to be addressed:

(i) How much slack should be provided?
(ii) How to distribute the total slack?

For the first problem, an intuitive approach can be adapted. In the standard SVM formulation, a functional margin of 1 is required. Since we have at least two pattern vectors per set, we prefer to provide a total free slack of n, which is a free slack of 1 per set on the average. It is intuitive that the smaller the value of the expression $y_i\langle \mathbf{w} \cdot \mathbf{x}_{i,k}\rangle + b$ is, the more likely it requires free slack.

We introduce two approaches for distributing the restricted free slack. In the first approach, there is a total slack of n, and each pattern vector is allowed to receive from this slack, irrespective of the set they belong to. We refer to this case as *pooled free slack* (PFS). In the second approach, however, each set of pattern vector receive an equal amount of 1 regardless. We refer to the second case as *free slack per set* (FSS).

In Fig. 3 and Fig. 4, examples of the application of PFS and FSS methods on 2-dimensional data are given, respectively. The free slack distributed by PFS and FSS are shown in the third dimension. The figures on the left are top views showing the original data. The figures on the right are front views, only showing the amount of slack assigned. PFS has less restriction on the distribution of free slack. Therefore, more free slack is assigned to those pattern vectors closer to the hyperplane or that are on the other side. On the other hand, FSS method assigns a maximum free slack of 1 to the pattern vectors that are closer to the hyperplane or

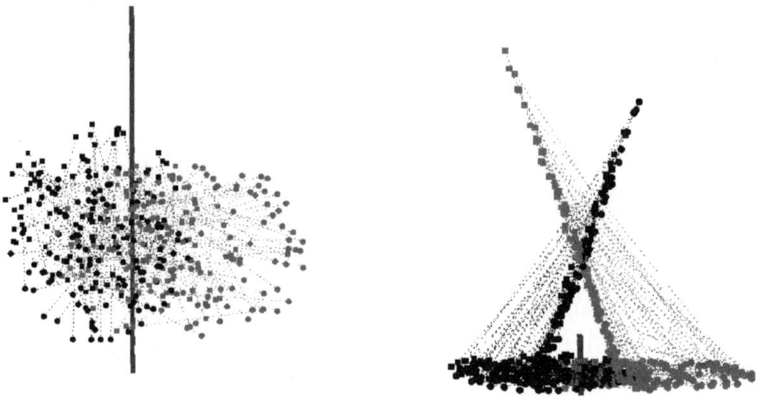

(a) (b)

FIGURE 3. Distribution of pooled free slack (PFS). (a) Top view, (b) Front view.

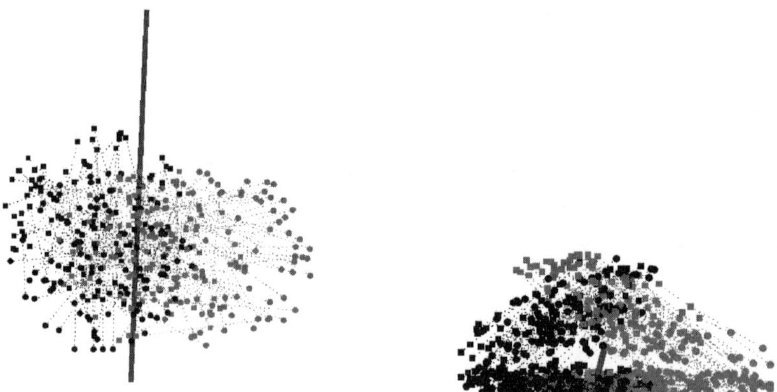

(a) (b)

FIGURE 4. Distribution of free slack per set (FSS). (a) Top view, (b) Front view.

on the other side. The remaining free slack is distributed over to the other pattern vectors, including the ones further away from the separating hyperplane.

3.2.1. *2-norm soft selection.* Assume that there are sets $\mathbf{X}_i$, $i = 1, \ldots, n$ of vectors, $\mathbf{x}_{i,k}$, $k = 1, \ldots, t$. The following optimization problem finds the maximal margin with PFS approach

$$\text{(3.3a)} \quad \min \quad \frac{1}{2}\|\mathbf{w}\|^2 + \frac{C}{2}\sum_{i=1}^{n}\xi_i^2$$

$$\text{(3.3b)} \quad \text{subject to } y_i(\langle \mathbf{w} \cdot \mathbf{x}_{i,k}\rangle + b) \geq 1 - \xi_i - v_{i,k}, \quad i = 1, \ldots, n;\ k = 1, \ldots, t$$

$$\text{(3.3c)} \quad \sum_{i=1}^{n}\sum_{k=1}^{t} v_{i,k} \leq n$$

$$\upsilon_{i,k} \geq 0, \quad i = 1, \ldots, n; \ k = 1, \ldots, t. \tag{3.3d}$$

The Lagrangian function for (3.3) is given as follows,

$$\begin{aligned} L(\mathbf{w}, b, \xi, \hat{\xi}) = {} & \frac{1}{2}\|\mathbf{w}\|^2 + \frac{C}{2}\sum_{i=1}^{n} \xi_i^2 \\ & - \sum_{i=1}^{n}\sum_{k=1}^{t} \alpha_{i,k}[y_i(\langle \mathbf{w} \cdot \mathbf{x}_{i,k}\rangle + b) - 1 + \xi_i + \upsilon_{i,k}] \\ & - \beta\left(n - \sum_{i=1}^{n}\sum_{k=1}^{t} \upsilon_{i,k}\right) \\ & - \sum_{i=1}^{n}\sum_{k=1}^{t} r_{i,k}\upsilon_{i,k}. \end{aligned} \tag{3.4}$$

In this equation, $\boldsymbol{\alpha}$, β and $\mathbf{r}$ are Lagrangian multipliers. Taking partial derivatives with respect to the primal variables $\mathbf{w}$, b, and each ξ_i and $\upsilon_{i,k}$, and imposing stationarity, we get the following expressions:

$$\begin{aligned} \partial L/\partial \mathbf{w} &= \mathbf{w} - \sum_{i=1}^{n}\sum_{k=1}^{t} y_i\alpha_{i,k}\mathbf{x}_{i,k} = 0; & \partial L/\partial b &= \sum_{i=1}^{n}\sum_{k=1}^{t} y_i\alpha_{i,k} = 0, \\ \partial L/\partial \xi_i &= C\xi_i - \sum_{k=1}^{t}\alpha_{i,k} = 0; & \partial L/\partial \upsilon_{i,k} &= \beta - \alpha_{i,k} - r_{i,k} = 0. \end{aligned} \tag{3.5}$$

Substituting the partial derivatives (3.5) in (3.4), we can write the following dual formulation:

$$\max \quad \sum_{i=1}^{n}\sum_{k=1}^{t}\alpha_{i,k} - \frac{1}{2}\sum_{i=1}^{n}\sum_{k=1}^{t}\sum_{j=1}^{n}\sum_{l=1}^{t} y_i y_j \alpha_{i,k}\alpha_{j,l}\langle \mathbf{x}_{i,k} \cdot \mathbf{x}_{j,l}\rangle - \frac{1}{2C}\sum_{i=1}^{n}\left(\sum_{k=1}^{t}\alpha_{i,k}\right)^2 - \beta n \tag{3.6a}$$

$$\text{subject to} \ \sum_{i=1}^{n}\sum_{k=1}^{t} y_i\alpha_{i,k} = 0 \tag{3.6b}$$

$$0 \leq \alpha_{i,k} \leq \beta, \quad i = 1, \ldots, n; \ k = 1, \ldots, t. \tag{3.6c}$$

Note that kernel induced maps to nonlinear feature spaces can be used by replacing the linear dot product $\langle \mathbf{x}_{i,k} \cdot \mathbf{x}_{j,l}\rangle$ in (3.6) with a kernel $\mathbf{K}(\mathbf{x}_{i,k}, \mathbf{x}_{j,l})$.

For the FSS case, the constraint (3.3c) becomes $\sum_{k=1}^{t}\upsilon_{i,k} \leq 1$ for each $i = 1, \ldots, n$. Consequently, this change introduces a separate Lagrangian multiplier β_i for each constraint, and therefore the corresponding component in the Lagrangian function changes to

$$-\sum_{i=1}^{n}\beta_i\left(1 - \sum_{k=1}^{t}\upsilon_{i,k}\right). \tag{3.7}$$

Following the differentiation of the Lagrange function, imposing stationarity and substituting the expressions back in the Lagrangian, we can obtain the dual formulation for the FSS case as,

$$\text{(3.8a)} \qquad \max \quad \sum_{i=1}^{n}\sum_{k=1}^{t}\alpha_{i,k} - \frac{1}{2}\sum_{i=1}^{n}\sum_{k=1}^{t}\sum_{j=1}^{n}\sum_{l=1}^{t} y_i y_j \alpha_{i,k}\alpha_{j,l}\langle \mathbf{x}_{i,k}\cdot \mathbf{x}_{j,l}\rangle - \frac{1}{2C}\sum_{i=1}^{n}\left(\sum_{k=1}^{t}\alpha_{i,k}\right)^2 - \sum_{i=1}^{n}\beta_i$$

$$\text{(3.8b)} \qquad \text{subject to} \quad \sum_{i=1}^{n}\sum_{k=1}^{t} y_i\alpha_{i,k} = 0$$

$$\text{(3.8c)} \qquad 0 \le \alpha_{i,k} \le \beta_i, \quad i = 1,\dots,n;\ k = 1,\dots,t.$$

3.2.2. *1-norm soft selection.* An alternative formulation for an SVM classifier is to penalize 1-norm penalty term, as shown in (2.3). Here we present the soft selection formulations for the 1-norm SVM. The derivation of the primal and dual optimization problem are similar to those discussed for the 2-norm case, therefore we briefly present these formulations. The primal for the FSS approach is given as follows:

$$\text{(3.9a)} \quad \min \quad \frac{1}{2}\|\mathbf{w}\|^2 + C\sum_{i=1}^{n}\xi_i$$

$$\text{(3.9b)} \quad \text{subject to } y_i(\langle \mathbf{w}\cdot \mathbf{x}_{i,k}\rangle + b) \ge 1 - \xi_i - \upsilon_{i,k}, \quad i = 1,\dots,n;\ k = 1,\dots,t$$

$$\text{(3.9c)} \quad \sum_{i=1}^{n}\sum_{k=1}^{t}\upsilon_{i,k} \le n$$

$$\text{(3.9d)} \quad \xi_i,\ \upsilon_{i,k} \ge 0, \quad i = 1,\dots,n;\ k = 1,\dots,t.$$

As in the 2-norm case, we derive the Lagrangian function for the primal problem, differentiate it with respect to the primal variables, impose stationarity and substitute the resulting expressions back to obtain the following dual formulation.

$$\text{(3.10a)} \quad \max \quad \sum_{i=1}^{n}\sum_{k=1}^{t}\alpha_{i,k} - \frac{1}{2}\sum_{i=1}^{n}\sum_{k=1}^{t}\sum_{j=1}^{n}\sum_{l=1}^{t} y_i y_j \alpha_{i,k}\alpha_{j,l}\langle \mathbf{x}_{i,k}\cdot \mathbf{x}_{j,l}\rangle - \beta n$$

$$\text{(3.10b)} \quad \text{subject to} \quad \sum_{i=1}^{n}\sum_{k=1}^{t} y_i\alpha_{i,k} = 0$$

$$\text{(3.10c)} \quad \sum_{k=1}^{t}\alpha_{i,k} \le C, \quad i = 1,\dots,n;$$

$$\text{(3.10d)} \quad 0 \le \alpha_{i,k} \le \beta, \quad i = 1,\dots,n;\ k = 1,\dots,t.$$

In (3.10), the dot product $\langle \mathbf{x}_{i,k}\cdot \mathbf{x}_{j,l}\rangle$ can be replaced with a kernel $\mathbf{K}(\mathbf{x}_{i,k}, \mathbf{x}_{j,l})$ for a kernel induced nonlinear feature space.

The primal and dual formulations for the 1-norm FSS case are very similar to those for the 1-norm PFS case, where the general free slack of n in constraint (3.9c) now implies that the free slack is restricted to 1 for each set in FSS. Consequently, in the dual formulation the term βn changes to $\sum_{i=1}^{n}\beta_i$ in the objective and the constraint $0 \le \alpha_{i,k} \le \beta$ changes to $0 \le \alpha_{i,k} \le \beta_i$.

4. Selection Methods

In the previous section, we developed linear and nonlinear mathematical formulations that can be solved efficiently. The solutions of these problems allow some pattern vectors to use free slack and provide more flexibility for the separating hyperplane. We can use the solution in various ways. The two main sources of information we can use are

(i) the amount of slack each pattern vector receives
(ii) the new hyperplane.

We propose two different approaches based on these sources of information: *Iterative Elimination* and *Direct Selection*. We introduce these approaches, as well as a naïve method for comparison.

4.1. Iterative elimination. The two main soft selection formulations are mainly developed to give the separating hyperplane more flexibility, but at the same time to identify those pattern vectors which are misclassified or very close to the hyperplane. We can obtain a more separated subset of positive and negative pattern vectors if we remove those pattern vectors with high free slack. So, our intuitive basic approach is: at each iteration, supply an incremental amount of free slack of n; solve the soft selection problem in order to identify the pattern vectors with the maximum free slack from each set and remove them; and repeat the iterations with the updated data until only one pattern vector per set remains. This approach is summarized in Algorithm 1, below.

Algorithm 1 Iterative elimination.

$\mathbf{X} \Leftarrow \mathbf{X}(0)$
$j \Leftarrow 1$
while $j < t$ **do**
  $\{\mathbf{w}, b, \upsilon\} \Leftarrow \text{SOFT SELECTION}(\mathbf{X}, \mathbf{y}, n)$
  $r \Leftarrow \emptyset$
  for $i = 1$ to n **do**
    $l = \arg\max_{k=1,\ldots,t-j+1}\{\upsilon_{i,k}\}$
    $r \Leftarrow r \cup \mathbf{x}_{i,l}$
  end for
  $\mathbf{X} \Leftarrow \mathbf{X} \setminus r$
  $j \Leftarrow j + 1$
end while
return $\mathbf{X}$

In Algorithm 1, $\mathbf{X}(0)$ is the original input of pattern vectors, $\mathbf{y}$ is the vector of labels for each set $\mathbf{X}_i$, n is total the free slack amount provided for the soft selection problem, $(\mathbf{w}, b)$ is the hyperplane and $\upsilon_{i,k}$ is the individual free slack value of the pattern vector $\mathbf{x}_{i,k}$ from the solution to the soft selection problem, and r is the set of pattern vectors to be removed at each iteration.

4.2. Direct selection. The alternative to iterative elimination is to use the information regarding the new hyperplane. However, this time, we provide more free slack to give more flexibility to the separating hyperplane. The intuition in this approach is to diminish the influence of $t - 1$ pattern vectors from each set in order to find the final hyperplane in a single iteration. The selection might be based on the free slack values, however depending on the location of individual pattern vectors with respect to the hyperplane, multiple pattern vectors with zero free slack per set might be observed. Instead, we use the distance of the pattern vectors to the hyperplane as our selection criterion. The direct selection algorithm can be summarized as solving the soft selection problem given $n(t-1)$ amount of total free slack (i.e., $n(t-1)$ for PFS or $t-1$ for FSS), and from each set, returning the pattern vector furthest from the resulting hyperplane. The direct selection is given in Algorithm 2.

Algorithm 2 Direct selection.

1: $\{\mathbf{w}, b, v\} \Leftarrow$ SOFT SELECTION$(\mathbf{X}(0), \mathbf{y}, n(t-1))$
2: $\mathbf{X} \Leftarrow \emptyset$
3: **for** $i = 1$ to n **do**
4: $\quad l = \arg\max_{k=1,\ldots,t}\{y_i(\langle \mathbf{w} \cdot \mathbf{x}_{i,k}\rangle + b)\}$
5: $\quad \mathbf{X} \Leftarrow \mathbf{X} \cup \mathbf{x}_{i,l}$
6: **end for**
7: **return** $\mathbf{X}$

In Algorithm 2, $\mathbf{X}(0)$ is the original input of pattern vectors, $\mathbf{y}$ is the vector of labels for each set $\mathbf{X}_i$, $n(t-1)$ is total the free slack amount provided for the soft selection problem, $(\mathbf{w}, b)$ is the hyperplane and $v_{i,k}$ is the individual free slack value of the pattern vector $\mathbf{x}_{i,k}$ from the solution to the soft selection problem, and the amount $y_i(\langle \mathbf{w} \cdot \mathbf{x}_{i,k}\rangle + b)$ is the distance of $\mathbf{x}_{i,k}$ from the hyperplane $(\mathbf{w}, b)$.

4.3. Naïve elimination. We also consider a naïve elimination method, which combines the iterative elimination concept and the distance approach from direct selection. The basic idea is to solve a standard SVM problem; for each set, find the pattern vectors that are closest to the separating hyperplane and remove them; repeat the same process $t-1$ times until only a single pattern vector remains in each set. The naïve elimination is included in our analysis in order to compare the performance of the iterative elimination and direct selection methods to an SVM based approach which does not depend on the soft selection formulations. A similar iterative method is used for the MI setting [1]. The naïve elimination method can be summarized in Algorithm 3.

In Algorithm 3, $\mathbf{X}(0)$ is the original input of pattern vectors, $\mathbf{y}$ is the vector of labels for each set $\mathbf{X}_i$, SVM is a standard SVM classifier, $(\mathbf{w}, b)$ is the hyperplane from the solution to the standard SVM formulation, and r is the set of pattern vectors to be removed at each iteration.

5. Computational Results

In this section, we show the computational results of the proposed methods developed in Section 4. We start with the description of how the data is generated and how the performances of the methods are compared. We present comparative results of the iterative elimination method, direct selection method and the naïve elimination method. We include the results for the iterative elimination and direct selection methods using PFS and FSS approaches with 1-norm and 2-norm penalization schemes. The majority of the results are demonstrated using the linear

Algorithm 3 Naïve elimination.

$\mathbf{X} \Leftarrow \mathbf{X}(0)$
$j \Leftarrow 1$
while $j < t$ **do**
  $\{\mathbf{w}, b\} \Leftarrow \text{SVM}(\mathbf{X}, \mathbf{y})$
  $r \Leftarrow \emptyset$
  for $i = 1$ to n **do**
    $l = \arg\min_{k=1,\ldots,t-j+1}\{y_i(\langle \mathbf{w} \cdot \mathbf{x}_{i,k}\rangle + b)\}$
    $r \Leftarrow r \cup \mathbf{x}_{i,l}$
  end for
  $\mathbf{X} \Leftarrow \mathbf{X} \setminus r$
  $j \Leftarrow j + 1$
end while
return $\mathbf{X}$

formulations of PFS and FSS. We also present the results using nonlinear formulations.

5.1. Simulated data and performance measure. The simulated data is generated depending on two parameters which determine the *dimensionality* and the *separability* of the pattern vectors. For each combination of the parameters 100 data sets are created and classified using PFS and FSS for both 2-norm and 1-norm methods. The simulated data is distributed uniformly in a hypersphere with radius r and center $\mathcal{C}$. $\mathcal{C}$ is also distributed uniformly in a hypersphere with radius c. Note that when r is kept constant, c determines the separability of the data. The dimension of the data, denoted by d, is another parameter. In Fig. 5 three instances with different separability values (a) $c = 0$ (b) $c = r/2$ and (c) $c = r$ are shown for $d = 2$.

In the following subsections, the measure for the performance of the method is the objective function value obtained by the standard SVM formulation for the final set of selected pattern vectors. For this purpose we set the restricted total free slack to zero for the respective formulation, in which case they are equivalent to the standard SVM formulation. The results are later normalized within each dimension d and separability c for the compared methods. The normalization is

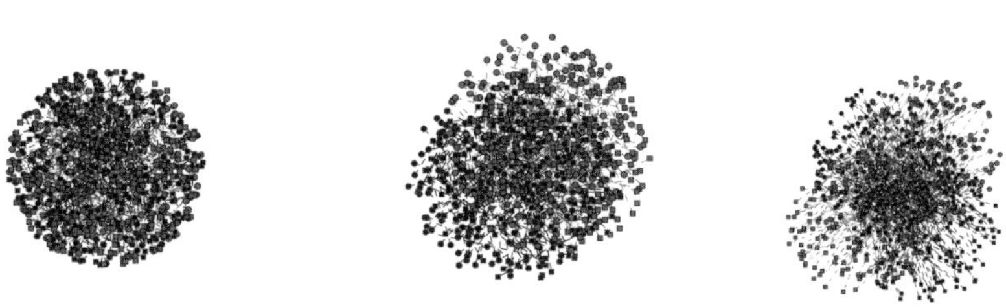

(a) (b) (c)

FIGURE 5. Data with separability (a) $c = 0$, (b) $c = r/2$, (c) $c = r$.

done by measuring the mean $\mu_{d,c}$ and the standard deviation $\sigma_{d,c}$ of all the objective function values obtained from all of the compared methods, and normalize each objective function value using the mean and the standard deviation.

5.2. Iterative elimination results. We compared the performance of the linear PFS and FSS methods for both 2-norm and 1-norm formulations on simulated data for $d = 2, 4, \ldots, 20$ and $c = 0, r/2, r$. We tested the performance of both methods on 100 instances of simulated data sets for each combination of d and c. The number of pattern vectors per set is set to 6. After 5 iterations, only one pattern vector per set remains and the remaining pattern vectors are used for measuring the performance using the standard SVM formulation. In Fig. 6, the average performances of the PFS vs FSS method are presented. The 1-norm and 2-norm formulations are compared separately. The final assessment is performed by the objective function value of 1-norm standard SVM with the selected pattern vectors from 1-norm PFS and FSS. Similarly, 2-norm standard SVM is used for 2-norm PFS and FSS.

For each combination of the two parameters, the objective function values from 100 instances of data sets for both PFS and FSS methods are normalized. The charts on the left and on the right shows the difference in the normalized objective function of FSS with respect to PFS for 1-norm and 2-norm, respectively. Let z_{PFS} and z_{FSS} denote the objective function values from standard SVM obtained by the PFS and FSS methods, respectively. Fig. 6 shows the values $z_{\text{FSS}} - z_{\text{PFS}}$ for both 1-norm and 2-norm cases for each combination of the parameters. The FSS method performs better for lower values of separability of the data set and PFS method performs better with higher values of separability of the data set. For both 1-norm and 2-norm cases, the increase in the dimensionality of the data (number of features) has a more significant effect on the dominance of PFS for higher values of separation compared to dominance of FSS for lower values of separation.

5.3. Iterative elimination vs. naïve elimination. In this set of results, we used 2-norm PFS for the elimination method. We note that this formulation reduces to 2-norm standard SVM formulation when the total free slack is set to zero. Simulated data is generated as explained in Section 5.1 for $d = 2, 4, \ldots, 20$ and $c = 0, r/2, r$. Note that $t = 6$ and free slack parameter $p = 1$. For each

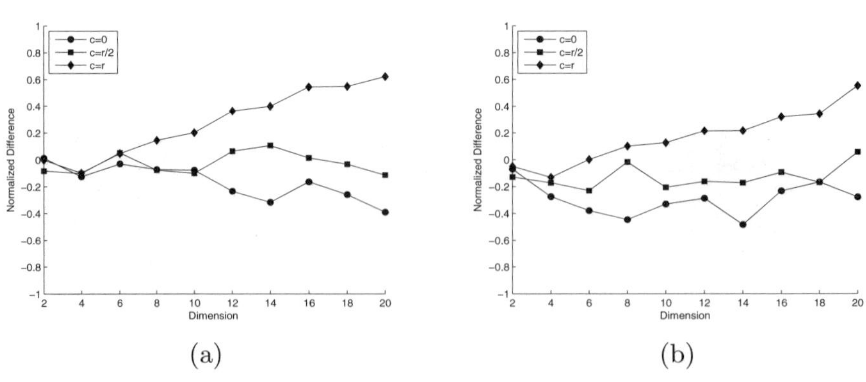

FIGURE 6. Comparison of PFS and FSS for (a) 1-norm, (b) 2-norm.

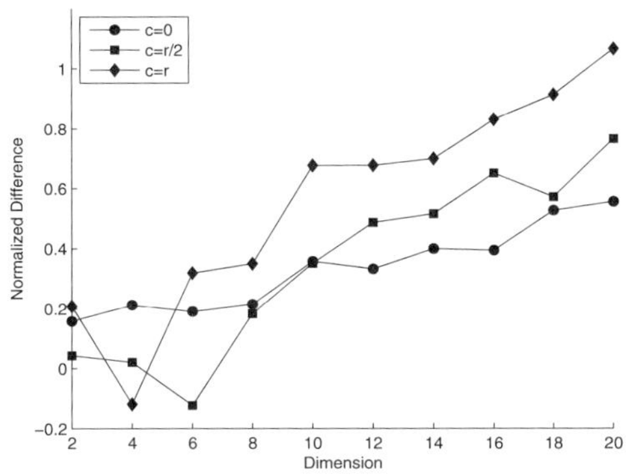

FIGURE 7. Normalized difference between iterative elimination and naïve elimination.

combination of the parameters 100 instances of simulated data sets are generated and tested using PFS elimination and standard SVM elimination. The results are normalized as explained in Section 5.1. Let z_{PFS} and z_N denote the final objective function values obtained from PFS elimination and standard elimination.

In Fig. 7, the values $z_N - z_{\text{PFS}}$, for $d = 2, 4, \ldots, 20$ are plotted for each c value. It is clear from the figure that as the number of dimensions increases the PFS elimination is significantly superior to the naïve elimination method. The difference becomes more apparent for higher levels of data separation. The increase amounts in the performance for both dimensionality and separability clearly mark the success of the proposed formulations due to the increased flexibility of the separating hyperplane.

5.4. Direct selection results. We generated and evaluated data as explained in Section 5.1 for $d = 2, 4, \ldots, 20$ and $c = 0,\ r/2,\ r$ for total slack parameter $p = 1, \ldots, 5$ with 100 instances each. Note that $t = 6$. In Fig. 8, the effect of the increase in total slack is shown. The three graphs in the figure are in the order of increasing separation in the data. In each graph, the objective function values for the highest total slack parameter $p = 5$ is assumed to be the base and the differences between the others and the base, $z_i - z_5$, $i = 1, \ldots, 4$ are charted. The amount of free slack does not contribute significantly for completely overlapping data in graph (a). However, it is clear from graph (b) that if there is some separability in the data, increasing amount of slack improves the performance of the method for higher dimensions. This difference is even more amplified in graph (c) for higher values of separability of the data. In graphs (b) and (c), the increase in the difference between $p = 5$ and others for higher dimensional data are also apparent. Based on these results, we can conclude that the best free slack parameter is $t - 1$ for a dataset with t pattern vectors in each set.

Next we compare the direct selection method with single elimination. For this comparison we use 2-norm PFS for both iterative elimination and direct selection methods. Note that $t = 6$ and free slack parameter $p = t - 1$. In Fig. 9, the performances of iterative elimination and direct selection are shown with the values $z_{\text{DS}} - z_{\text{IE}}$, where z_{IE} and z_{DS} are the objective function values obtained from 2-norm

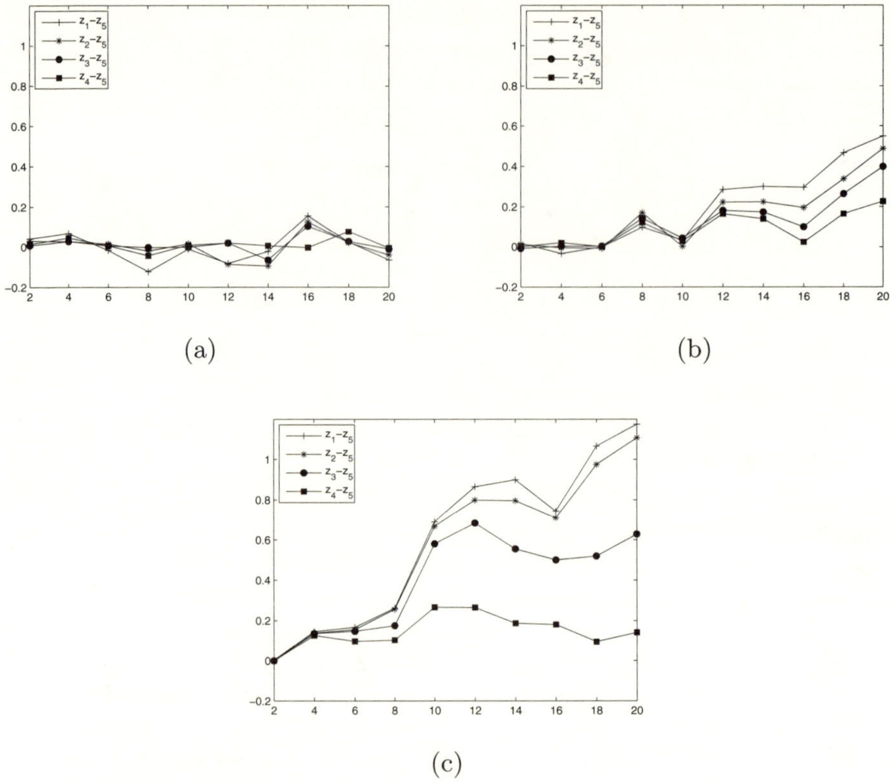

FIGURE 8. Effect of the amount of free slack on data with separability (a) $c = 0$, (b) $c = r/2$, (c) $c = r$.

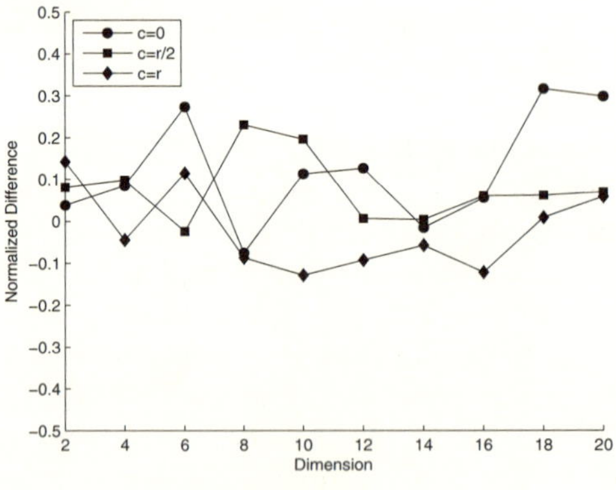

FIGURE 9. Comparison of iterative elimination and direct selection methods.

standard SVM for the pattern vectors found by iterative elimination and direct selection respectively. The results fluctuate and there is no significant dominance of one method over the other. However we observe from the figure that, on the average, the iterative elimination method performs slightly better than the direct elimination method.

5.5. Results on nonlinear data. We compared the iterative elimination method and the naïve elimination method using the nonlinear formulation. In Fig. 10, data instance is generated for $t = 6$ as explained in Section 5.1. It is clear from the figure that the solution found by the naïve elimination method is inferior to the solution found by iterative elimination method, where the pattern vectors are significantly separated.

The naïve elimination method is compared with PFS iterative elimination method on nonlinear data. 100 instances of data sets are generated as explained in Section 5.1 with 6 pattern vectors per set for 20 dimensional data and for three levels of separation $c = 0$, $r/2$, r. The results obtained from the two methods are significantly different. Therefore, in Fig. 11, we give the ratio of the mean values from the naïve method to the mean values from the iterative elimination method on the left. The graph on the right is the ratio of the standard deviation values of the

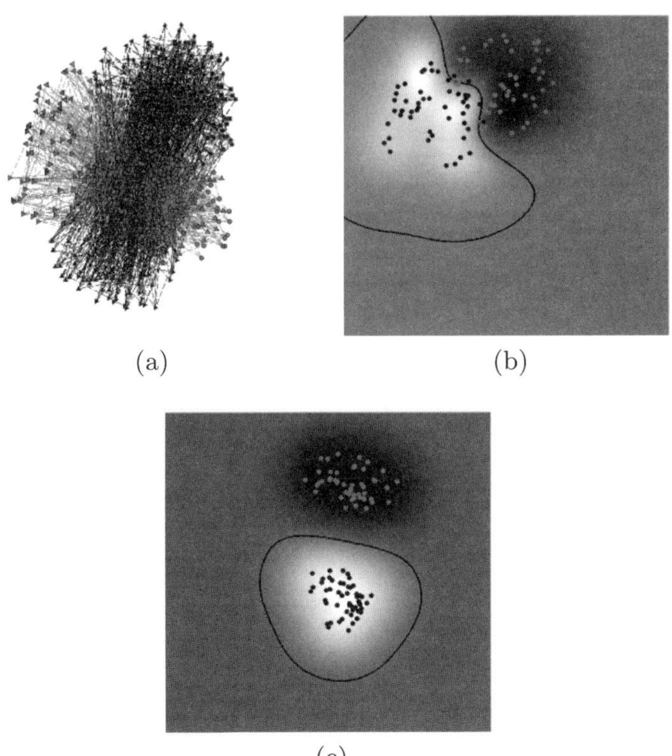

FIGURE 10. Comparison of iterative elimination and naïve elimination on nonlinear data (a) original data, (b) naïve elimination, (c) iterative elimination.

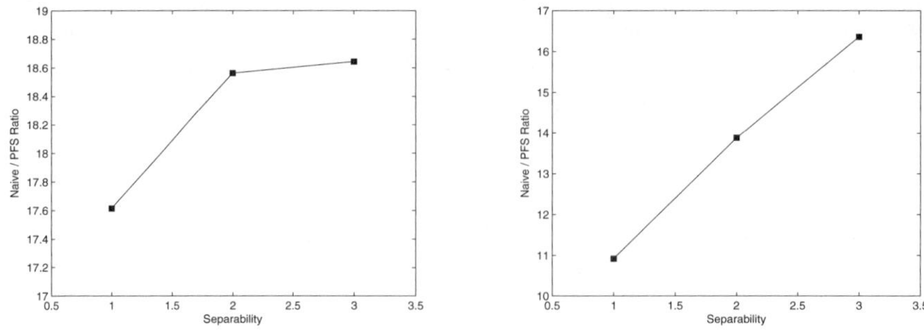

(a) (b)

FIGURE 11. The results for naïve elimination and iterative elimination on nonlinear data for different levels of separability: (a) the ratio of the mean (b) the ratio of the standard deviation.

two methods. This result confirms the significant superiority of the proposed iterative elimination method over the naïve elimination method. Therefore, we use this method in the next section where we use PFS based iterative elimination method on a high dimensional nonlinear neural data set.

6. Selective Classification Applied to Neural Data

We applied the selective classification methods developed to a real life neuroscience problem. The neural data we study is the local field potentials (LFP) collected from multiple channels implanted in different cortical areas of a macaque monkey during a visual discrimination task. This task involves recognizing a visual GO stimuli which is followed by a motor response. The visuomotor task is repeated multiple times for the same experiment with different stimuli-response combinations. These differences are grouped as different classes of data for classification. The main objective is to be able to detect these differences over the time course of the task, which requires extensive computational effort to achieve robust results from the multi-dimensional and highly nonlinear neural data.

The visual stimuli are designed to create *lines* and *diamonds*. The go stimuli is chosen to be either lines or diamonds from one session to another. We are interested in detecting different cognitive stages of the visual discrimination task over the time line. We distinguish different sets of labels for each cognitive stage. Three different stages are anticipated:

(i) the detection of the visual stimulus
(ii) the categorical discrimination of the stimulus
(iii) the motor response.

The first and the third stages are relatively easy to detect and has been reported in previous studies. However the second stage has not ever been detected to our knowledge. This stage involves a complex cognitive process whose onset and length vary over time. The details of this study can be found in [24].

The motivation behind the development of selective classification methods is to perform classification while accounting for the temporal variations in the underlying complex cognitive processes. Let $\mathbf{S}_j$ be the matrix where each row i of n rows is a vector of recordings from trial i at time point, j. Now consider the expanded matrix $|\mathbf{S}_j^T \mathbf{S}_{j+1}^T, \ldots, \mathbf{S}_{j+t-1}|$, which involves $n \times t$ rows of vector patterns, such that the corresponding rows of $\mathbf{S}_j$ and $\mathbf{S}_{j+1}$ are two consecutive recordings over the time line from the same trial. This matrix is composed of all pattern vectors within a window of size t from j to $j+t-1$. Our initial objective is to provide more recordings for classification. However, the standard SVM classifier is influenced by the noisy recordings. We assume that one recording among the t recordings from each trial comes from the underlying cognitive process we want to detect. Selecting the most distinguishable recording from each trial at a given time window is a hard problem. Therefore we use the *pooled free slack* (PFS) method with iterative elimination to detect and remove noisy recordings iteratively to achieve better recordings for the given time window.

The data we study consists of around 4000 trials. Because of the computational limitations of the optimization software[1] we use, the entire data could not be processed simultaneously. The strategy followed is to consider 200 trials at a time with equal numbers of positive and negative recordings. Nonlinear PFS method is applied with a window of 3 recordings from each trial for each time point, which correspond to 15 milliseconds. The recordings with the highest slack is eliminated from each set at each iteration. This is repeated until there is only one pattern vector remains from each trial. Each independent batch of 200 trials resulted in

[1]The optimizer used for this study is CPLEX 10.1.0.

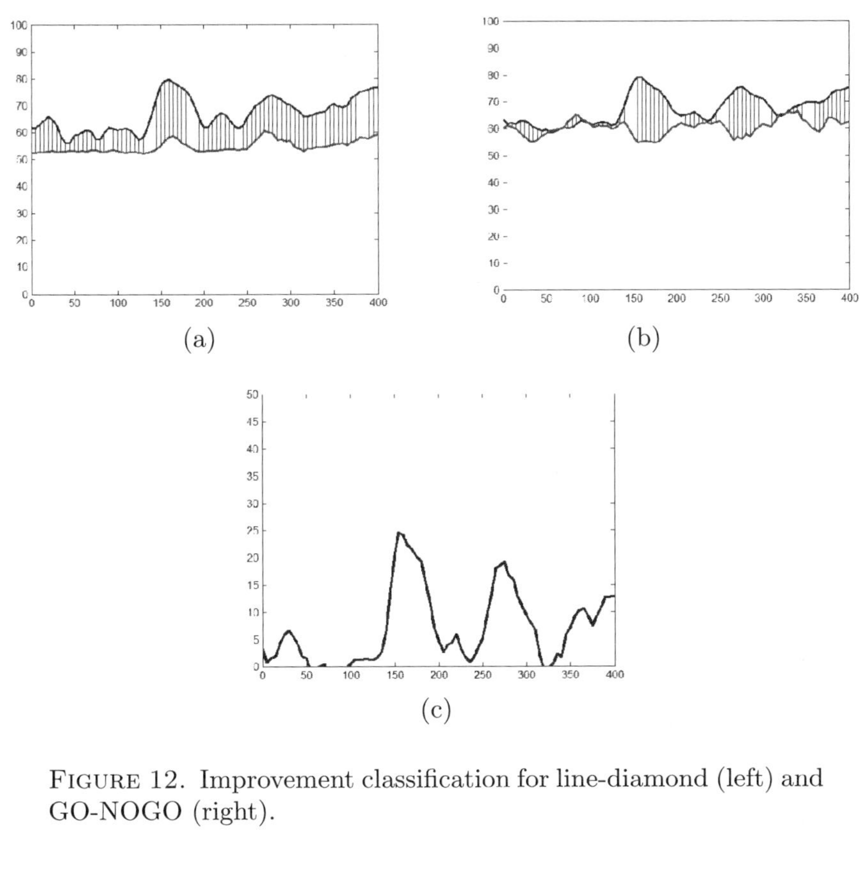

FIGURE 12. Improvement classification for line-diamond (left) and GO-NOGO (right).

a consistently separated cumulative set of selected recordings. The classification accuracy of the selected recordings from each time window was evaluated with the standard SVM classifier using 10-fold classification. In Fig. 12, the graph on the left is the comparison of the classification accuracy results from PFS method with iterative elimination and the results from the standard SVM classification. The iterative elimination shows a consistent increase of 10% in the accuracy over the time line. This increase can be adjusted by the *baseline* approach. In order to create a baseline, we randomly assign class labels to pattern vectors and apply the iterative elimination methods, so that we can detect the increase in the accuracy for random data. In Fig. 12 (b), the time intervals with the actual improvement are very clear after the baseline approach. Interestingly, there is a slight decrease in the baseline where there is a prominent increase in the accuracy. The results are further highlighted in the graph on the right which shows the difference between the accuracy obtained from the iterative elimination and the baseline. The categorical difference is clearly detected with a sharp peak around 160 milliseconds after the stimulus onset, which has not been detected by any previous study. The second peak around 275 milliseconds is too late for the categorical differentiation, however would probably be related to post processing of the categorical difference.

7. Conclusion

In this chapter we introduce novel selective linear and nonlinear classification methods which are generalizations of the standard SVM classifiers. Sets of pattern vectors sharing the same label are given as input. One pattern vector is selected from each set in order to maximize the classification margin with respect to the selected positive and negative pattern vectors. The problem of selecting the best pattern vectors is referred to as the *hard selection* problem. Kernelized hard selection problems are also developed for classification. However, these combinatorial problems are hard to solve. We propose alternative approaches with tractable formulations. The selective nature of the these formulations is maintained through the restricted free slack concept. The intuition behind this concept is to reverse the combinatorial selection problem by detecting influential pattern vectors which require free slack to decrease their effect on the classification functions. Iteratively removing such problematic pattern vectors, we can find those with better classification results.

Two variations of the free slack method, namely pooled free slack (PFS), and free slack per set (FSS), are introduced for selective linear classification together with kernelized dual formulations for selective nonlinear classification. These methods are further extended to direct separation by increasing the total free slack to diminish the effect of multiple pattern vectors per set and provide more flexibility for the hyperplane to reorient itself with respect to well separated pattern vectors. The performance of iterative elimination and direct selection algorithms are compared with each other, as well as with a naïve elimination algorithm that uses standard SVM method and ideas from the proposed methods. Results are reported for linear and nonlinear simulated data.

The motivation for the development of selective classification methods comes from the classification of cognitive states in a neural data set. The neural data is intracranial local field potential (LFP) recordings acquired from a macaque monkey during a visual discrimination task. SVM methods, in general, are very well suited for neural data due to the non-linearity and multi-dimensionality of the recordings. However, due to the temporal noise in the data, the results are significantly hindered. In the context of the selective methods, a sliding small time window of recordings are considered over all recordings. Well separated recordings are selected by the iterative elimination method. The selected recordings are evaluated with standard SVM methods which result in a significant increase over the time line. The increase is adjusted by a baseline method which isolates the actual improvement peaks. These peaks mark the onset and length of the categorical discrimination stage of the visuomotor task, which involves a complex cognitive process that has not been detected by previous studies. This result suggest that the proposed selective classification methods would provide promising solutions for other neuroscience problems.

References

1. S. Andrews, I. Tsochantaridis, and T. Hofmann, *Support vector machines for multiple-instance learning*, Advances in Neural Information Processing Systems, vol. 15, MIT Press, Cambridge, MA, 2003, pp. 561–568.
2. P. Auer, *On learning from multi-instance examples: empirical evaluation of a theoretical approach*, 14th Internat. Conf. on Machine Learning, Morgan Kaufmann, San Francisco, CA, 1997, pp. 21–29.
3. K. Bennet and C. Campbell, *Support vector machines: hype or hallelujah?*, SIGKDD Explorations **2** (2000), no. 2, 1–13.
4. M. Brown, W. Grundy, D. Lin, N. Cristianini, C. Sugne, T. Furey, M. Ares, and D. Haussler, *Knowledge-base analysis of microarray gene expression data by using support vector machines*, PNAS **97** (2000), no. 1, 262–267.
5. C. Cifarelli and G. Patrizi, *Solving large protein folding problem by a linear complementarity algorithm with* $0-1$ *variables*, Optim. Methods Softw., 2005 (submitted).
6. N. Cristianini and J. Shawe-Taylor, *An introduction to support vector machines*, Cambridge University Press, Cambridge, UK, 2000.
7. T. G. Dietterich, R. H. Lathrop, and T. Lozano-Perez, *Solving the multiple instance problem with axis-parallel rectangles*, Artificial Intelligence **89** (1997), 31–71.
8. G. N. Garcia, T. Ebrahimi, and J. M. Vesin, *Joint time-frequency-space classification of* EEG *in a brain-computer interface application*, EURASIP J. Appl. Signal Processing **2003** (2003), no. 7, 713–729.
9. T. Gartner, P. A. Flach, A. Kowalczyk, and Alex J. Smola, *Multi instance kernels*, 19th Internat. Conf. on Machine Learning (San Francisco, 2002), Morgan Kaufmann Publishers, San Francisco, CA, 2002, pp. 179–186.
10. Z. Huang and H. Chen and C. J. Hsu and W. H. Chenb and S. Wuc, *Credit rating analysis with support vector machines and neural networks: a market comparative study*, Decision Support Systems **37** (2004), 543–558.
11. C. W. Hsu, C. C. Chang, and C. J. Lin, *A practical guide to support vector classification*, Tech. Report, National Taiwan University, 2004; `http://www.csie.ntu.edu.tw/~cjlin/papers/guide/guide.pdf`,
12. T. Joachims, *Text categorization with support vector machines: learning with many relevant features*, European Conference on Machine Learning (C. Ndellec and C. Rouveirol, eds.), Springer, Berlin, 1998, pp. 137–142.
13. ———, *Making large-scale* SVM *learning practical*, Advances in Kernel Methods—Support Vector Learning (B. Schölkopf and C. Burges and A. Smola, eds.), MIT Press, Cambridge, MA, 1999, pp. 69–184.
14. T. N. Lal, M. Schroeder, T. Hinterberger, J. Weston, M. Bogdan, N. Birbaumer, and B. Schlkopf, *Support vector channel selection in* BCI, IEEE Trans. Biomed. Engr. **51** (2004), no. 6, 1003–1010.
15. S. Lee and A. Verri (eds.), *Pattern recognition with support vector machines*, SVM 2002 (Niagara Falls, Canada), LNCS 2388, 2002, Springer, Berlin Heidelberg, Germany, pp. 213–236.

16. P. M. Long and L. Tan, PAC *learning axis aligned rectangles with respect to product distributions from multiple instance examples*, Machine Learning **30** (1998), no. 1, 7–22.
17. O. L. Mangasarian, *Multi-surface method of pattern separation*, IEEE Trans. Inform. Theory **IT-14** (1968), 801–807.
18. O. L. Mangasarian and E. W. Wild, *Multisurface proximal support vector classification via generalized eigenvalues*, Data Mining Institute, Tech. Report, no. 04-03, 2004.
19. W. S. Noble, *Support vector machine applications in computational biology*, Kernel Methods in Computational Biology (B. Schölkopf, K. Tsuda, and J.-P. Vert, eds.), MIT Press, Cambridge, MA, 2004, pp. 71–92.
20. E. Osuna and F. Girosi, *An improved training algorithm for support vector machines*, Neural Networks for Signal Processing VII—Proc. of the 1997 IEEE Workshop (J. Principe, L. Gile, N. Morgan, and E. Wilson), IEEE, New York, NY, 1997, pp. 276–285.
21. J. Platt, *Fast training of* SVMs *using sequential minimal optimization*, Advances in Kernel Methods: Support Vector Learning (B. Schölkopf, C. Burges and A. Smola, eds.) MIT Press, Cambridge, MA, 1999, pp. 185–208.
22. S. Ray and M. Craven, *Supervised versus multiple instance learnining: an emperical comparison*, 22nd Internat. Conf. on Machine Learning (Bonn, Germany), ACM Press, New York, NY, 2005, pp. 697–704.
23. B. Schölkopf and A. J. Smola, *Learning with Kernels*, MIT Press, Cambridge, 2002.
24. O. Seref, C. Cifarelli, E. Kundakcioglu, P. M. Pardalos, and M. Ding, *Detecting categorical discrimination in a visuomotor task using selective support vector machines*, (H. R. Arabnia, M. Q. Yang, J. Y. Yang, eds.), BIOCOMP '07, 580–590 (to appear).
25. J. Shawe-Taylor and N. Cristiani, *Kernel methods for pattern analysis*, Cambridge University Press, Cambridge, UK, 2004.
26. Q. Tao, S. D. Scott, and N. V. Vinodchandran, SVM-*based generalized multiple instance learning via approximate box counting*, 21st Internat. Conf. on Machine Learning (C. E. Brodley, ed.), vol. 69, ACM, New York, NY, 2004, pp. 779–806.
27. T. B. Trafalis and H. Ince, *Support vector machine for regression and applications to financial forecasting*, Internat. Joint Conf. on Neural Networks (IJCNN'02) (Italy), 2002, IEEE, Washington, DC, vol. 6, pp. 348–353.
28. V. Vapnik, *The nature of statistical learning theory*, Springer-Verlag, New York, Inc., NY, 1995.
29. ______, *Statistical learning theory*, John Wiley, NY, 1998.
30. V. Vapnik and A. Chervonenkis, *The necessary and sufficient conditions for consistency in the empirical risk minimization method*, Pattern Recognition and Image Analysis **1** (1991), no. 3, 283–305.
31. J. Wang and J.-D. Zucker, *Solving the multiple instance problem: a lazy learning approach*, 7th Internat. Conf. on Machine Learning (ICML-2000), Morgan Kaufmann, San Francisco, CA, 2000, pp. 1119–1125.
32. Q. Zhang and S. A. Goldman, EM-DD: *an improved multiple-instance learning technique*, Neural Information Processing Systems 2001, vol. 14, The MIT Press, Cambridge, MA, 2001, pp. 1073–1080.

Department of Industrial and Systems Engineering, University of Florida, 303 Weil Hall, P.O. Box 116595, Gainesville, FL 32611-6595, USA

E-mail address, O. Seref: `seref@ufl.edu`

E-mail address, O. E. Kundakcioglu: `erhun@ufl.edu`

E-mail address, P. M. Pardalos: `pardalos@ufl.edu`

Titles in This Series

17 **Olga Kharlampovich, Editor,** Summer school in group theory in Banff, 1996, 1998
16 **Alain Vincent, Editor,** Numerical methods in fluid mechanics, 1998
15 **François Lalonde, Editor,** Geometry, topology, and dynamics, 1998
14 **John Harnad and Alex Kasman, Editors,** The bispectral problem, 1998
13 **Michel C. Delfour, Editor,** Boundaries, interfaces, and transitions, 1998
12 **Peter C. Greiner, Victor Ivrii, Luis A. Seco, and Catherine Sulem, Editors,** Partial differential equations and their applications, 1997
11 **Luc Vinet, Editor,** Advances in mathematical sciences: CRM's 25 years, 1997
10 **Donald E. Knuth,** Stable marriage and its relation to other combinatorial problems: An introduction to the mathematical analysis of algorithms, 1997
9 **D. Levi, L. Vinet, and P. Winternitz, Editors,** Symmetries and integrability of difference equations, 1996
8 **J. Feldman, R. Froese, and L. M. Rosen, Editors,** Mathematical quantum theory II: Schrödinger operators, 1995
7 **J. Feldman, R. Froese, and L. M. Rosen, Editors,** Mathematical quantum theory I: Field theory and many-body theory, 1994
6 **Guido Mislin, Editor,** The Hilton Symposium 1993: Topics in topology and group theory, 1994
5 **D. A. Dawson, Editor,** Measure-valued processes, stochastic partial differential equations, and interacting systems, 1994
4 **Hershy Kisilevsky and M. Ram Murty, Editors,** Elliptic curves and related topics, 1994
3 **Rémi Vaillancourt and Andrei L. Smirnov, Editors,** Asymptotic methods in mechanics, 1993
2 **Philip D. Loewen,** Optimal control via nonsmooth analysis, 1993
1 **M. Ram Murty, Editor,** Theta functions: From the classical to the modern, 1993